I TATTI STUDIES IN
ITALIAN RENAISSANCE HISTORY

Sponsored by Villa I Tatti

Harvard University Center for Italian Renaissance Studies

Florence, Italy

THE DUKE
AND THE STARS

*Astrology and Politics
in Renaissance Milan*

Monica Azzolini

HARVARD UNIVERSITY PRESS
Cambridge, Massachusetts
London, England
2013

Library of Congress Cataloging-in-Publication Data

Azzolini, Monica, 1971–

The duke and the stars : astrology and politics in Renaissance Milan /
Monica Azzolini.

p. cm.

Includes bibliographical references (p. 321) and index.

ISBN 978-0-674-06663-2   1. Astrology and
politics—History.   2. Gian Galeazzo Sforza, Duke of Milan, 1469–1494.
3. Milan (Italy)—History—To 1535.   I. Title.

BF1729.P6A99 2012

133.50945'2109024—dc23   2012015760

To A. C., S. K., K. P., and N. G. S.

Contents

Preface   *ix*

Illustrations   *xi*

Introduction   *1*

1 The Science of the Stars:
 Learning Astrology at the University of Pavia   *22*

2 The Making of a Dynasty:
 Astrology under Bianca Maria Visconti and Francesco Sforza   *65*

3 Astrology Is Destiny:
 Galeazzo Maria Sforza and the Political Uses of Astrology   *100*

4 The Star-Crossed Duke:
 Gian Galeazzo Sforza and Medical Astrology   *135*

5 The Viper and the Eagle:
 The Rise and Fall of Astrology under Ludovico Sforza   *167*

 Epilogue   *210*

Abbreviations   *215*

Notes   *219*

Bibliography   *321*

Acknowledgments   *355*

Index   *361*

Preface

This book is a study of the uses and function of astrological prediction in one of the most sumptuous courts of the Italian Renaissance, the Sforza of Milan. The research that eventually led to this book started with the casual discovery of a large body of documents in the Archivio di Stato di Milano on a sultry summer day back in 2002. A few weeks later—armed with little teaching experience and a bag full of hope and goodwill—I left Europe for the United States to take up my first academic post. Copies of the documents travelled with me to Seattle, but only to lay there, virtually forgotten, for another few years. Busy with teaching and life on a new continent, I worked toward the publication of a couple of articles stemming from my doctoral dissertation on the intellectual and social context of Leonardo da Vinci's anatomies in Milan. The plan, then, was to revise my doctoral dissertation to turn it into a book. It was only when I moved to Sydney, Australia, in the fall of 2004 that my plans changed and my research took an unexpected turn. The nearest copies of Leonardo's anatomical sheets were in Canberra, a full one-hour flight from where I taught. The archives that I used when writing my doctorate were a good twenty-one hours away. Conceiving of a plan B became desirable and, increasingly, a necessity. I looked over my notes and the thousand copies of documents collected over the years, and there it was, a stack of documents on astrology that I could barely decipher. There is no name for the strange disease that takes over a scholar's mind when intrigued by what one cannot fully grasp (or at least no name exists that is truly flattering to that scholar!). No sooner had I landed my second job on another continent—this one further still from my archives—that I was caught in the grip of this nameless disease: I could barely decipher the symbols on many of those pages and yet I wanted to make sense of those documents and tell their story.

The rehabilitation of astrology—one of the infamous "wretched subjects" of pre-modern societies—in Renaissance and Early Modern Studies is now almost complete. Astrology is no longer shunned by serious academics. Astrologers have been the subject of entire monographs, and in recent decades there has been a steady flow of works exploring the intellectual tenets of the discipline. Astrology's relationship with political power, especially in a courtly context, however, has yet to receive organic treatment, and the role of the court astrologer still awaits a major study. In particular, very little attention has been paid to the middling practitioners that populated Renaissance courts and did not rise to fame through publishing. My documents are crowded with precisely such figures: people who did not publish even a single work, whose names are now almost completely forgotten, but who, at times, held pride of place in advising powerful rulers. The discovery of these documents—and the characters that populated them—prompted a series of questions: What were these documents doing in the ducal diplomatic correspondence, and to whom were they addressed? Who were these astrologers who offered their services, and were they successful in seeking employment? Were they in stable employment, or did they offer their services only occasionally? Many more questions emerged as I progressed through my research. Not all my questions found an answer, but the answers that I found are in the following pages.

Illustrations

Figure 1. *Lamento del Duca Galeazo, Duca di Milano, quando fu morto in Sancto Stephano da Gioanandrea da Lampognano* (Florence: Bernardo Zucchetta, for Piero Pacini da Pescia, October 24, 1505).   *6*

Figure 2. *Homo signorum,* or zodiac man, from Johannes Ketham, *Fasciculo de medicina in vulgare* (Venice: Giovanni and Gregorio de Gregori da Forlì, February 5, 1493/94).   *14*

Figure 3. *Homo venorum,* or vein man, from Johannes Ketham, *Fasciculo de medicina in vulgare* (Venice: Giovanni and Gregorio de Gregori da Forlì, February 5, 1493/94).   *15*

Figure 4. Diagram of the phases of the Moon, from Johannes Sacrobosco, *Computus* in British Library, MS Arundel 88, fol. 38r.   *31*

Figure 5. Diagram of the phases of the Moon, from Johannes Sacrobosco, *Computus* in Cambridge University Library, MS li.III.3, fol. 46v.   *32*

Figure 6. Frontispiece woodcut of the celestial spheres and the four elements, from *Judicium cum tractatibus planetariis* (Milan: Filippo Mantegazza, December 20, 1496).   *35*

Figure 7. Table of contents of Giovanni Battista Boerio's transcription of
           eleven out of the twelve original chapters of John of Bruges's
           De veritate astronomie, from British Library, MS Arundel 88,
           fol. 15r. *37*

Figure 8. Table of contents of the *Judicium cum tractatibus planetariis* (1496),
           which contains eleven out of twelve chapters of John of Bruges's *De
           veritate astronomie.* *38*

Figure 9. Ornate initial from Alcabitius, *Liber introductorius* (Venice: Erhard
           Ratdolt, 1485), sig. aa2r. *43*

Figure 10. Ornate initial within a dense two-column page in tironian type from
           Ptolemy's *Quadripartitum,* in *Opera astrologica* (Venice: Bonetto
           Locatelli for Ottaviano Scoto, December 20, 1493), sig. A2r. *44*

Figure 11. Diagram of the aspects of the planets from Alcabitius, *Liber
           introductorius* (Venice: Erhard Ratdolt, 1485), sig. aa5r. *55*

Figure 12. Ptolemaic celestial sphere with the zodiac and the ecliptic, from
           Johannes Sacrobosco, *Sphaera* (Venice: Simone Bevilacqua da Pavia,
           1499), sig. a2v. *56*

Figure 13. *Figura coeli,* or square celestial chart with the house division, from
           Alcabitius, *Liber introductorius* (Venice: Erhard Ratdolt, 1485),
           sig. bb1v. *59*

Figure 14. Geniture of Ludovico Maria Sforza, from Girolamo Cardano, *De
           exemplo centum geniturarum,* in *Opera Omnia,* 10 vols. (Lyon: Jean
           Antoine Huguetan & Marc Antoine Ravaud, 1663), vol. 5, 463. *61*

Figure 15. Raffaele Vimercati donating his *iudicium* to Francesco Sforza, fourth
           Duke of Milan and father of Galeazzo Maria Sforza, from Biblioteca
           Trivulziana, MS Triv. 1329, *Liber iudiciorum in nativitate Comitis
           GaleazMarie Vicecomitis Lugurum futuri ducis* (1461), fol. 2r. *99*

Figure 16. Geniture of Galeazzo Maria Sforza, from Biblioteca Trivulziana, MS
           Triv. 1329, *Liber iudiciorum in nativitate Comitis GaleazMarie
           Vicecomitis Lugurum futuri ducis* (1461), fol. 21r. *106*

Figure 17. Annius of Viterbo's interrogation about the possible death of the
           King of Naples, Ferrante of Aragon (chart), from ASMi, *Sforzesco,
           Miscellanea* 1569, Annius of Viterbo to Galeazzo Maria Sforza,
           Genoa, November 24, 1475. *124*

Figure 18. Reconstructed chart from Ambrogio Varesi da Rosate's
            interrogation about the possible death of Innocent VIII dated July
            18, 1492, derived from Varesi's letter to Ludovico Sforza in ASMi,
            Autografi, Medici 219, Ambrogio Varese da Rosate to Ludovico,
            Milan, July 20, 1492.   *195*

Figure 19. Reconstructed chart of the Sun's entry into the first degree of Aries
            (Spring Equinox), derived from Varesi's letter to Ludovico Sforza in
            ASMi, *Autografi,* Medici 219, Ambrogio Varese da Rosate to
            Ludovico, Milan, July 20, 1492.   *196*

THE DUKE AND THE STARS

Introduction

A number of obscure physicians, astrologers, and physician-astrologers loom largely in this book. Mostly unknown to present-day historians, some of them were, nonetheless, important personalities in their time. The reason many of them have fallen victim to oblivion is rather simple: they were professional astrologers who wrote or published little, and when they did it was often in the form of private correspondence with their prospective or actual clients, or in that of annual prognostications that were posted on university boards in the faculty of arts and medicine. This book is about the Sforza dukes and their use of astrology, but it is also about these "minor" characters: men who are barely remembered in the histories written about their more famous clients, people who were written out of the history of great men.[1] My argument, simply put, is that these historical actors are far from insignificant. I argue, instead, that the study of these minor professional figures contributes greatly to our understanding of Renaissance cultural, social, and political history. It does so in two important ways: first, by providing a corrective to the idea that Renaissance politics was driven to a large extent by a type of political pragmatism devoid of many of those cultural elements that were characteristic of its time; and secondly, by providing a variety of illuminating examples of how astrological theory was put into practice in Renaissance daily life.

Regarding the first point: Italian Renaissance leaders may have been cunning, calculating men driven by personal and dynastic ambitions, but this did not make them immune from embracing the worldview of their contemporaries. Such a worldview encompassed, among other things, celestial influence, which stipulated that the movement of the celestial spheres brought about changes in the sublunary world. This was not limited to seasons and natural events, however; rather, it went as far as influencing individuals, kingdoms, and even religions.[2] As this book will demonstrate, people adhered to the principles of astrology to different degrees. The vast majority of people were happy to admit that celestial bodies exerted an influence on Earth; a good part of them believed that the nature of this influence could be determined and interpreted by the professional astrologer. A smaller number among this last group held a more deterministic view—this one not shared even by all Renaissance astrologers in equal measure—that went as far as to argue that one could choose the best moment to attempt an action on the basis of the configuration of the skies. Not all Renaissance political leaders abided by this last principle—which saw its application in the astrological technique of elections—but there were certainly some areas of political and civic life where these principles were applied with more consistency than others in this period. War was one of them (travel was another). One example is the passing of the baton of command to the captain-general of an army. To be propitious, the event had to happen "per puncto d'astrologia," namely at a precise time of day, as determined by one or more astrologers. Although met with skepticism even by some contemporaries, this practice was well documented in the Florentine republic, whose officials followed it on a number of occasions in the course of the fifteenth century.[3] Likewise, we now know that Ludovico Maria Sforza—to whom Chapter 5 of this book is dedicated—applied this very same protocol in appointing his own generals. While preparing himself to face the French army and defend his duchy from foreign occupation at the end of the fifteenth century, Ludovico chose to appoint his generals "per puncto d'astrologia," in the belief, no doubt, that this would guarantee him a more favorable outcome. This allows us to speculate that the practice may have been more common than it has been generally thought. Far from making Ludovico Maria Sforza less cunning and calculating, moreover, this example suggests that the Duke of Milan resorted to *all* possible means to ensure success. This included astrological counseling, which was deemed conjectural but firmly rooted in the legitimate and reliable art of astrology.

In the Renaissance, astrology was far from the discredited art that populates contemporary newspaper columns. As a university discipline in the degree of arts and medicine, it was imparted rigorously. As the sister discipline of astronomy—its ancillary discipline in what made up the "science of the stars"—it was considered not only legitimate, but also based on true "scientific principles." As such, there is nothing surprising in seeing Renaissance princes seek astrological advice in all spheres of Renaissance life, war included. While we may not share Ludovico's worldview any longer, therefore, we should refrain from selectively sanitizing political history of all those aspects that seem to us "superstitious" and thus unpleasant (even when these were sometimes contested in their own time, as astrology certainly was).

The second point—and the second major contribution of this work to current scholarship—regards our present understanding of Renaissance astrology. While much has been done to restore astrology to the intellectual and cultural place it occupied in pre-modern societies, to date historians have tended to pay much less systematic attention to astrological practice in its political and social contexts.[4] We know relatively little of astrologers' favored techniques and their relationships with their clients.[5] Similarly, we know little of how these relationships were initiated and later negotiated. By exploring astrologers' private correspondence and other ephemeral writings, this book aims to expand and refine historians' present understanding of the ways in which astrological theory met the demands of Renaissance men and women and was put, literally, in practice. It does so by examining one particular set of clients—Renaissance Italian lords and princes—and focusing on one particular family, the Sforza. While the angle is seemingly narrow, this has allowed me to write a series of microhistories of the Sforza dukes and their astrologers in the specific political and social contexts in which these figures operated. This microhistorical dimension has added texture to my analysis, allowing me to weave astrological practice tightly into the political and social context in which these astrologers operated. This important dimension would have been largely lost had I decided to widen the scope and chronology of my study.[6] Therefore, while earlier historians have often concentrated on the intellectual and theoretical bases of astrology in the Renaissance, this book strives to bring theory and practice together, moving back and forth between astrological techniques and the ways in which these were practically applied in specific contexts. *Professory*

Some caveats may be in order, however. As H. C. Erik Midelfort said of his mad German princes, "these records document only the lives and the miseries of the powerful."[7] This is true of this book as well. My research documents the use of astrology of one particular elite group in one particular part of Italy and its results cannot be taken as representative of the general attitude of Renaissance people toward astrology. My choice was guided, at least to some extent, by an element of serendipity—the fortuitous encounter of a large number of astrological documents in the Milanese archives in the summer of 2002 (when I was looking for very different type of material). But it was also dictated by methodological problems. While Italian courts kept meticulous records of their daily activities, thus providing contemporary historians with extremely rich evidence of their social, cultural, and political interactions, similar sources are rarely available for the merchant classes or the lower classes. In the case of astrology, moreover, the problem may be compounded further by the fact that fifteenth-century astrology was mostly practiced by learned physicians. This may have put astrological consultation out of the reach of a large sector of the population. On occasion, however, we are fortunate enough to be able to explore some of the attitudes that common people held toward astrology from the account of courtiers who commented on the circulation of astrological prognostications. While this glance is generally brief and episodic, it allows us to say that Renaissance people from all walks of life paid attention to astrological forecasting. We know for certain, for instance, that Renaissance lords were often concerned about the power of astrological predictions to stir the populace into action. It seems certain, therefore, that while the prime consumers of astrological counsel were Renaissance elites, the populace was not immune to the powers of astrology.

In the Renaissance prognostications circulated both privately and publicly. Their nature and form changed accordingly. In their letters, astrologers and astrologer-physicians often provided highly personalized advice to their clients. Given the proximity of some of them to their own lords, moreover, there can be little doubt that on occasion they delivered their advice orally. When writing their annual prognostications (or *iudicia*), instead, astrologers strove to provide general forecasts: these included the weather, the harvest, and the likely outcome of conflict in the forthcoming year. They also ventured, however, to express predictions about various professions or categories of people,

and, more importantly, specific European and Italian lords and kings. Some
of this advice, both private and public, was undoubtedly sensitive. While
control was exerted over personal correspondence, annual prognostications
had indisputably a public dimension and circulated widely among Renaissance elites and within Northern Italian courts. Some of these had enough
currency to be registered by contemporary chroniclers (themselves, sometimes, members of Italian Renaissance courts), who often commented on the
accuracy (or otherwise) of the events that were forecast. Such was the case of
the alarming predictions that circulated soon before the dramatic assassination of Galeazzo Maria Sforza, fifth Duke of Milan, on Boxing Day of 1476
(Figure 1). In recording the event, the two Ferrarese courtiers Bernardino
Zambotti and Girolamo Ferrarini did not miss the opportunity to celebrate
one of the brightest talents of their native Ferrara, the university professor
and court astrologer Pietro Buono Avogario. A few years before the event,
Avogario had correctly forecast Galeazzo's death. "A great lord," Avogario had
written in his prognostication (*iudicio*) for 1475, "will die this year either
by sword or poison." Both Bernardino Zambotti and Girolamo Ferrarini
quoted the relevant passage almost verbatim from Avogario's prognostication,
Ferrarini adding that, "people say openly that these words in his prognostication came true in the person of the Duke of Milan, who is now dead. And
this is very true, as I said before."[8] Astrological prognostications obviously
had people talking, and this factor was well known to Italian rulers who,
as we shall see, attempted to control this type of information. Indeed, this
trafficking of astrological "intelligence" was deemed relevant to Renaissance
politics—Galeazzo Maria Sforza himself requested that annual prognostications published in other cities be collected and subject to control. Tracing the
circulation of this type of "intelligence," therefore, both tells us more about
the circulation of knowledge and helps us illuminate Renaissance political
praxis.

The evidence on which this book is based comes from a vast array of documentation for the period of Sforza domination, c. 1450–1500. Its interest lies
primarily in the political use of astrology at the Milanese court of the Sforza,
where physicians and astrologers who had trained at the local university of
Pavia often found employment. Perhaps surprisingly, however, it relies heavily
on diplomatic and other archival sources that are not, generally speaking, the
bread and butter of the historian of science. Few historians of science and
medicine have ventured to use these sources to write their histories, and yet

O Sacra et senza macula Maria
madre delbuõ iesu figliuola e sposa
fonte di charita humile et pia
Vergine bella et misericordiosa
refugio de gliafflicti albergo et pace
splendor del sole stella luminosa
Per me priegha il tuo figliuolo se tipiace
che ase raccoglia questa anima tapina
che lassa il mondo misero et fallace
O coronata in cielo alta regina
soccorrimi allostremo di mia guerra
siche moblii della infernal sucina
Et uoi chel corpo mio uedete in terra
& laltrui ferro nel mio sangue tinto
dirouiilnome mio et chitanto erra

Galeazo Maria son duca quinto
di Milano hor udite idolor miei
cõgliocchi iluolto di lacrime dipinto
Nel mille quatrocento septanzei
del mese di dicembre poi natale
el di sancto sstephano auenzei
Co mei andando al culto diuinale
catholico et deuoto a udir la messa
saprete chi agran torto massale
Nella chiesa del martyr doue e messa,
pura bambagia atorno a una croce
per certa cerimonia iui concessa
Et un gridando largo ad alta uoce
uẽne uerso di me cõ uolto humano
& col cor tristo spietato et feroce

FIGURE 1. *Lamento del Duca Galeazo, Duca di Milano, quando fu morto in Sancto
Stephano da Gioanandrea da Lampognano* (Florence: Bernardo Zucchetta, for Piero
Pacini da Pescia, October 24, 1505). By kind permission of the Archivio Storico Civico,
Biblioteca Trivulziana, Milan. © Comune di Milano.

much can be gleaned from diplomatic correspondence about the daily practice of Renaissance physicians and astrologers, their relationships with their clients, and the type of services requested. A notable exception to this trend is represented by the work of French historian Marilyn Nicoud, whose studies on medicine and medical care at the Sforza court represent a fine companion to the present work.[9] Much like Nicoud's work, the present study allows us to explore the delicate relationships between Renaissance professionals and their clients and patrons. In the case of astrologers, the nature of these relationships could vary greatly from permanent service to occasional consultations. In many cases, moreover, this relationship was negotiated on the basis of proven knowledge and competence. We can presume that some of these astrologers built their reputations over time and were therefore renown for their predictive skills. This may have been the case of Annius of Viterbo, the Dominican preacher who was consulted by Galeazzo Maria Sforza to find out if his enemy Ferrante of Aragon was going to die or not. Others, less famous, may have practiced locally and may have been known to the duke or his courtiers in this way. Others again, like the physician-astrologers Ambrogio Varesi da Rosate and Gabriele Pirovano, may have held positions at the University of Pavia. Be that as it may, it was not unusual for a client to question or test the practitioner's knowledge by asking for a second opinion. Galeazzo Maria Sforza, for instance, ordered his trusted secretary Cicco Simonetta to request an annual prognostication from three different astrologers, none of whom was to know that others had been consulted. The same happened only weeks before Ludovico Maria Sforza lost the Duchy of Milan to the French. Weary of the soundness of Ambrogio Varesi's predictions, Ludovico consulted other astrologers, who expressed a different opinion. What these examples demonstrate is that patrons did not accept astrological advice uncritically, but often exerted quality control over this information, thus testing the knowledge of the astrologers they employed.[10]

If historians of medicine and science have often failed to appreciate the richness of these archival resources, this is true to an extent also of political historians. Despite the presence of astrological material in diplomatic archives, political historians on the whole have regarded any evidence of astrological consultation as marginal to their own discipline, often considering astrological counsel as an aberration, a regrettable form of superstition of little or no consequence for European political history.[11] As a result, within the narrative of political history, astrologers and physicians have fallen into the margins of courtly life and diplomacy and have largely disappeared. Much remains to be done, therefore, to shed light on the political role of astrology (and indeed, medicine) in European history.

Finally, I could have chosen to write about another city and another court. Why Milan? Not all archives have been catalogued with criteria that facilitate the research of historians of science or medicine, and records that are abundant in one archive may be completely lacking in another. In writing history, therefore, there are limits to what one can do that are not solely intellectual, but also practical. As those who have worked in Italian archives will testify, each archive has something unique and different (without denying, of course, the similarities). The difference in the case of Milan is that in the eighteenth century some of the archival documentation was ordered by subject. This order by subject ("ordinamento per materia") was first devised and implemented by the archivist Luca Peroni in Milan in the eighteenth century and later exported to other Italian archives. While generally berated by modern archivists as painfully inadequate to the needs of archival preservation and modern archival criteria, this system had the undoubted merit of collecting documents according to subject, thus sometimes making life easier for the historian. There are, of course, limits to such a classification system: only a fraction of the relevant material of a given subject was separated and classified in this way while other relevant sources remained scattered in the rest of the archive. But this classification provides a solid starting point for any thematic research. In Milan, these subjects, at least initially, included medicine and physicians and astrology and astrologers. It was on this basis that Ferdinando Gabotto was able to write two pioneering articles on astrology at the Sforza court, drawing almost exclusively from a *fondo Astrologi*. Sadly, however, this *fondo* no longer exists.[12] While the *fondo Medici e Medicina* remains seemingly intact, for some reason at some point in the late eighteenth or early nineteenth century, somebody decided to dismantle its astrological counterpart. Some letters were then inserted in their original chronological series (according to provenance and date), while others were arbitrarily put aside in a *fondo Miscellaneo* encompassing astrological and other miscellaneous papers. Other documents cited by Gabotto, instead, are now much harder to locate; some are possibly no longer extant. The reasons for dismantling the *fondo Astrologi* remain unclear, but at every stage of the process the arbitrary decisions of the archivists and the lack of an efficient system of internal referencing may have introduced errors and increased the risks of dispersion of this precious material. (It is clear, for instance, that some of the undated documents in the *fondo Miscellaneo* were originally accompanied by dated letters with which they were sent, but no trace of these letters remains in the present division.) Even if now dispersed, the *fondo Astrologi* used by Gabotto represents the foundation of this study. Without those two pioneering articles—with all the limits of nineteenth-century research—this work would have been hardly possible. Building on it, I have striven to reconstruct the uses of astrology at the court of the Sforza of Milan.

two important sources

Much important work has been done in the past fifty years and, no doubt, more studies will continue to appear that clearly demonstrate the importance of astrology within early modern society and its relevance to the history of science, intellectual history, and social and cultural history. However, one area that so far has not been explored systematically, especially in the Italian context, is court astrology.[13] While virtually every aspect of Italian court life has come under the scrutiny of art, cultural, and intellectual historians over the past decade, astrology on the whole has been granted very little attention and still awaits sustained research.[14] Yet, as this book demonstrates, astrologers seemingly gravitated naturally around Italian Renaissance princes, often offering their services, and at times becoming their privileged advisors. It is therefore worth pondering why so much material about astrology can be related to courts. Were Renaissance courts hotbeds of astrological practice? Answering this question with confidence would require extensive studies of other important Italian courts, not least the Roman curia. But we can already say that astrologers populated Italian courts in healthy numbers. Why so? There are a number of possible explanations, all probably, to an extent, valid. I have already mentioned issues of preservation of the documentation; we know more about Renaissance elites than any other social class. There is also a clear economic factor—if the casting of a horoscope was the subject of a monetary transaction, regular, even daily, consultation with an astrologer may have required a different type of remuneration. This could have taken the form of land, privileges, or a university job. Only princes and nobles could dispense this type of remuneration. I have also mentioned that astrology held high status in the university curriculum; this provided it with legitimacy and a place at court, where all major university disciplines (law, medicine, and even theology) received some form of patronage. (Renaissance patronage

historians tend to avoid astrology

was not limited to art, architecture, and literature—it encompassed all sort of "scientific" disciplines, something that is worth remembering.) The relationship between the court and the university was indeed very close, and many of the figures employed at court at some point or another maintained posts at the *Studium* of Pavia. As the book reveals, the "science of the stars" that was taught at Pavia—which embraced both the "science of movements," or astronomy, and the "science of judgments," or astrology—was clearly applied in the political arena and skillfully used by the dukes of Milan and their entourage for a wide variety of purposes.

There may also be other reasons. There was something eminently attractive in the use of astrological counsel in political activities. It is not a coincidence that astrology rose in prominence in times of crisis, when uncertainty mounted, and decision-making on the basis of the information provided through diplomatic channels faltered. As much as other predictive disciplines such as prophecy, astrology attempted to provide answers when these were most difficult to obtain.[15] As political uncertainty grew in the fifteenth century, so did the circulation of prophecies, astrological prognostications, and accounts of prodigies and omens. Whether it was phrased in the language of religion, in that of celestial influence, or in that of Nature, the explanation was often one and the same: that God's hand was behind the social and political turmoil that characterized much of the fifteenth century and early sixteenth century. As political leaders grew more anxious about their future, therefore, they paid increasing attention to the words of prophets, prophetesses, and astrologers.[16]

One other aspect should be considered, this one in relation to the broader intellectual movement of Renaissance humanism and the discovery of Roman history. The fact that astrology was sought after by emperors and kings in Roman times does not only show its attractiveness among powerful rulers, but may also have been an additional factor that made it attractive to more humanistic-inclined Renaissance princes.[17] This is an aspect that would certainly deserve reflection and further study. Finally, there was the connection with learned medicine. Not all physicians embraced astrology with enthusiasm, but some did, and if they reached a position of prominence within the court, they may have used it to foster their own intellectual agenda and promote astrological medicine.

Astrology at Court:
Applications, Concepts, and Techniques

[margin note: 2 very Popular forms used in court]

It is worth pondering briefly on the ways in which astrology was employed within some Italian courts. Court astrology could take many forms, but two "types" were the most prominent: medical astrology and political astrology. Both are very well represented in Milan. Put briefly, medical astrology represents that area of Renaissance medicine where the influence of the stars accounted for the state of health of one's patient. This was sometimes considered distinct from judicial astrology, which attempted to foretell the destinies of individuals and nations, and was seen instead to be part of natural astrology, which, together with medical astrology (or astrological medicine), included also weather prediction.[18] The theory that the heavenly bodies influenced human affairs and particularly the state of the human body and its four humors had an ancient pedigree. Within the Hippocratic-Galenic tradition, health was believed to be the result of the balance of the four humors (sanguine, phlegmatic, melancholic, and choleric) in an individual, while disease ensued when this balance was lost. The fluctuation of the humors within the body could be ascribed to a number of factors: these included air (and thus climate), food and drink, sleep and vigil, motion and rest, evacuation and repletion, and the "passiones animi"—what was generally called the six non-naturals.[19] Each disease, moreover, was believed to have a lifecycle, and each humor of the body was believed to have its own rhythm that, in a state of illness, culminated in a paroxysm (*paroxismus*).[20] Medieval and Renaissance physicians drew from a rich ancient medical tradition to compile dedicated health treatises (*regimen sanitatis*) that provided advice on how to maintain a state of health and thus prevent illness. In the case of princes and their families, these treatises could also be personalized, making the genre of the *regimen sanitatis* very popular within Renaissance courts.[21] It was also believed, however, that external causes could contribute to throwing the normal balance of the humors out of kilter, and these included celestial influence. Disease could ensue, therefore, *ex radice inferiori,* namely from factors related to the sublunary world (and thus to the six non-naturals), and *ex radice superiori,* namely from the movement of the stars, which were thus defined as "causes" of the illness.[22]

[margin note: 4 humors theory]

Within this framework of celestial and terrestrial correspondences, various theories found their place. One of them was the Hippocratic-Galenic theory of the critical days expounded in Galen's *De diebus criticis,* whereby he elaborated how certain days after the onset of an illness, called *dies indicativi* (days in which the patient will present symptoms that allow the doctor to predict when the crisis would happen), could be used to formulate a prognosis of death or recovery for the patient.[23] This was based largely on the regular motions of the two luminaries, the Moon and the Sun (which were seen as both causes and signs), and involved a set of calculations used to establish the progression of a patient's illness.[24] In particular, the chartable cycle of the Moon through the twelve signs of the zodiac (which the Moon takes roughly twenty-eight days to complete) was seen as considerably important in determining the course of a patient's illness.[25]

Similar principles, and particularly that of the Moon's movement through the various signs of the zodiac, also guided the kind of astrological medicine elucidated in another text seemingly compiled in late antiquity, the ps.-Hippocratic *Astronomia* or *Astrologia Ypocratis,* which had large circulation from the fourteenth to the sixteenth century.[26] Moreover, the theory of lunar phases at the core of medical astrology often intersected with the theory of fevers, one of the most common "signs" of illness on which medieval physicians based their diagnosis and prognosis.[27] A clear example of this is offered by the anonymous author of the *Astronomia Ypocratis,* who explained how "when illness betakes anyone and the Moon is in Aries with Mars and the Sun, the patient will suffer a malady of the head because of the lack of heat [. . .] yet he will have hot fevers which will not subside."[28] The suggested treatment generally involved bloodletting and purging. Much like purging, drawing blood was used to restore the balance of the four humors in the patient's body, but this practice, too, was often closely linked to planetary movements and to the ancient theory of *melothesia,* whereby different parts of the body were associated with different signs of the zodiac and planets.[29] Accordingly, bloodletting was often practiced following a set of rules related to the position of the planets (particularly the Moon) in the various signs of the zodiac. The rules were simply formulated and easy to memorize. On bloodletting, for instance, the author of the extremely popular *Fasiculo de medicina,* a Renaissance medical work attributed to Johannes de Ketham, addressed the reader as follows:

If you want to know when it is a good time to draw blood [. . .] first follow the general rules, and first and foremost, that at the time of new and full Moon it is not helpful to draw blood even if the Moon is placed in a

favorable sign. Avoid also cutting any member of the body with iron when
the Moon is in the sign that governs that part of the body. When the Moon
is in an airy or fiery sign, however, the operation is more effective than if it
were in an earthy or watery sign. Moreover, the young need to have their
blood let when the Moon is rising, while the old when it is waning. And
again, in spring and summer bloodletting should be performed on the
right part of the body, while in autumn and winter on the left side.[30]

[margin note: rules for 'blood letting']

These general instructions were followed by more precise ones: the medical
practitioner who wanted to let a patient's blood was to look at the sign in
which the Moon was at the time, and if at all possible, also at the Moon's relation to other planets and the ascendant (in other words, the practitioner had
to consider the Moon's "aspects") at the time when the patient fell ill before
going ahead with the operation. Accordingly, the patient could have their
blood let when the Moon was in Taurus, Cancer, Libra, or Sagittarius, but not
when it was in Scorpio or Capricorn; and never if the Moon was conjunct with
or in a negative aspect to Saturn or Mars, even if placed in a favorable sign.
Similarly, the author advised against treating diseases of the neck, eyes, throat,
and nails when the Moon was in Taurus; the shoulders, arms, and nails when
in Gemini; the chest, lungs, and bile in Cancer; the stomach, heart, chest,
hips, liver, and intestines in Leo; the belly and all internal organs in Virgo; the
intestines, kidneys, vesica, and all other members down to the genital organs
in Libra; the genital organs and anus in Scorpio; the thighs, legs, and joints in
Sagittarius; the knees and nerves in Capricorn; the legs and everything else
down to the ankles in Aquarius; and the feet in Pisces. Finally, a last set of
instructions gave specific days of the month in which to avoid or practice
phlebotomy.[31] To help memorizing these basic principles, these two theories—
melothesia and phlebotomy—were often visualized in both manuscript and
print in the shape of a man, with the various signs of the zodiac superimposed
on his body in the case of the zodiac man (*homo signorum*), with rubrics connected to specific points on the body in the case of the vein man (*homo
venorum*) (Figures 2 and 3). That the two images and the principles that guided
them were seen as connected can be evinced also by the close proximity of the
two figures and the texts that accompany them in the *Fasiculo*.

Within medical astrology, diagnosis, prognosis, and therapy were thus formulated on the basis of a number of elements: these could include the chart
that was cast at the moment when the patient fell ill (his decumbiture), the

[margin note: reliance on positioning & Moons & Planets]

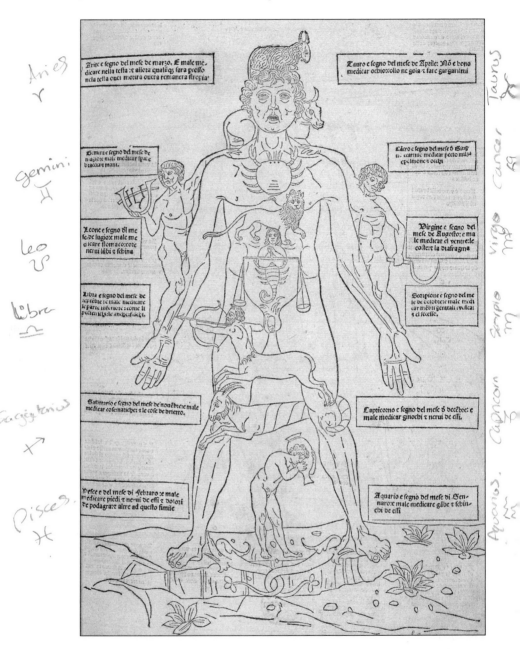

FIGURE 2. *Homo signorum,* or zodiac man, from Johannes Ketham, *Fasciculo de medicina in vulgare* (Venice: Giovanni and Gregorio de Gregori da Forlì, February 5, 1493/94). By kind permission of the British Library, London.

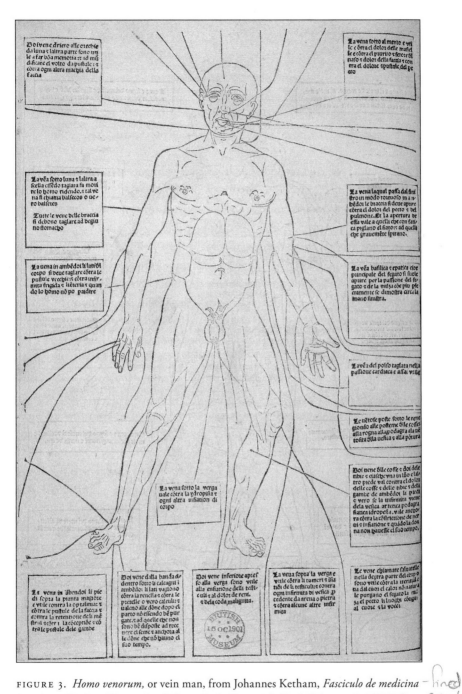

FIGURE 3. *Homo venorum,* or vein man, from Johannes Ketham, *Fasciculo de medicina in vulgare* (Venice: Giovanni and Gregorio de Gregori da Forlì, February 5, 1493/94). By kind permission of the British Library, London.

patient's birth chart (his geniture or nativity), the position of the stars at various stages in the disease (these could be determined quickly by consulting astronomical tables), and the astrological prediction of the patient's urine without examining it (and, therefore, called "unseen"). Prognosis and treatment depended on where certain planets, and particularly the Moon, were situated at a certain point in the illness.

Although medical astrology had both its proponents and opponents, there is no doubt that it played an important role within the curricula of Italian Renaissance universities, seemingly growing in influence during the fourteenth and fifteenth centuries.[32] Did most Italian fifteenth-century physicians abide by the principles of astrological medicine? Giving a confident answer to this question in the present state of research is difficult.[33] It seems that some did and others did not, but a clear pattern has yet to emerge. We know that Galen's works were part of the university curriculum of Italian universities and an abbreviation of his *De diebus criticis* was read by Giovanni Battista Boerio, later physician of Kings Henry VII and Henry VIII of England, while he was a student at Pavia.[34] We also know that this Galenic theory of the critical days was employed by those court physicians who attended to the care of Gian Galeazzo Sforza, sixth Duke of Milan, during his lengthy illness, and that a comparison between the malefic configuration of his natal chart and the planetary positions at the moment of his death had convinced the physicians that his death was inevitable.[35]

While we are fortunate to possess extensive correspondence regarding Gian Galeazzo's illness and death, few medical historians have dedicated their efforts to investigating the role of physicians at the bedsides of fifteenth-century Italian rulers and their families through the diplomatic correspondence housed in Italian archives.[36] When compared with the few other documented instances of illness that occurred at the Sforza court, the picture becomes complicated and blurry, since astrological medicine is not often explicitly adduced as the medical theory behind the treatment of other Sforza family members.[37] One might speculate, therefore, that during the 1480s and 1490s medical astrology progressively rose to higher prominence within the practice of courtly medicine.[38] One of the reasons could be that this corresponded to a similar rise in prominence of astrology as a political discipline within the court of Ludovico Maria Sforza, called il Moro, then *de facto* ruler of Milan in place of his nephew Gian Galeazzo. As Ludovico himself was miraculously saved from death by the physician-astrologer Ambrogio Varesi da Rosate,[39] he may have privileged physicians well versed in astrological medicine within his court. More research about Renaissance court medicine would be necessary, however, to establish possible trends and fashions within the medical establishment of Renaissance courts.

In its turn, the term "political astrology" encompasses a diverse set of practices that rulers and other political figures used to advance their political agendas. This application of astrology to public life could take many forms: astrology could be employed to construct narratives of legitimacy, to predict the death of one's enemies, to sign alliances, and to forge dynastic marriages. A variety of astrological techniques were used to put astrology at the service of politics. These generally included casting natal horoscopes, but also making astrological interrogations, elections, and revolutions, namely short-term and medium-term prognostications often related to specific actions or contexts that would help the ruler make political decisions. All of these techniques required the casting and interpretation of celestial figures.

Drawing a celestial chart is certainly the first step in the astrologer's art. Natal charts were the staple of most astrologers' practices, and the casting and interpretation of natal charts—or their annual revolutions—was most certainly part of a court astrologer's life. During their lifetime most princes and members of the Renaissance elites would either request their horoscopes (or those of their family members) to be cast or receive them as gifts. These sorts of gifts could either come from friends as tokens of *amicizia,* or, more often, from prospective astrologers looking for patronage.[40] Either way, genitures had an important function: while at the personal level they could provide the client with self-knowledge and personal guidance, at the public level they could function either as self-aggrandizing advertisement or as forms of political propaganda, both positive and negative. In the case of a leader, this personal information was often considered politically sensitive: genitures could exalt the qualities of a leader, but also reveal weaknesses or predispositions that could be exploited by unscrupulous enemies to their advantage. For this reason, a leader's horoscope possessed considerable political value, and its interpretation could be regarded as a confidential matter. In cases in which the horoscope was particularly positive, however, it could be used to one's advantage, often in the form of political propaganda.[41]

As the births of the heirs of princes and rulers were often public events accompanied by lavish display and celebrations, the information requested to cast a horoscope was often of public knowledge (although the all-essential time of birth was sometimes known only to a few). The casting of the horoscope and its interpretation, however, required specialized competence. For this and other similar tasks, therefore, fifteenth- and sixteenth-century rulers resorted to one or more astrologers. By the fifteenth century, a number of astrologers started collecting famous and less famous genitures in great numbers. The existence of these collections incontestably indicates that the horoscopes of the European elites—political and military rulers, men of letters and intellectuals, popes and cardinals, wives and children of powerful men—were often of public domain.[42] Astrologers collected these charts for various reasons, but one was almost certainly that of creating a numerically significant sample of charts that could be scrutinized against the life events of the people in question. There was nothing better than the horoscope of a public figure to verify the true nature of astrology's predictive value. This could be done using either a natal chart or the annual revolution of the same (and sometimes comparing the two). With their lives in the public eye, the life events of political leaders could be constantly compared to their charts, and this allowed the astrologer to test his own prediction against real life and explain further certain events that had possibly escaped his attention (or, alternatively, explain away things that he had predicted incorrectly).[43] In the sixteenth century some of these collections were published as self-contained works. The most famous of these collections are certainly those of Girolamo Cardano and Luca Gaurico, but these are by no means the only ones extant. To consider only those which are most relevant to this study, those of the Sforza family are best known in their printed version in Girolamo Cardano's *Liber de exemplis centum geniturarum,* but had been circulating in manuscript since the fourteenth century.[44]

Astrological interrogations and elections, instead, were used to choose a person's course of action. Interrogations could address politically sensitive questions such as the likely death of an enemy or the birth of an enemy's heir.[45] Elections, on the other hand, were used to decide when it was best to commence an action and for this reason could include choosing the best moment to marry or travel, to nominate the general of an army, to make one's most faithful men swear allegiance, or to engage in war.[46]

We often seem to know more about the clients' needs than we do about the astrologers' motives in writing their judgments and prognostications. It is clear why rulers and Renaissance elites may have wanted to resort to astrology as a form of political counsel. It is not always clear, however, why astrologers wrote what they wrote, what they hoped to obtain, and what kind of pressure they were under to oblige their lords. As a number of examples in this book will reveal, there were times when a court astrologer could provide a dispassionate judgment of a client's nativity or an interrogation, but there were also times when such an exercise was colored by expectations and the astrologer may have felt the need to shift the emphasis of his judgment on one aspect of the chart rather than another to please his lord. Without this implying that the astrologer was fraudulent, we have to admit the possibility that not all advice was dispassionate and disinterested. Whatever the case, this does not change the fact that astrologers were taken seriously and considered valuable. Likewise, this does not diminish in any way, I would argue, the influence that these men may have exerted on their lords' political decisions.

Court Professions: A Note on Terminology

Cohorts of physicians swarmed the Sforza court at times of illness. Considering the importance of the health of a ruler for the stability of his state, it is not difficult to imagine why Renaissance ruling elites surrounded themselves with not just one doctor, but a team of doctors.[47] In the documents of the time, these professional figures are generally referred to with the term "magister." When signing their letters, moreover, the term "physicus" or "doctor artium et medicine" often followed their names. Unlike those practitioners who offered only political advice to the Sforza dukes with their astrological interrogations, elections, or prognostications, these medical professionals never define themselves with the term "astrologus." In this book, therefore, I have chosen to use the following three terms to define these professional categories and their areas of competence: I shall use *physician* or *doctor* to indicate those professional figures who provided healthcare to the Sforza dukes and their families. Even in those instances where astrological medicine is applied, I will still refer to these professional figures with the term that indicates their primary role of healthcare practitioner. I shall reserve the term *astrologer* for those professionals who provided astrological consultations to the dukes that did not relate immediately to their health or someone else's. Some of these figures, as we shall see, were both astrologers and astronomers, others both astrologers and physicians. In contrast, I shall adopt the term *physician-astrologer* to refer to those professionals who claimed expertise in both of these areas and firmly operated at court both as physicians and as astrologers, sometimes by using astrological medicine, sometimes by offering astrological counsel of a political nature.

These distinctions are, of course, somewhat artificial. The Dominican preacher and historian Annius of Viterbo (1432?–1502), for instance, offered to Galeazzo Maria Sforza, fifth Duke of Milan, both a series of annual prognostications and an interrogation. Should we call him an astrologer? He was certainly considered one by the French Renaissance physician Simon de Phares, and yet historians nowadays remember him chiefly as a preacher and a masterful forger.[48] As this example shows, Renaissance professional boundaries do not always closely fit the historical characters with which this book is concerned. Many of these figures, given their high social and professional status, operated in multiple intellectual spheres and occupied different social roles that included, for instance, holding positions within the Privy Council, the major organ of government at the Sforza court, or being invested with diplomatic missions. Since it is important to appreciate the complexities of the various personalities involved, I have striven to provide an indicative "label" to describe each historical character's main area of competence as it appears from the surviving documents.

As noted before, to date there is no book dedicated exclusively to the practice of astrology within an Italian Renaissance court. In particular, little attention has been paid to the political dimension of this "predictive art." This book attempts to remedy this situation by looking at one single case, that of the Sforza court in Milan. While also extending my research to other Italian archives in Mantua, Modena, and Florence, I have preferred to maintain a focused narrative on Milan. The richness of the material and the complexities of its political and social significance, I believe, require a dedicated study. Yet, in recounting the story of the Sforza's use of astrology at court, I have come across documents that pertain to other Northern Italian courts—especially that of the Gonzaga in Mantua—that lead me to believe that there is great merit in casting the net wider in future investigations and in attempting a comparative approach between Milan and other Italian courts.[49]

This book is about the ways in which medicine, politics, weddings, and wars found a common interpreter in the astrologer at the court of Milan. It opens with an exploration of how astrology was imparted in the *Studium* of Pavia and circulated in the Duchy of Milan. In the absence of firm evidence of what the curriculum at Pavia may have looked like, I attempted to determine what kind of astrological texts were popular in the second half of the fifteenth century. I looked at one particular student notebook in some detail and suggested that many more texts than those listed in the Bologna curriculum—many by Arabic and medieval authors—may have been imparted to students and read by university professors. The rest of the book is made up of a series of case studies, or microhistories, of how astrology was employed by the Sforza dukes. My first case study (Chapter 2) focuses on the newly elected Duke of Milan, Francesco Sforza (1401–1466), and his wife, Bianca Maria Visconti (1425–1468). In this chapter I illustrate the rich variety of astrological counselors who knocked at Francesco Sforza's door offering their services. This chapter illustrates how, as heir to the Visconti, the Sforza were naturally perceived as patrons of astrology and how Bianca Maria herself used her influence to support one of her father's favorite astrologers. Chapter 3 moves on to consider how astrology was put to use by Galeazzo Maria Sforza (1444–1476), Francesco's eldest son. Galeazzo was an avid collector of astrological intelligence, and he used it on more than one occasion to guide him in his political praxis. Astrology was also used against him, in this case to announce his death across the breadth of Northern Italy by means of astrological prognostications. Chapter 4 moves in quite a different direction, looking at how astrology was used in medicine. Gian Galeazzo Sforza (1469–1494) represents an ideal case study of how astrological medicine was employed to treat him but also to justify his untimely death. Finally, Chapter 5 centers on the figure of the seventh Duke of Milan, Ludovico Maria Sforza (1452–1508), the keenest consumer of astrological counsel among the Sforza. Ludovico resorted to astrology on a daily basis, planning trips, weddings, the appointment of generals, and the entry into battles with the use of astrology. From 1487 he started favoring one astrologer-physician among all others, Ambrogio Varesi da Rosate. Varesi's story of ascent and fall can be mapped onto that of his lord.

The escalating uncertainty of the times was clearly a key factor in making astrology so relevant to Italian politics and allowing it to take center stage in political decision-making under Ludovico. Navigating the murky waters of Renaissance political diplomacy was never easy, even less so at the end of the fifteenth century. Ludovico Maria Sforza, with his ill-veiled double game, must be held at least partly responsible for that. Bigger forces were involved in his dangerous game, and Ludovico's personal ambitions may have rendered him too blind to allow him to face them. A number of factors undid the Duchy of Milan during the last quarter of the fifteenth century, acting as prelude to the crisis of the Italian Wars of the following century. One of them, I would argue, was Ludovico's over-reliance on the advice of his astrologers, and Varesi's in particular. The political legacy of the Sforza rulers, therefore, cannot be properly assessed without taking astrology into consideration.

I

✦

The Science of the Stars

Learning Astrology at the University of Pavia

In the dramatic days that followed the occupation of Milan by the troops of the King of France, Charles VIII of Valois, in 1499, Leonardo da Vinci, who had served Ludovico Maria Sforza for over twenty years, left the Lombard capital with all of his belongings. The year that followed saw the Florentine artist travel across the peninsula in search of more stable employment: first to Mantua, then to Venice, finally to Florence. Leonardo, it was reported, was neglecting painting in order to dedicate himself almost obsessively to geometry.[1] It may not be surprising, therefore, to discover two copies of Euclid's *Elementa* (one in the vernacular) among the books listed in one of the artist's library inventories tentatively dated to 1503–1504.[2] It is clear that Leonardo was still under the influence of his friend and teacher Luca Pacioli (c. 1445–1517), the Franciscan monk who had been invited to Milan by Duke Ludovico Maria Sforza to teach mathematics and who had imparted Leonardo lessons in Euclidean geometry. Leonardo's list of books, however, offers some genuine surprises. Together with the two copies of Euclid's *Elementa*—a text that, as we shall see, was the staple of the mathematical curriculum of the degree in arts and medicine at most Italian universities—Leonardo also possessed a book entitled *Sphera mundi* (most likely a printed edition of Sacrobosco's

Sphaera),[3] a copy of Regiomontanus's *Kalendar,* an unidentified book on the quadrant, a text by Albumasar (most likely either the *Introductorium maius* or *De magnis coniunctionibus*), and a vernacular copy of Alcabitius's *Introductorius* that he had borrowed from the Florentine astronomer-astrologer Francesco Serigatti.[4] While the first three books fell into the broad remit of geometry and astronomy, the other two belonged unequivocally to the field of judicial astrology, this despite the fact that on at least one occasion Leonardo expressed his reservations about the validity of the latter.[5] To these works we could also add two copies of a *Chiromantia,* and a copy of Michael Scot's *Physiognomia,* two texts that, as we shall see in the following pages, appeared often in the collections of students of medicine and astrology and were often present in fifteenth-century Italian scientific collections.[6]

In the context of this book it is not necessary to establish whether Leonardo believed in astrology, or indeed whether his apparent aversion toward judicial astrology was more simply motivated by his dislike of Ludovico's chief astrologer, Ambrogio Varesi da Rosate, as one scholar has suggested.[7] Leonardo certainly knew Varesi, as both were present at a "scientific duel" staged by Ludovico Maria Sforza in February 1498.[8] In the *Divina proportione,* where Pacioli describes the event, the Franciscan mathematician praised Varesi as an "expert investigator of the celestial bodies and interpreter of future events."[9] In the same work, however, he argued that geometry was superior to astrology and astronomy, much like Leonardo did in his *Paragone.*[10] As I have argued elsewhere, disciplinary rivalry was rife at the Sforza court, and both Pacioli and Leonardo may have tried to advance their social status at court through debate and writing.[11] That said, there is no way of telling whether Leonardo's objections to judicial astrology were motivated by his personal dislike of Varesi or by his dislike of the discipline more generally. It is possible that, like some contemporaries, Leonardo saw in Varesi the person responsible for Ludovico's fall in 1499, but if this was the case, the evidence does not allow us to prove it conclusively. What we can say, however, is that Leonardo's scientific collection bears some revealing similarities with that of many of his contemporaries who would have studied for a degree in arts and medicine at a university like Pavia. As this chapter will illustrate, the *Sphaera* of Sacrobosco and a text on the quadrant were two works that were taught regularly in courses of spherical astronomy at Italian universities, while Alcabitius's *Introductorius* and the works of Albumasar were regular readings for students in astrology. This is in itself relevant, as it highlights how works of spherical astronomy as well as judicial astrology were reasonably affordable and had wide circulation among learned men in Milan.

The purpose of this chapter is therefore to reconstruct, to the degree possible, the *curriculum studiorum* of students of astrology at the University of Pavia, the *Studium* of the Duchy of Milan. What books of astronomy and astrology were most commonly studied at university? In a seminal study on the teaching of astronomy, Olaf Pedersen located a series of texts that were taught regularly at late-medieval universities. He called this group of texts a *corpus astronomicum* and argued that this *corpus* first developed when Sacrobosco himself taught in Paris and was expanded later to contain other astronomical treatises.[12] Can we similarly locate a *corpus* of astrological texts— a Renaissance *corpus astrologicum*—that constituted the staple of fifteenth-century teaching in astrology at Italian universities? Can we establish what kind of texts were taught at Pavia? The task is not easy—we do not possess official documents for Pavia—but the existing evidence goes some way in suggesting what Pavian students of astrology may have read. Our knowledge of astrological manuscripts, their contents, provenance, and ownership, is still very limited. Scores of them remain unstudied in Italian and other libraries. How many of these were student notebooks, for instance, is impossible to say. How many had Lombard origin even more so. My attempt to establish what texts were likely to be read in Pavia, therefore, is *par force* summary and imperfect. It is just the start of a process that will require the collaborative and cumulative work of many more historians and philologists.[13]

While a full study of university teaching in arts and medicine at Pavia goes beyond the scope of this work, the *Studium* of Pavia was the most common choice for Lombard students wanting to pursue a career as physicians and astrologers in the fifteenth century, and for this reason determining to the degree possible what kind of astronomy and astrology was studied in Pavia is clearly relevant to the study of Sforza court astrology with which this book is concerned. Not only did many of the physicians and astrologers associated with the Sforza court study at Pavia, but many of them also taught there. The contacts between the university and the court, therefore, were considerable, and the boundaries between court and university often porous. The appointment of university professors at Pavia was officially administered by the Sforza Privy Council, the core organ of Sforza administration, but was often heavily influenced by the Duke of Milan himself.[14]

The early history of Italian universities is notoriously poorly documented and Pavia is no exception: little evidence exists about the way in which teaching was imparted and what texts were studied. The documentation for Pavia, unsurprisingly, has some serious lacunae: the archives of the college of arts and medicine, which must have existed at some point, have not been preserved, and for this reason we similarly lack matriculation lists for the period under consideration. Furthermore, only some of the rolls, or *rotuli,* namely the lists of professors teaching at Pavia (together with their disciplines), have come down to us.[15] The best sources of information, therefore, are the graduation papers of those students who completed their courses at Pavia. These documents include the name of the student, the person who "promoted" him for the degree, the names of other representatives of the university, and those of other participants at exams, including other students.[16] As Paul Grendler highlighted, however, not all of the promoters also taught at the university, and this makes it harder to establish who held a chair and who did not.[17] These documents are complemented by correspondence regarding the *Studium* among the Sforza papers at the Archivio di Stato in Milan. This correspondence is quite varied, ranging from salary payments to issues concerning students. Very little, however, is said about the kind of texts that were taught. Even in the case where official information about the *curriculum studiorum* exists, moreover, the list of texts indicated should only be taken as prescriptive, and is probably only partly representative of what was studied at Italian *Studia* from year to year. As Nancy Siraisi has argued in her exemplary study of the University of Padua, a sounder approach is one based on a reconstruction of the interests of those associated, whether directly or indirectly, with the various universities.[18] Such work remains to be done for Pavia. Answering the questions outlined above, therefore, requires a multidirectional approach.

A first modest attempt in this direction is provided in the following pages, which try to trace the most important aspects of university teaching of astronomy and astrology in Pavia during the Sforza period. The evidence to be presented here comes from a variety of documents: given the dearth of information as to what was taught at Pavia, the personal inventories of university professors who taught there are particularly valuable. Likewise, the libraries of Milanese physicians and other members of the elite may provide an indication of which astrology and astronomy books circulated in Milan in manuscript or print. Some of these books, as we shall see, were deemed essential to the study of *astrologia.* To these one should add the library of the Sforza family housed in Pavia, which served in many ways as a reference library for courtiers and university professors alike. Other scant evidence comes from the astrology and astronomy books printed in the Duchy of Milan, but this evidence should not be taken as fully representative of what circulated in print in Milan as we know that the Milanese elite had easy access to books printed in other Italian cities, particularly Venice, which by the 1480s had come to dominate the Italian printing industry (and, indeed, for a time, the European printing industry as well).[19]

Unfortunately, I have thus far been unable to trace many astrology and astronomy manuscripts that could be firmly attributed to Pavian professors or students. The lamentable dispersion of this kind of material and the scant information included in most library catalogs about any notes of ownership or provenance makes such a task particularly arduous. Yet, as we shall see, the one actual Pavian manuscript I have located provides much insight into the interests and type of knowledge available to one particular fifteenth-century Pavian student and allows us to make more general statements as to the kinds of books that may have been read by other fifteenth-century Pavian students.[20]

The Astrologer's World: a *Corpus Astrologicum*?

By the second half of the thirteenth century the study of astronomy and astrology occupied a firm place in the curricula of Italian universities.[21] There was not, however, a degree in astronomy/astrology proper. Rather, astronomy and astrology were part of the training imparted to students in arts and medicine at most European universities as astrology was closely linked to medical theory and practice.[22] The fact that the two Latin terms *astrologia* and *astronomia* were used interchangeably in the period with which this book is concerned (as indeed in an earlier period) indicates that the two discplines were not thought to be different and irreconcilable, but part of the same realm of knowledge concerned with celestial motion and its effects. Then as now, people were able to distinguish what constituted astronomy and astrology, and there is ample evidence that the two disciplines were not considered to be equivalent. What is most striking to a modern reader, however, is the fact that astronomy (i.e., the study of celestial motion) was often seen as propaedeutic and subservient to astrology and astrological medicine, which focused on the predictable effects of this motion on Earth and the human body, and that for this reason, at least initially, greater emphasis and importance was given to astrology over astronomy.[23] Together with the study of the movements of the planets in the sky, therefore, students also learned about their effects on Earth. Studied within three distinctive scientific disciplines—mathematics, natural philosophy, and medicine—that formed the traditional four-year cycle of academic studies, astrology was, therefore, at the core of the arts and medicine degree.[24] Most Italian universities had one to two professors of astrology/astronomy: often at least one of them also imparted some other teaching in medicine, be it theoretical or more often practical, thus confirming the strong link between astrology and medicine in the period.[25]

In the first half of the fifteenth century, the *Studium* of Pavia followed this trend, with the appointment of one, often two, and rarely three professors of astrology within the faculty of arts and medicine.[26] The trend is confirmed also for the second half of the century, when at least one professor was employed.[27] As noted earlier, the statutes of the University of Pavia are no longer extant; for this reason we can only make conjectures as to what exactly was studied there. Given the high mobility of both students and professors in the Italian Renaissance, we can assume, however, that it was not remarkably different from what was imparted at other Italian universities, especially when it came to the foundational texts of the first and second year.[28] The most complete evidence of an Italian *curriculum studiorum* in the "science of the stars" is that provided by the University Statutes of Bologna in 1405.[29] According to these statutes, during their course of studies in arts and medicine students were instructed in Euclidian mathematics, Ptolemaic spherical astronomy, and classical as well as Arabic astrology. Expertise in arithmetic and geometry served equally for astronomical measurements and for astrological and astro-medical calculations, and was imparted in the first year of studies through the reading of such texts as the *Algorismus* of Johannes de Sacrobosco (or possibly the ps.-Sacrobosco's *Algorismus de minutiis*) and Euclid's *Elementa geometriae* (in Bologna's case, in the version with Campanus of Novara's commentary). Then came instruction on how to use astronomical tables, which in the fifteenth century meant essentially the Alphonsine tables and their canons by John of Saxony,[30] followed by basic planetary theory, which was studied through the popular textbook *Theorica planetarum*.[31]

The second year progressed in much the same fashion with the reading of the second book of Euclid's *Elementa,* the study of Jean de Linières's canons on the Alphonsine tables, and Sacrobosco's *Sphaera* for further instruction in spherical astronomy.[32] To this was added instruction on how to construct and use an astrolabe. The text indicated in the statutes is Messahallah's *De astro-labio,* but it is possible that others were adopted as well.[33] Astrology started to feature more prominently in the third year of study: together with the third book of Euclid's *Elementa* and a treatise on the quadrant that explained the construction and use of another important astronomical instrument, students started to familiarize themselves with the basic principles of astrology by reading Alcabitius's *Introductorius ad iudicia astrorum* (also simply called *Introductorius*), as well as ps.-Ptolemy's *Centiloquium,* with Haly's commentary.[34] As we have seen, together with Euclid's *Elementa,* Leonardo possessed both a copy of Alcabitius's *Introductorius* and one of a treatise on the quadrant, two texts that were deemed appropriate for advanced students in arts and medicine. The fourth and final year saw the introduction of medical astrology proper with the teaching of William of England's *De urina non visa,* a text that imparted rules for the examination of urine, not by inspecting it, but simply by casting a horoscope instead.[35] This was accompanied by the study of Ptolemy's *Quadripartitum* (*Tetrabiblos*) and the third book of the *Almagest.* More medical astrology was imparted via the medical curriculum proper, with the study of Galen's *De diebus creticis* (or, as we shall see, possibly its abridged medieval version) in the first three years of the course of study in theoretical medicine.[36]

Paul Grendler's survey of the teaching of astrology and astronomy at all the major Italian universities confirms that Euclid, Sacrobosco, and the *Theorica planetarum* were statutory texts taught regularly in Renaissance universities up to the sixteenth century. These texts were at the core of what Olaf Pedersen called the *corpus astronomicum* of late-medieval universities, which first developed when Sacrobosco himself taught in Paris and later grew to contain other astronomical treatises.[37] Of course innovation happened, as with the introduction of the *Theoricae novae planetarum* of the Vienna astronomer Georg Peurbach or the *Sphaera* of the French astronomer Oronce Finé at the University of Pisa, a clear sign that universities absorbed new texts and theories and provided innovative teaching.[38] Moreover, even when the texts were the same, how knowledge was imparted to the students could differ dramatically. This is best exemplified by examining Lynn Thorndike's edition of the commentaries of Sacrobosco's *Sphaera,* which reveals how the medieval professor Cecco d'Ascoli could impart a great deal of astrology while purportedly teaching spherical astronomy. Now as much as then, therefore, we can assume that university professors' personal interests and preferences shaped the curriculum at least to some degree.[39]

Can we also speak of a *corpus astrologicum,* then? Unfortunately, much less is known about the teaching of proper astrological texts: given the extreme popularity of Ptolemy's *Quadripartium,* ps.-Ptolemy's *Centiloquium,* and the *Centiloquium Hermetis* in manuscript and print—the first the key astrological text of antiquity, the others two very popular texts of astrological aphorisms— we can be reasonably sure that these texts constituted core parts of the curriculum. But a history of the teaching of astrology at Italian universities remains largely to be written.[40] Crucial evidence of the type of texts that could have been studied at Pavia, however, is provided by a single manuscript that once belonged to one of its students, the Genoese physician-astrologer Giovanni Battista Boerio. Since the author of this manuscript, which is now at the British Library, has not been previously identified, this invaluable source has remained almost completely untapped by historians.[41] While admittedly being only one example of the kind of texts read by Pavian students, this manuscript is significant in more ways than one. First, it includes some texts that we also find in the Bolognese curriculum, such as Johannes de Lineris's *Canons* on the Alphonsine tables (*Canones primi mobilis Johannis de Lineriis*) and William of England's *De urina non visa* (*De urina non visa et de concordia astrologiae et medicinae et caeterae*). Secondly, and possibly of greater interest, it includes a number of texts that do not appear in the university statutes.[42] These other texts seem to fall into three categories: 1) texts that appeared in print in the years immediately following Boerio's transcription of them (thus suggesting that they may have become must reads for aspiring astrologers); 2) texts relevant to Boerio and his contemporaries because they treated celestial phenomena occurring around the time when he was writing; and, finally, 3) ephemeral works such as prognostications, recipes, and horoscopes that were linked more intimately with Boerio's own life and practice.

But who was Giovanni Battista Boerio, and what did he study at Pavia? A fascinating insight into the life of this Renaissance astrologer is provided by his own notes in the notebook. At various points in the manuscript, our Pavian student left notes that tell us something more about his background and interests: we know, for instance, that he was a student of arts and medicine who audited courses in astrology, he was the son of a physician from Taggia (a small town near Genoa) called Bartolomeo who died in 1483, and that on August 5,

1484, he finished transcribing sections of Sacrobosco's *Computus* about the phases of the Moon.[43] Boerio's transcription was accurate and professional: not only did he copy the relevant passage from Sacrobosco on the illumination of the Moon, but also the accompanying diagrams he must have found in his original source (Figures 4 and 5).[44] A few weeks later, moreover, he diligently transcribed the text of ps.-Aristotle's *Chiromantia,* thus showing the breadth of interests that may have characterized fifteenth-century scientific learning.[45] (A copy of this or another text of the *Chiromantia,* as noted, was also part of Leonardo da Vinci's library.) As indicated by a note at the end of the text, at this time Boerio was not in Pavia. Indeed, he had been forced into temporary exile by the plague epidemic that hit the university town in 1484. In this note he defines himself as an arts and medicine student taking courses in natural philosophy and astrology, adding that he was from Taggia but that he resided in Valenza, near Alessandria, presumably because of the plague.[46] A very similar note also accompanies the text of Messahallah's *De revolutionibus annorum mundi,* which Boerio finished transcribing just after the *Chiromantia,* on September 24, 1484,[47] and another gloss of this type follows the annual prognostication authored by the ducal physician-astrologer Gabriele Pirovano for the year 1484, which, Boerio tells us, he had copied down from the original.[48]

Little else is known of Boerio's student days in Pavia. The Genoese student must have graduated from the university, and for a time he probably returned to his native Taggia to practice. Sometime after 1498, at a time when political and social life in Genoa was severely disrupted by the war between the French king and the Duchy of Milan, Boerio travelled to England, probably with his two brothers, in search of better prospects. By November 1504, we find him employed as chief physician to King Henry VII and later to his son Henry VIII.[49] By then Boerio must have been an experienced physician and astrologer, or it is unlikely he would have attracted the attention of such an impressive patron. His nationality may have made him an attractive candidate for the job. In London he came into contact with a group of English humanists and intellectuals, some of whom, like Thomas Linacre, were Italophiles and had trained in medicine in Italy. Indeed, it was Linacre who later took up Boerio's post as chief physician, taking upon himself the task of raising the status of medicine in England by founding the Royal College of Physicians that still exists today in London.[50] To this group of Italophiles one can add the name of Erasmus of Rotterdam, who took charge of accompanying Boerio's sons during their travels to Italy in the summer of 1506. For reasons

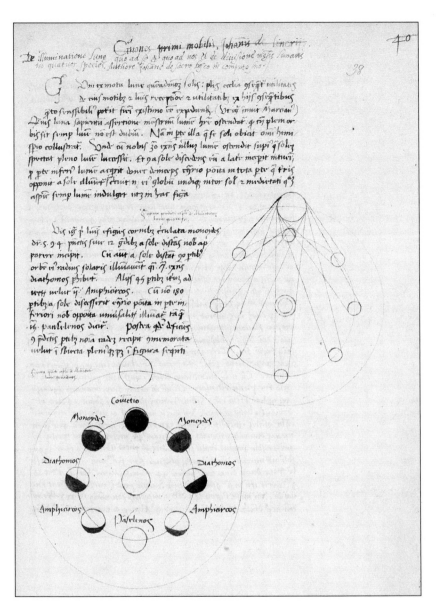

FIGURE 4. Diagram of the phases of the Moon, from Johannes Sacrobosco, *Computus* in British Library, MS Arundel 88, fol. 38r. By kind permission of the British Library, London.

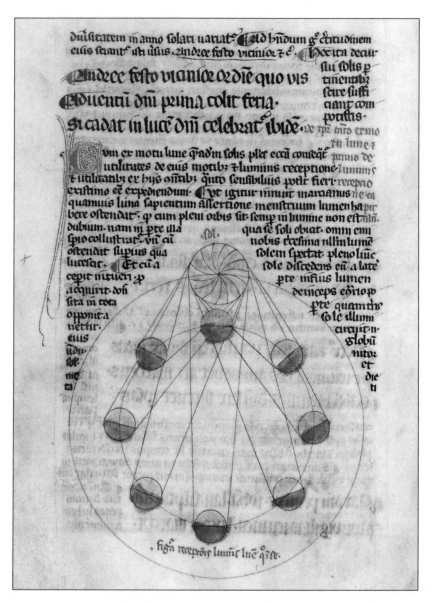

FIGURE 5. Diagram of the phases of the Moon, from Johannes Sacrobosco, *Computus* in Cambridge University Library, MS li.III.3, fol. 46v. By kind permission of the Syndics of Cambridge University Library.

that remain obscure, while Erasmus was in Italy the relationship between him and Boerio deteriorated, and the young Erasmus abandoned his appointment. Nevertheless, their relationship flourished again on Erasmus's return to England, and it is possibly on this occasion that the Dutch humanist dedicated his translation of Lucian's *De astrologia* to his long-time Italian friend. The friendship, however, did not last very long, and, from that moment on, Erasmus's comments to his correspondent William Gonnel regarding Boerio were anything but flattering.[51] Despite his troubled friendship with Erasmus and having now been all but totally forgotten by most historians, Boerio must have been an accomplished practitioner if at his death he could make generous donations to his native town of Taggia.[52]

Boerio's notebook was probably not the only one he composed during his student days, but so far no other manuscript has surfaced that can be attributed to him with any certainty. The manuscript has all the traits of a student notebook: like most students, Boerio was not always methodical and precise in copying his texts, sometimes leaving them unfinished, sometimes interrupting them only to continue a few pages later when there was space. For instance, the copy of Zael's *Quinquaginta praecepta*—a work made up of snappy astrological aphorisms—is broken up by a single sheet that contains an interrogation entitled *Judicium de quaestione quadam*. (Boerio, however, appended a note to the bottom of fol. 94v of the *praecepta* reminding himself of where the text continued.) The question is related to an ecclesiastical benefice. It is not clear, however, if this was a hypothetical question used as a student exercise, a question posed by Boerio himself, or cast for a client.[53] To cite another example, a fragment on the celestial configurations bringing about the plague appears between an anonymous treatise on physiognomy and a series of aphorisms on astrological embryology taken from ps.-Ptolemy's *Centiloquium* and the *Centiloquium Hermetis*, two other extremely popular books of astrological aphorisms.[54]

The manuscript obviously has a very complex textual history. Indeed, with its composite nature, this fascinating notebook raises as many questions as it answers with respect to the complex transmission of astronomical and astrological texts at Italian universities. In many ways, however, this notebook constitutes a precious source of information as it reveals the breadth of interests of fifteenth-century Pavian students and the rich variety of texts available to them. Four sets of considerations seem particularly relevant in the context of university learning: first, the manuscript contains a great deal more Arabic astrology than indicated in the Bolognese curriculum; second, the book contains a great deal of medical astrology; third, it contains a series of shorter anonymous texts (prognostications as well as medical recipes) that can help us understand the type of practical knowledge that was not imparted through lengthy books but that must have circulated among a community of students and court practitioners; fourth, the manuscript contains a series of charts and their interpretations, some of which related to politically sensitive issues such as the production of a healthy male heir.

Medieval and Arabic astrology comprises a substantial element of the manuscript: MS Arundel 88 opens with a chapter entitled *De natura Scorpionis et de Saturni et Jovis in Scorpio coniunctionis significatione* addressing the topic of the great conjunction of Saturn and Jupiter. The conjunction of the superior planets was believed to herald large-scale events on Earth and was therefore particularly worthy of study. The text copied by Boerio has its origin in a medieval text attributed to Johannes of Bruges entitled *De veritate astronomie,* but it is likely that Boerio copied it from an incomplete, anonymous source circulating in the Duchy of Milan at the time.[55] Little is known of the circulation of *De veritate astronomie* in Italy, but an anonymous work with the same incipit and explicit as that of MS Arundel 88 was published in Milan by the presses of Filippo Mantegazza in 1496.[56] The examination of this rare Milanese incunabulum—which circulated with the title *Judicium cum tractatibus planetariis compositum per quendam hominem sanctissimum et prophetam anno Christi 1096,* contained a table of contents, and had a very primitive woodcut of the celestial spheres (Figure 6)—reveals that this was indeed the text copied by Boerio, which eventually found its way to the Milanese publisher Filippo Mantegazza and was printed on December 20, 1496, when Milan was already involved in a disastrous war against France. The publication of this text clearly indicates that the work circulated in the Duchy of Milan, and Boerio's transcription seems to attest to an earlier popularity among Pavian students and professors.[57]

This work draws largely on Albumasar's *De magnis coniunctionibus,* as well as ps.-Ptolomy's *Centiloquium,* Alcabitius's *Introductorius,* and Abraham Ibn Ezra's *Liber de nativitatibus* and *De mundo vel seculo* (respectively, a work of natal astrology and one of conjunctionist astrology) and follows a chronological order where various conjunctions are mentioned and analyzed according to the triplicity in which they appear and the historical events that accompanied them. (In this specific case, Bruges limits his analysis to con-

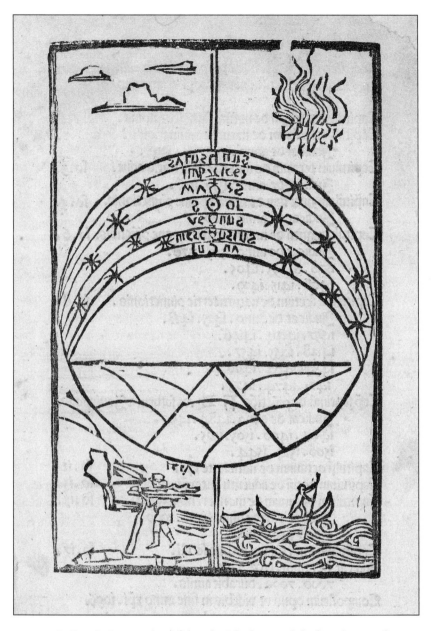

FIGURE 6. Frontispiece woodcut of the celestial spheres and the four elements, from *Judicium cum tractatibus planetariis* (Milan: Filippo Mantegazza, December 20, 1496). By kind permission of the British Library, London. © British Library Board.

junctions in the sign of Scorpio, in the watery triplicity.)[58] A number of Arabic and medieval authors dedicated their works to the treatment of the large-scale effects of planetary conjunctions, and particularly the conjunction of the outer planets Saturn, Jupiter, and Mars in the twelve signs and the four triplicities in which they occurred.[59] These conjunctions, it was believed, heralded the rise and fall of religious sects and dynasties.[60] The conjunction of 1484, the year in which Boerio transcribed this text, for instance, received much attention from astrologers and was later associated with the rise of Protestantism and the birth of Luther.[61] From the *De veritate astronomie,* Boerio transcribed eleven out of the twelve chapters that comprised the original work—exactly those parts of Bruges's work that were later printed in the Milanese incunabulum in 1496 (Figures 7 and 8)—after which point he stopped to transcribe a text of mundane astrology that also addresses the rise and fall of religious sects and dynasties, namely Messahallah's *De revolutionibus annorum mundi.*[62]

It is very likely that Boerio transcribed all he had in front of him and that, therefore, in Northern Italy the *De veritate astronomie* circulated in incomplete form. Be that as it may, Boerio's interest in this passage of the *De veritate astronomie* certainly lay in the fact that it concerned itself with the conjunction of the superior planets Saturn and Jupiter in Scorpio, which was due to occur on November 18, 1484, the year in which Boerio transcribed this passage. This is confirmed by the marginal notes that accompany the text, some of which are autobiographical. On the first page of the manuscript, a note singles out for special consideration the conjunctions of Saturn and Jupiter of 1464 and 1484. A subsequent note next to the discussion of the 1464 conjunction records that Boerio was born on the morning of January 7 of that year, before dawn, in the Genoese town of Taggia, while another one reports how during the 1484 conjunction, the whole of Italy, and Lombardy in particular, had been struck by a plague epidemic.[63]

In a manner not uncommon to late-medieval texts, the *De veritate astronomie* also has an interesting prophetic and apocalyptic tone, which includes a chapter on the arrival of the Antichrist, another on the renewal of the world, one on the end of the world, and—in Boerio's case—an interesting marginal reference to the writings of the French theologian cum astrologer Pierre d'Ailly.[64] The astrological-prophetic nature of this text is further amplified by the fact that in Boerio's notebook this is interspersed with an anonymous sybilline prophecy rich in astrological elements revolving around Lombardy

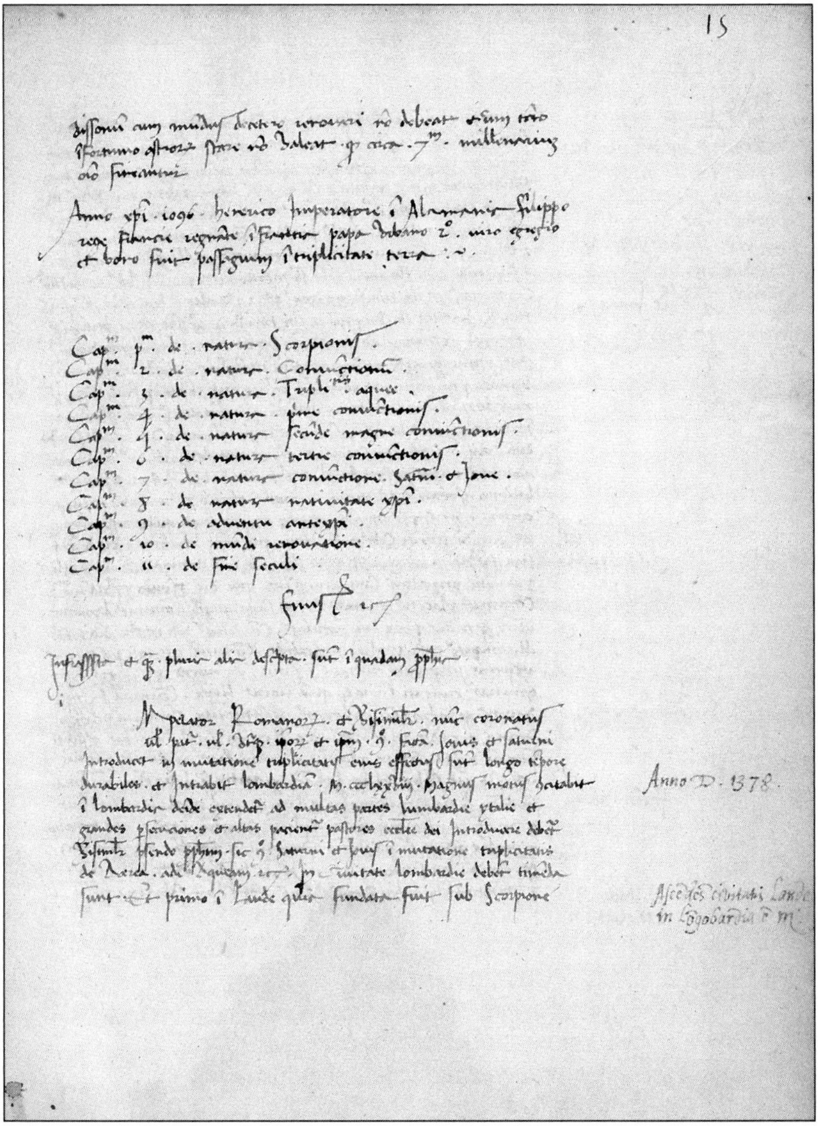

FIGURE 7. Table of contents of Giovanni Battista Boerio's transcription of eleven out of the twelve original chapters of John of Bruges's *De veritate astronomie,* from British Library, MS Arundel 88, fol. 15r. By kind permission of the British Library, London. © British Library Board.

Judicium cum tractatibus planetariis cõpofitu3 per quen
dam hominem fanctiffimum ꝛ pꝛophetam anno xꝓi . 1096 .

Capitulum pꝛimum de natura figni fcoꝛpionis .' a folie . 1 .
Capitulum fecũdum de natura coniunctionis . a fo . 2 .
     Judicat de anno . 1447 . . 1998 .
Capitulum tertium de natura triplicitatis aque . fo . 3 .
     Judicat de anno . 1336 . 1356 .
Capitulum quartum de natura pꝛime ꝯiunctionis . fo . 5 .
     Judicat de anno . 1360 .
Capitulum qntũ de natura fecũde magne ꝯiũctiõis . fo . 6 .
     Judicat de anno . 1365 . 1376 .
     1378 . 85 . 95 . 1405 .
     1406 . 1415 . 1429 .
Capitulum fextum de natura tertie ꝯiunctionis . fo . 8 .
     Judicat de anno . 1425 . 1428 .
     1437 . 1444 . 1446 .
     1448 . 1453 . 1457 .
     1454 . 1464 . 1466
     1473 . 1474 . 1475 .
Cap feptimũ de ꝯiũctiõe . ♄ ♃ . v3 faturni ꝛ Jouis . fo . 10 .
     Judicat de anno . 1484 . 1490 .
     1494 . 1496 . 1503 . 1505 .
     1506 . 1524 . 1544 .
Capitulũ octauum de natiuitate xꝓi . fo . 12
Capitulũ nonũ de aduentu antaxꝓi . fo . 13 .
Capitulum decimum de mondi renouatione . fo . 15 .
     Judicat vfq3 de anno . 1750 . 2000 .
     3500 . 5250 . 7000 .
Capitulum vndecimum de fine feculi . fo . 17 .
Judicat de annis . 1702 . 1762 .
     7000 . 7000 . durabit annis .
Compofitum opus vt videbis in fine anno xꝓi . 1096 .

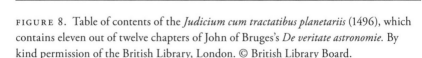

FIGURE 8. Table of contents of the *Judicium cum tractatibus planetariis* (1496), which contains eleven out of twelve chapters of John of Bruges's *De veritate astronomie*. By kind permission of the British Library, London. © British Library Board.

and the Holy Roman Emperor, both texts demonstrating how prophetic writing and conjunctionist astrology were often intermingled.[65] This was hardly unusual in this period, as testified by the great popularity of Johannes Lichtenberger's *Prognosticatio,* certainly the most representative example of how astrology and prophecy could successfully mingle together to create a text rife with political and religious meaning.[66] Our case is no different. The figures of the Antichrist and a Last World Emperor were common to most ps.-Joachite prophecies in circulation in the late Middle Ages; depending on the political leanings of its author, this last figure was either identified with the Holy Roman Emperor (as in our example) or with a French king to be elected emperor by the pope. Within this rich prophetic and astrological tradition, the sybilline prophecy about the Holy Roman Emperor added to the *De veritate astronomie* may have provided, therefore, a fine filo-imperial counterpoint to a series of very popular Francophile prophecies that circulated widely in the decades preceding the French descent into Italy.[67]

After this brief astro-prophetic section, Boerio's manuscript continues with the transcription of Messahallah's *De revolutionibus annorum mundi,* another important text of mundane astrology.[68] Other Arabic sources follow: the *Sententiae Almansoris* and the *Quinquaginta praecepta Zaelis,* two other common sources of astrological aphorisms together with ps.-Ptolemy's *Centiloquium* and the *Centiloquium Hermetis.*[69] Boerio both transcribed and re-read the *Sententiae* and the *Praecepta* carefully, underlining sentences, making brief annotations, and indicating on at least one occasion when the two sources of aphorisms he had copied agreed.[70]

Why did Boerio transcribe these texts? Some of them resonated strongly with the celestial events that were taking place at the time of his exile from Pavia, but others have less immediate associations with planetary conjunctions. It is plausible to imagine how some of them, so relevant to contemporaries' lives, were particularly attractive to students of astrology keen to refine their understanding of the effects of celestial phenomena on Earth; but were they exceptional, or were they a regular feature of university instruction in astrology at the time? In other words, were they part of a fifteenth-century *corpus astrologicum,* albeit probably broader and more varied than the *corpus astronomicum*? Of course, answering this question on the basis of one manuscript is virtually impossible. After all, Boerio may well have been an eclectic student and his readings hardly representative of what fifteenth-century university students learned. The latter hypothesis, however, is immediately cast into doubt when we compare the contents of MS Arundel 88 with the fifteenth-century miscellaneous manuscript of the Paduan university *prototipografo* Bartolomeo Valdizocco studied by Tiziana Pesenti Marangon.[71]

The interest in this manuscript lies in the fact that it contains a small number of astrological texts that we also find in Boerio's notebook, namely William of England's *De urina non visa* and the *Capitula* (or *Sententiae*) *Almansoris*. It also includes the *Centiloquium Hermetis*, Albumasar's *De revolutionibus annorum mundi,* and a work of astrological medicine entitled *Compendium medicinalis astrologie* of the medieval physician-astrologer Nicolò Paganica.[72] These other three texts, while not exactly the same as those we find in Boerio, bear similarities with others in MS Arundel 88. Valdizocco includes Albumasar's *De revolutionibus annorum mundi* where Boerio had Messahallah's work by the same title, and Nicolò Paganica's *Compendium* could have provided a modern alternative to Boerio's text on the critical days, on which more will be said shortly. Valdizocco's manuscript has certainly a contemporary Venetian feel—it includes two works written by two of Valdizocco's contemporaries, Prosdocimo de' Beldomandi's *De electionibus* and Nicolò de' Conti's *De motu octave sphere*—but it is not dissimilar in overall content from Boerio, including some medical astrology, some aphoristic works, and a work of mundane astrology.[73] In short, a bit of everything that students of astrology presumably learned at Italian universities. Not too differently from Boerio's example, Valdizocco's manuscript, with its copious marginal notes, has rightly been defined as "a specialized library": the manuscript seems to have brought together ancient and Arabic texts with more recent "local" authors into a unified scientific *compendium*.

The fact that little Arabic astrology appears in the Bolognese curriculum of 1405 (Alcabitius and Haly's commentary to the *Centiloquium* being the exceptions) may make the presence of a number of Arabic works in these two manuscripts rather unusual. The fact that many of these texts were printed around the time in which Boerio attended university at Pavia, however, indicates that Boerio's interest in these Arabic texts was far from unique and allows us to speculate with some reason that by the 1480s these works may have become part of a broad *corpus astrologicum* imparted at Italian universities. While we know from the number of extant manuscripts that these texts circulated widely in the Middle Ages, the only way to appreciate what students really read when attending their four-year degrees is through notebooks like Boerio's or early printed texts that were clearly marketed to students.

The hypothesis that at some point these Arabic astrological texts became part of the *corpus astrologicum* imparted to Italian students in arts and medicine becomes more probable when we also examine an interesting Venetian incunabulum published by the Bergamasque printer and priest Bonetto Locatelli with the help of the Lombard publisher and bookseller Ottaviano Scoto in 1493/94, which contains some of the texts included in the two manuscripts just discussed. This remarkable incunabulum includes Ptolemy's *Quadripartitum;* ps.-Ptolemy's *Centiloquium;* the *Centiloquium Hermetis;* the *Centum quinquaginta propositiones Almansoris* (also called *Capitula* or *Sententiae*); Zael's *De interrogationibus* and *De electionibus;* Messahallah's *De receptionibus planetarum;* his *Epistola in rebus eclipsis;* and the *De revolutionibus annorum mundi.*[74] Particularly significant is the fact that the contents of this incunabulum are relatively similar in content or genre to those works included in MS Lat. 7307 now at the Bibliothèque Nationale de France, a manuscript that once belonged to the Sforza Library of Pavia.[75] We know that at the time of its printing, the Lombard Scoto maintained a commercial interest and a bookshop (*bottega*) in Pavia and that in the late fifteenth century, Venetian printers and publishers started targeting the university market; this may allow us to speculate that the texts included in the Venetian incunabulum responded to the needs of the university curriculum.[76]

There are other elements that support this hypothesis. This Venetian miscellanea is an intriguing example of economically effective Renaissance printing. Unlike other astrological texts printed in Venice by the press of Erhard Ratdolt (Figure 9), the major publisher of astronomical and astrological texts in Italy in the period 1476–1486, this Venetian incunabulum is remarkably simple and unadorned (Figure 10). The text is printed in two columns in a small font—up to sixty-six lines of small type to a full column, to be precise—in an obvious attempt to maximize the amount of text that could be fitted onto a single page and thus reduce costs. Its table of contents is very basic, lacking, for instance, any page reference or decorative element. The economy of printing compensates for the size of the work, which counts 150 folios. According to Martin Lowry's estimates, around this time scientific and medical books could cost between 1 *lira* 12 *soldi* and a bit more than 3 *lire,* with the imperial *lira* being worth circa 20 *soldi.*[77] Even admitting that this incunabulum may have been more expensive than the average scientific work, such a book remained relatively affordable for Renaissance elites. As way of comparison, the salary of a Pavian university professor of medicine at this time could reach up to 750 *fiorini,* or the staggering amount of 24,000 *soldi* per year (or about 300 *ducati,* one Pavese *fiorino* equaling 32 *soldi*), while the cost of feeding a *cameriere* or servant at court was estimated around 1,920 *soldi* per year (or 24 *ducati*). To make another comparison, a good horse would cost more or less 40 *ducati* (or about 3,200 *soldi*), the same as about fifty books, while eight weeks' worth of bread would have equaled the cost of a book worth a bit more than 3 *lire.*[78] By being relatively affordable and by providing texts that had become part of the *corpus astrologicum* of Renaissance physician-astrologers, this incunabulum was probably marketed for a university audience of professors and wealthy students.[79] Two aspects seem to confirm this hypothesis: first, the text opened with a defense of astrology by the physician-astrologer Girolamo Salio of Faenza addressed to Domenico Maria Novara, Nicolaus Copernicus's professor in Bologna. We know that Girolamo Salio specialized mainly in medical and astrological books and texts edited by university professors or targeted for a university audience, and the copy of this very work now preserved at the Biblioteca Universitaria in Bologna contains Novara's own signature.[80] Secondly, Scoto's interest in tapping into the student market is well documented, as testified by his *bottega* in Pavia and his financing of the publication of Cristoforo Barzizza's *Introductorium ad opus practicum medicinae,* a book aimed at Pavian and Paduan medical students, the very same year (to this was appended Rhasis's *Liber IX Almansoris,* a classic of the medical curriculum, with Barzizza's own commentary).[81]

Further evidence of a Renaissance *corpus astrologicum* comes from the library inventories of fifteenth-century physicians. The remarkable inventory of the Paduan physician Alessandro Pellati is exemplary in this respect.[82] The inventory of Pellati's books at his death in 1487 is remarkable both for the number of books it contains (over 160) and for the astronomical and astrological texts it lists. The books that can be counted as part of the *corpus astronomicum* include Sacrobosco's classic *Sphaera,* ps.-Sacrobosco *Algorismus de minutiis,* the *Algorismi* of Prosdocimo de' Beldomandi, a *Computo in astrologia* (possibly Sacrobosco's), multiple copies of the *Theorica planetarum* (at least one in Campanus's version), possibly copies of Peurbach's *Theorica nova planetarum,* and Thebit Bencora's *Tractatus de motu octavae sphaerae.* To these, one can add a series of computational tables, both in print and manuscript: copies of the Alphonsine tables, the astronomical tables by Giovanni Bianchini, those of Johannes of Lübeck, and those of the near-contemporary Nicolò de' Conti.[83] The library also contained Johannes Regiomontanus's

LIBELLVS YSAGOGICVS ABDILAZI. ID EST SER/
VI GLORIOSI DEI:QVI DICITVR ALCHABITIVS
AD MAGISTERIVM IVDITIORVM ASTRORVM:
INTERPRETATVS A IOANNE HISPALENSI.SCRI
PTVMQVE IN EVNDEM A IOHANNE SAXONIE
EDITVM VTILI SERIE CONNEXVM INCIPIVNT.

Oſtulata a domino prolixitate uitẹ
ceypbadala.i.gladii regni & durabi
litate ſui bonor̨cuſtodia quocꝫ ope
rum eius ſiue bonorũ atcꝫ extenſio
ne ſui ĩperii:exordiamur id qᵬ uo/
lumus narrare. Cum uidiſſem cõ
uentũ quorundã antiquorũ ex au/
ctoribus magiſterii iuditiorũ aſtro/
rum edidiſſe libros quoſ uocauerũt
itroductorios huiꝰ magiſterii id eſt
iuditiorũ aſtrorũ: ſed quoſdã ex eis
nõ fuiſſe ſcrutatos diligenter uniuerſa q̃ neceſſaria ſũt in eodeꝫ
magiſterio de his quẹ cõueniunt introductorio . quoſdã uero
ea quẹ neceſſaria ſũt ‸ptuliſſe ‸plixe: & quia qᵬ n̨eceſſariũ eſt in
eo periiſſe cernerẽ.Quoſdã quocꝫ in ordinatione eorũ quẹ ‸p/
tulerunt nõ inceſſiſſe itinere diſciplinẹ cõſpexiſſem cẹpi edere
hunc librũ:& poſui eũ introductoriũ:atcꝫ collegi ĩ eo ex dictis
antiquorũ quicqd neceſſariũ ẽhuic magiſterio ͭm modũ intro
ductoriũ:& nõ introduxi ratiocinatiões diſputationi ſiue de/
fenſioni eorũ quẹ‸ptulimus n̨eceſſarias cũ ſint ĩ libro Ptolemẹi
qui appellaͭ alarbamacaleͭ id eſt quattuor tractatuũ:& in libro
meo quẽ edidi in cõfirmatione magiſterii iuditiorũ aſtrorum:
& ĩ deſtructiõe epiſtolẹ haiſſebenhali in anullatiõe eius ex ra
tiocĩnatione quẹ ad hoc poſſint ſufficere:& diuiſi eũ in quincꝫ
differentias.Prima differentia eſt ĩ eſſe circuli ſignoꝝ eſſentiali

FIGURE 9. Ornate initial from Alcabitius, *Liber introductorius* (Venice: Erhard Ratdolt, 1485), sig. aa2r. The page has ample margins and the text is in elegant, but less space-efficient, roman type. The book was printed in 4to format. By kind permission of the Crawford Library, Royal Observatory, Edinburgh.

FIGURE 10. Ornate initial within a dense two-column page in tironian type from Ptolemy's *Quadripartitum,* in *Opera astrologica* (Venice: Bonetto Locatelli for Ottaviano Scoto, December 20, 1493), fol. A2r. The book was published in folio format. By kind permission of the Crawford Library, Royal Observatory, Edinburgh.

Kalendar with a *quadrans*,[84] two armillary spheres, and a quadrant. As noted, all these texts and instruments were essential tools for instruction in spherical astronomy and for mapping the position of the planets on the celestial sphere. A good number of them, moreover, were set texts for university instruction.

Many other books from Pellati's library, however, dealt with astrology proper. This *corpus astrologicum* contained introductory books on astrology, such as ps.-Ptolemy's *Centiloquium* (bound with Ptolemy's *Quadripartitum*) and Alcabitius's *Introductorius* (multiple copies both in manuscript and print), but also more advanced texts of Arabic and Hebrew astrology. These included Albumasar's *Introductorium maius,* Zael's *De interrogationibus* (together with another unidentified work by him), Abraham Ibn Ezra's *De nativitatibus,* two manuscripts vaguely indicated as *Ali con altri tratadi* (at least one of which probably was Haly Abenragel's *De iudiciis astrorum,* published in Venice as early as 1485 or the *Astronomia Ypocratis* with Haly's prologue that Pellati edited in 1483),[85] and possibly a work by Messahallah. To these Arabic and Hebrew texts, more recent ones written by late-medieval authors were added, many of which are hard to identify with any certainty. The *Suma strologica* listed in the inventory is likely to be the very popular work of John of Ashenden's *Summa astrologiae iudicialis de accidentibus mundi,*[86] but the vague description of a *manuale in astrologia* or a series of *libri, quaderni,* or *tratadi de astrologia* make it impossible to identify these texts conclusively. The same is true for a *Libro de astrologia con alguni tondi e tavole,* which could be either a work of astrology or of astronomy with some tables and "volvelles" (movable parts). To this list we can add recent works printed by Venetian presses, like Werner Rolewinck's *Fasciculus temporum* and Hyginus's *Poeticon astronomicon,*[87] and other works that we have encountered in Boerio's manuscript, namely an unidentified *Chiromantia* and Michael Scot's *Physiognomia.* The fact that these texts of chiromancy and physiognomy appear in Boerio's manuscript and also in Pellati's and Leonardo da Vinci's book lists suggests that by the late fifteenth century they had become mainstream texts of university instruction and part of the corpus of texts that were deemed useful to natural philosophers and aspiring doctors in arts and medicine.[88]

Other examples can be added to those discussed so far. For example, the library of the famous Florentine physician Antonio Benivieni (1443–1502)—friend of Marsilio Ficino and also close to the stern critic of astrology Girolamo Savonarola—indicates that even a physician more famous for his anatomical studies than for dabbling in astrology possessed a good number of astrological and astronomical books.[89] His 1487 inventory was divided into subjects, and the entry astronomy/astrology counted a dozen books: among those that belonged to the *corpus astronomicum* we can count Sacrobosco's *Sphaera* with two commentaries as well as the *Computus,* another work by Sacrobosco with other works (possibly one of the incunabula which also included the *Theorica planetarum?*), and a book on perspective with the rules of composition of the quadrant. Benivieni's tables included the Alphonsine tables and the tables of Prosdocimo de' Beldomandi, while the *corpus astrologicum* encompassed Haly Abenragel's *De iudiciis astrorum,* an undefined book with many works of astrology (including the *Tabule Toletane* and Alcabitius's *Introductorius*), another copy of Alcabitius with a commentary, Ptolemy's *Quadripartitum,* and other unspecified works.[90] To these, one should add ps.-Ptolemy's *Centiloquium* and another unspecified book of astronomical calculations.[91] Given the relatively small representation of astronomy and astrology books in Benivieni's collection—a dozen as opposed to over seventy books of medicine—it seems very likely that most of these books dated back to Benivieni's student days. Alcabitius, the *Centiloquium,* and the *Quadripartitum* were most certainly statutory texts, while learning how to use the Alphonsine tables and the quadrant were essential skills imparted at university. Less obvious, however, is the fact that Benivieni owned Haly Abenragel's *De iudiciis astrorum,* one of the most extensive compilations of the Middle Ages that contained all branches of astrology, including revolutions.

How does this compare with Pavian libraries? Unfortunately for us very few inventories of Milanese medical libraries have surfaced so far, and none as exhaustive and representative as those of Pellati or Benivieni.[92] One of these inventories, albeit very incomplete, is that of the Milanese physician-astrologer Antonio Bernareggi, who taught natural philosophy, medicine, and astrology at the University of Pavia during Filippo Maria Visconti's reign and became court astrologer and physician to the duke's family. More will be said about Bernareggi's activities and astrological practice in Chapter 2. Suffice it to say here that, with the generosity proper to a dedicated teacher, in his last will before dying, Bernareggi expressly requested that part of his library remain to his heirs and part be given to the library of the convent of Saint Thomas of Pavia so as to be made accessible to poor students who were unable to purchase those texts. The choice of Saint Thomas was significant: the university of Pavia did not have its own library, and for this reason, as we shall see, professors occasionally borrowed from the ducal library at the Castle of

Pavia while students seem to have resorted to college or conventual libraries in the city or borrowed directly from their professors.[93] Similarly, the university did not possess its own buildings, and up to 1489 many of its lectures were given within the premises of the convent of Saint Thomas.

The convent was at the core of university teaching. Apparently, the university seal, the Statutes of Pavia and those of Bologna, and the book with the names of all the students who had obtained degrees and of the professors who taught there were preserved in the church.[94] Unfortunately, we do not have the complete list of the books Bernareggi owned, but only those books he donated to the convent. Yet even such an incomplete list yields some precious information about Bernareggi's teaching and the needs of Pavian students. Together with a hefty copy of Guido Bonatti's *Liber astronomiae*—one of the most popular medieval *summae* of classical, Arabic, and Hebrew astrology— Bernareggi donated to the convent Ptolemy's *Almagest,* Haly Abenragel's *De iudiciis astrorum,* Alfraganus's *De aggregationibus scientiae stellarum,* and Giorgio Anselmi's *Astronomia*.[95] We can assume that the convent library had copies of most of the statutory texts adopted at the university and thus that, with the possible exception of the *Almagest,* Bernareggi's donation simply supplied a few novelties that were probably not present at the time. Bernareggi's donation, therefore, may be seen to reflect what we could call "new additions," namely Arabic texts that may have become more fashionable in the second half of the fifteenth century.

The examples discussed so far may provide an impressionistic idea of what fifteenth-century students actually read, but it still allows us to draw some valuable conclusions. The evidence clearly indicates that Arabic astrology was a strong presence among the readings of Pavian students. Far from being limited to Alcabitius and Haly's commentary, university teaching of astrology seemed to have made use of a range of other texts, often *summae* that addressed various types of astrology (elections and interrogations, judicial and mundane astrology), such as Haly Abenragel's *De iudiciis astrorum,* and sometimes also more specialized, shorter texts, like Messahallah's *De revolutionibus annorum mundi* or Zael's *De interrogationibus.* Similarly, Italian university professors did not shun medieval *summae,* such as those of Ashenden, Bonatti, and Anselmi. This evidence seems to suggest also that a *corpus astrologicum* existed, that it was probably both broader and more fluid than the *corpus astronomicum,* and that it comprised a number of ancient and Arabic texts of various degrees of complexity. These texts were often complemented by the works of late-medieval and Renaissance authors who sometimes had academic links to the *Studia* in which these texts were studied. As we shall see later in this chapter, some of these works are strongly reflected in the ducal library of Pavia, thus helping us add an important piece to the puzzle.

But now I would like to return briefly to the other aspects that character-ized MS Arundel 88 and that illuminate some of the texts that could have been studied at Pavia. As suggested, this manuscript contains a good deal of medical astrology. I have already mentioned how both Boerio's and Valdizocco's manuscripts contained one of the statutory texts indicated in the Bologna curriculum, namely William of England's *De urina non visa*. This was not, however, the only text of medical astrology to be found in MS Arundel 88. Our manuscript also contains the ps.-Galenic *Aggregationes de crisi et creticis diebus* often attributed to Bartholomew of Bruges, which pro-vided a nice alternative to Galen's lengthier *De diebus criticis* indicated in the Bologna University statutes.[96] This text, which has received some critical attention in recent years, was composed sometime in the later part of the thirteenth century by an anonymous author and concerns the art of astro-logical prognostication.[97] While almost entirely based on Book III of Galen's *De diebus criticis,* the *Aggregationes* constituted a handy shortcut to the more convoluted ancient text and thus was particularly suitable for student instruc-tion. Both texts are based on the principle that a skilled astrologer-physician can calculate the critical days of an illness according to the length of the phases of the Moon. The *Aggregationes* pays particular attention to the influ-ence of the Moon in its relation to the other planets and the signs of the zodiac. Together with the ps.-Hippocratic *Astronomia* or *Astrologia Ypocratis,* a spurious medico-astrological work of Greek or Byzantine origin that circu-lated widely in late-medieval Italy,[98] Galen's *De diebus criticis,* its abbrevia-tion *Aggregationes,* and William of England's *De urina non visa* constituted a staple of late-medieval medical astrology. Of the three main medical func-tions of late-medieval medicine, namely diagnosis, prescription, and prog-nosis, all these texts belonged to the latter. Indeed, they provided a useful series of rules to help physicians formulate a prognosis for their patients on the basis of celestial influences. As we shall see, similar theories were occa-sionally adopted at the Sforza court, particularly under the reign of Ludovico Maria Sforza, when his court physicians treated, albeit unsuccessfully, the recurrent illnesses of Ludovico's nephew, Gian Galeazzo Sforza, according to some of these principles. That the *Aggregationes* appears in Boerio's manuscript, therefore, finds a nice counterpart in the practical application of the very same theories in the treatment of Sforza patients.

A further intriguing aspect of MS Arundel 88 is the presence of a series of medical recipes concerning the treatment of wounds and ulcers (which include recipes used at the court of the Duke of Ferrara, Ercole I d'Este [1431–1505], one of which was explicitly credited to his court physician, Francesco Benzi); preventive measures to eliminate bed bugs; magic formulas; and love potions.[99] This, together with the annual prognostication of the Milanese court physician-astrologer Gabriele Pirovano clearly indicates that the culture of the *Studium* was not clearly separated from that of the court.

Finally, our manuscript takes us beyond the limits of Pavian student life, both to Boerio's birthplace and to England, where he was to spend almost his entire career. The manuscript includes a significant number of genitures and other charts, often with lengthy interpretations. Some have a local flavor, like those of Battista II Campofregoso, Doge of Genoa from 1478–1483,[100] or that of a certain Giovanni Battista Colis of Cremona, made by the Cremonese astrologer Battista Piasio.[101] These are followed by notes on the date of birth of various individuals, including Francis I, King of France, and then by lengthier horoscopes of the English royal family, starting with that of Henry VIII, copied by Boerio from an original produced at the University of Oxford in 1513, which Boerio believed contained a series of errors.[102]

Boerio's student notebook has taken us on a journey of exploration: it has served as a prism through which I examined the texts that constituted the *corpus astronomicum* and the *corpus astrologicum* of fifteenth-century students in astrology and astrological medicine at Pavia and other Northern Italian universities. Admittedly, Boerio's notebook is only one example of what a student might have read, and it is not fully representative of Boerio's interests, which may have encompassed many more works. What these other works may have been, however, emerges from the analysis of other manuscripts written in the same period, from their comparison with book inventories of contemporary physicians, and from the study of scientific book printing at the end of the fifteenth century. Such a comparative approach has allowed me to somewhat overcome the limitations of my source and has provided a vivid image of the kinds of books read and studied by aspiring physician-astrologers at Italian universities and of the books that guided their practice when they entered the profession. What is clear even from this preliminary work is that students of astrology often went well beyond the study of the statutory texts indicated in the Bolognese curriculum. Their knowledge of Arabic works of astrology—some including all techniques and genres, others only some—was probably more vast than we have assumed so far by looking at the medical curriculum of Italian universities. This knowledge, moreover, was possibly dictated by personal taste, but also almost certainly by general trends and interests associated with specific celestial and terrestrial events.

The Ducal Library at Pavia

One last element may be added to this picture that can help us once more connect university and court: the ducal library housed at the Sforza castle in Pavia. Among the first items listed in the 1488 inventory of the library of Pavia were a large clock (*horologium*), two globes (one celestial, one terrestrial), and a terrestrial map on a wooden surface. To these objects one could add a fine copy of Ptolomy's *Cosmographia* in large format and with images and tables. These were only some—probably the most valuable—scientific objects housed in the ducal library.[103]

With five inventories covering the period 1426–1490, we are in a privileged position to gauge the nature and content of the books housed in the ducal library at Pavia, which, with well over nine hundred items, constituted an exceptional collection for the time.[104] The majority of texts that made up the library included classical authors, literature, theology and other devotional texts, but the library also included numerous books of law, natural philosophy, medicine, astronomy, and astrology. Unfortunately, due to its lamentable dispersion in the aftermath of the French descent, it is not always possible to locate the manuscripts that originally comprised the ducal library. Nevertheless, Elizabeth Pellegrin's thorough study of the detailed inventory dated 1426 and her researches in the major European libraries have allowed her to identify many of the astrological and astronomical manuscripts that are now housed in the Bibliothèque Nationale de France.

The ducal library contained a number of texts that belonged to the *corpus astronomicum* and the *corpus astrologicum* of fifteenth-century universities, many of which were bound together. A close look at the astronomy books listed reveals close affinities with the private libraries examined above. Some of the astrology books, moreover, were bound together with works of astronomy, clearly indicating the continuity existing between one branch of knowledge and the other. Among the books that fall broadly under the heading of astronomy, we can count multiple copies of Johannes Sacrobosco's *Sphaera* (one of which was bound with a copy of Robertus Anglicus's *Tractatus quadrantis,* the *Canones in motibus super celestium corpora,* an anonymous *Liber de iudiciis in astrologia,* and Alfraganus's *Liber de aggregationibus scientiae stellarum*),[105] a copy of Johannes Sacrobosco's *Algoritmus de minutiis et integris,*[106] two copies of Campanus's commentary on Euclid's *Elementa* (one of which was bound with Thebit Bencora's *De motu octavae sphaerae,* Ptolemy's *Planispherium,* and the astrological work of Messahallah, *De interrogationibus*),[107] and a copy of Ptolemy's *Almagest* in Gerard of Cremona's translation.[108] Of course no proper astronomical and astrological library could do without its set of tables: that of Pavia had at least one copy of the Toledan tables and the *Canones in motibus super celestium corpora* mentioned above (possibly de Lineris' *Canon,* in which case the library probably contained at least one copy of the Alphonsine tables that go with it).[109]

Classical, Arabic, and medieval astrology featured prominently: the inventories list a commentary on Alcabitius's *Introductorius,*[110] a copy of ps.-Ptolemy's *Centiloquium* with the *Capitula Almansoris,* Michael Scot's *Liber particularis,* or *Astronomia,*[111] Albumasar's *De magnis coniunctionibus, annorum revolutionibus ac eorum profectionibus,* Zael's *De electionibus* and *Liber temporum,* and Messahallah's *Epistula de eclipsi lunae.*[112] To these we should add Messahallah's *De interrogationibus,* and the anonymous *Liber de iudiciis in astrologia* mentioned above.

Among the texts that expounded the theory of the critical days in the ducal library we can count an *Isagoge Ioannini et de diebus cresitis* [sic][113] and a text by Galen entitled *Super de crisis* that could have contained Galen's *De crisibus* or the *De diebus criticis* or possibly the *Aggregationes.*[114] Additional works relevant to the practice of medical astrology include a small *Tabula medicorum ad inveniendam lunam in signis gradibus et minutis,*[115] some *Tabule pulcherrime pro contemptu* [sic] *de sole et luna,*[116] and a *Tabula quedam de luna,*[117] all of which would have made the task of tracing the luminaries' movements faster and easier. To these at some point were added other astrological books; Albumasar's *Flores astrologiae,* two copies of Guido Bonatti's popular medieval compendium *Liber astronomiae* (here with the title *Ad iudicia stellarum,* one in parchment, one paper), another book of astrology attributed to Michael Scot (possibly his *Liber introductorius* or parts thereof, which contains a great deal of astrology), Alfodhol's work of divination *Liber iudiciorum et consiliorum,* and an anonymous *De naturis signorum et iudiciis eorum in astrologia.*[118]

By any standard the private library of the dukes of Milan encompassed an impressive collection of astrology books: these spanned many centuries, from late antiquity to the Middle Ages, and covered all genres of astrology. Such a library obviously constitutes a privileged source of information about the texts that court physicians and astrologers may have consulted in their daily practice. Indeed, we know with certainty that courtiers and physicians used the library regularly and occasionally borrowed books, thus allowing us to infer that its contents would have constituted a common source of information for members of the intellectual courtly community.[119] As noted, some of the texts available in the library were also works that appeared in MS Arundel 88, thus suggesting that their presence in the ducal library may have served the needs of professors teaching at Pavia. Although we do not know what else Boerio may have read during his course of studies at Pavia, what MS Arundel 88 tells us is sufficient to indicate that Arabic and medieval astrology played a larger role than it is often assumed in the training of Pavian physician-astrologers, and probably in that of most Italian students who aspired to practice this profession. Whatever else Boerio may have read can be similarly inferred by the readings of his contemporaries.

Of course, not all Renaissance physicians were well versed in astrology, and some may have been more enthusiastic than others in applying astrology to medicine.[120] Medicine itself was a complex body of knowledge with competing traditions and different schools of thought. While astrological medicine was certainly one important aspect of Renaissance medical practice, it would be mistaken to believe that its application was ubiquitous.[121] With this caveat in mind, one cannot fail to notice that some of the most famous physicians of the late Middle Ages and the Renaissance—including Pietro d'Abano, Biagio Pelacani da Parma, Marsilio da Santa Sophia, Michele Savonarola, and Girolamo Manfredi—were strong proponents of astrology and that this discipline flourished in Northern Italian universities such as Bologna, Padua, and Pavia.[122] We can safely say, therefore, that astrology played an important part in Italian university teaching within the curriculum of arts and medicine and that Pavia had a long-standing tradition of astrological teaching. As we shall see in subsequent chapters, what was learned at university—be it elections and interrogations or judicial, mundane, or medical astrology—was clearly put into practice at court.

From Theory to Practice:
Astrological Principles, Techniques, and Vocabulary

After having examined the kind of astrological books that circulated widely in fifteenth-century Italy and were taught, at least in part, at Italian universities, it is now time to turn to practice. Why were these books copied, purchased, and studied? What kind of principles and techniques did they explain? What practical needs did they meet? This book will not be able to provide lengthy explanations about all the astrological techniques employed by Renaissance astrologers, but since I will be referring regularly to certain astrological principles, terms, and techniques, it seems only fair to explain them here. Specialists in the field of Renaissance astronomy and astrology may well be familiar with many of the terms and techniques I will be discussing, but this is not necessarily true of all historians of science, let alone of students and political and cultural historians to whom this book is also addressed.

In the Italian Renaissance, virtually every university student in arts and medicine would acquire some basic knowledge of the heavenly movements; many of them, moreover, progressed far enough to study the influence of these movements on Earth below (what was called the sublunary world). That celestial motion exerted an influence on the world below was hardly questioned.[123] The principle of celestial influence at the core of astrology established that this influence varied with the position and mutual relationship of the planets and the luminaries (the Moon and the Sun) with respect to a given time and place on Earth. What needed to be established, at least for astrologers, was how to predict these effects to minimize those that were negative and possibly amplify those that were positive and, in general, how to know what to expect for the future in terms of things as varied as someone's life, the weather and the status of crops, future wars, and religious turmoil.

Each of the planets possessed particular characteristics that made it either "fortunate" or "unfortunate": cold and dry Saturn and hot and dry Mars were considered generally malevolent, and yet Mars was also the planet of war, and thus, if prominent in the birth chart of a ruler, for instance, could herald military and political success.[124] Jupiter's warm and moist nature, in contrast, was considered positive, the planet being fortunate and a source of fecundity and wealth. Temperate and moist Venus was equally fortunate and usually associated with voluptuousness and pleasure, while Mercury, although considered of changeable nature, was generally associated with intelligence and communication. The two luminaries, the Sun and the Moon, also had properties, the Moon being cold and moist and the Sun hot and dry; on the whole they were considered fortunate, with the Moon more fortunate than the Sun.[125]

As noted, planets are in mutual relationships with each other. The significant angular relationships that characterize this dynamic are known as "aspects." According to this principle, if two planets are about 0° apart, they are in conjunction; if they are 180° apart, they are in opposition; if they are at a 90° angle, they are in square or quartile aspect; if they are placed at a 120° angle, they are in trine; at 60°, in sextile (Figure 11). Some of these aspects are considered positive, like the sextile and trine aspects, others negative, like the square and the opposition. The evaluation of conjunctions depends on the qualities of the planets involved.

Various factors can influence a planet's quality: together with aspects, the sign of the zodiac through which a planet transits or the (mundane) house it occupies influences its character. The zodiac is that portion of the celestial sphere extending about 6° on each side of the ecliptic (i.e., the apparent path of the Sun) that includes in itself the path of the five planets known in the Renaissance (Saturn, Jupiter, Mars, Venus, and Mercury), plus the Moon and the Sun. This band is divided into twelve portions of 30°, each corresponding to one of the signs of the zodiac (Figure 12). The first degree of Aries (the first sign of the zodiac) coincides with the vernal equinox, the point at which the ecliptic intersects the celestial equator; signs are counted counter-clockwise from there (because of the precession of the equinoxes, however, this point has now slowly moved backward into Pisces).

Each individual planet, moreover, is believed to have a special relationship with a particular sign or signs of the zodiac and less favorable relations with others. Saturn, for instance, "rules" Capricorn and Aquarius, but it is in its "detriment" when in Cancer and Leo, the two signs diametrically opposed to Capricorn and Aquarius. Similarly, Mars rules Aries and Scorpio, but it is in its detriment when placed in Libra and Taurus. The luminaries rule one sign each: Cancer, the Moon, and Leo, the Sun. Luminaries and planets, furthermore, can also be "exalted" (or "in exaltation") when placed in exceptionally favorable positions: Mars, for instance, is in its exaltation when at 28° Capricorn; on the contrary, it is in its "fall" at 28° Cancer (Table 1). Two further astronomical points, the lunar nodes, namely the two points where the lunar path crosses the ecliptic, called respectively the *caput* and *cauda draconis,* are also considered important and are generally mapped on the chart.[126] Their

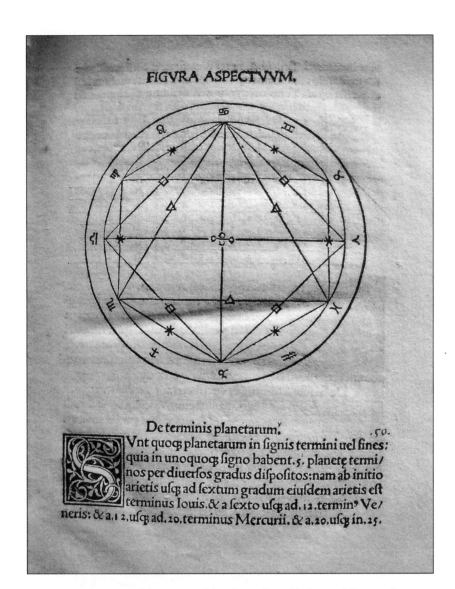

FIGVRA ASPECTVVM.

De terminis planetarum.   .50.

Vnt quoqʒ planetarum in fignis termini uel fines:
quia in unoquoqʒ figno habent.5. planetę termi/
nos per diuerfos gradus difpofitos:nam ab initio
arietis ufcʒ ad fextum gradum eiufdem arietis eft
terminus Iouis.& a fexto ufcʒ ad.12.termin⁹ Ve/
neris: & a.12.ufcʒ ad.20.terminus Mercurii. & a.20.ufcʒ in.25.

FIGURE 11. Diagram of the aspects of the planets from Alcabitius, *Liber introductorius* (Venice: Erhard Ratdolt, 1485), sig. aa5r. By kind permission of the Crawford Library, Royal Observatory, Edinburgh.

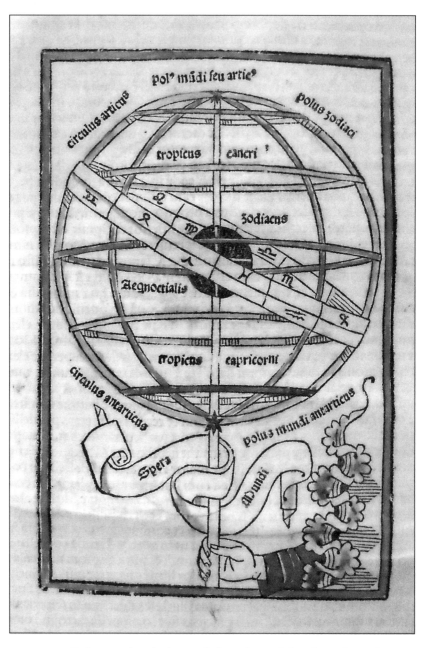

FIGURE 12. Ptolemaic celestial sphere with the zodiac and the ecliptic, from Johannes Sacrobosco, *Sphaera* (Venice: Simone Bevilacqua da Pavia, 1499), sig. a2v. By kind permission of the Crawford Library, Royal Observatory, Edinburgh.

Table 1: Planets' dignities and debilities

Planet	Mansion/Sign	Detriment	Exaltation	Fall
Moon	Cancer	Capricorn	Taurus 3°	Scorpio
Mercury	Gemini and Virgo	Sagittarius and Pisces	Virgo 15°	Pisces
Venus	Taurus and Libra	Scorpio and Aries	Pisces 27°	Virgo
Sun	Leo	Aquarius	Aries 19°	Libra
Mars	Aries and Scorpio	Libra and Taurus	Capricorn 28°	Cancer
Jupiter	Pisces and Sagittarius	Virgo and Gemini	Cancer 15°	Capricorn
Saturn	Capricorn and Aquarius	Cancer and Leo	Libra 21°	Aries

significance lies in the fact that eclipses, celestial phenomena considered of great interpretative significance for the astrologer, occur only near the lunar nodes.

To add complexity to this taxonomy, the signs of the zodiac are further divided into four "triplicities," namely groups of signs with different characteristics, each of which are ruled by different planets or luminaries (as we just saw). These triplicities are associated with the four elements and are the fiery triplicity (Aries, Leo, and Sagittarius), the earthy triplicity (Taurus, Virgo, and Capricorn), the airy triplicity (Gemini, Libra, and Aquarius), and the watery triplicity (Cancer, Scorpio, and Pisces). In a clear correspondence between macrocosm and microcosm, and thus man, these are linked also with the four Hippocratic humors, the choleric (fiery), the melancholic (earthy), the sanguine (airy), and the watery (phlegmatic), which are thus sometimes used in astrological medicine.

As indicated, the quality of planets is also affected by the (mundane) houses that they occupy at a given moment in time. Houses are divisions of the ecliptic plane that depend on time and location; they are numbered counter-clockwise starting from the position of the eastern horizon, called the ascendant (which is taken as the cusp of the first house), moving to the nadir (the cusp of the fourth house), the western horizon (the cusp of the seventh house), the midheaven or *medium coeli* (the cusp of the tenth house

on the meridian), and back to the ascendant. Astrologers traditionally asso-
ciate each of these houses with an element of a person's life: the first house
relates to life in general, the second to business and wealth, the third to
brothers, the fourth to parents, and so forth (Figure 13).[127] Renaissance astrologers adopted various techniques of house division, the most common of
which was that attributed to Alcabitius. The late historian of astronomy, John
D. North, called this the "standard method" because it was the most widely
adopted in the fifteenth and early sixteenth centuries.[128] The complex relationship of the planets and the luminaries with the signs of the zodiac and the
houses is at the core of any astrological chart interpretation. Certain combinations are believed to be particularly momentous: for instance, malevolent
Saturn in the eighth house, the house of death, may not bode well for the
future of the person who has it in his horoscope, while Saturn in the second
house indicates possible difficulties with money and possessions. Benevolent
Jupiter in the second house, on the other hand, is seen as positive, and thus
promises financial success.[129]

As we can see from the brief explanation provided here, astrologers had to
consider a bewildering number of factors and their interactions when interpreting a chart: this welter of combinations and permutations both added
greater complexity to the interpretation and allowed the astrologer an element of freedom. Within a single chart, the weight to be given to each of
these factors—which sometimes were in conflict, sometimes in agreement
with each other—was not guided by fixed rules, and this left the astrologer
some room to maneuver. No astrologer, therefore, when faced with the same
chart, would propose exactly the same interpretation of it. Astrology, as many
of its practitioners emphasized, was a conjectural art like medicine: it could
be used to predict events but would never offer certainties. More importantly,
despite critiques to the contrary, it was never overly deterministic, leaving
room for free will and the will of God to intervene.[130]

To construct a horoscope an astrologer would use a table of houses, such
as that of Alcabitius, to draw up the positions of the ascendant, Midheaven
and other house cusps for the time and place required. From an ephemeris he
could then determine the positions of the planets at that time and enter them
into the skeleton chart.[131] Charts could be constructed for a variety of reasons,
the most common of which was the occasion of a person's birth. In this case
the chart was often called *figura nativitatis* or *genitura* (Figure 14). This chart
plotted the sky at the moment of birth onto a square figure that contained the

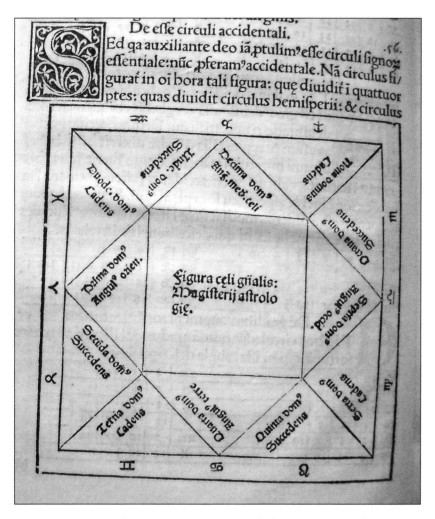

FIGURE 13. *Figura coeli,* or square celestial chart with the house division, from Alcabitius, *Liber introductorius* (Venice: Erhard Ratdolt, 1485), sig. bb1v. Counter-clockwise: first house: life / second house: riches / third house: brothers and relatives / fourth house: parents / fifth house: children / sixth house: health and sickness / seventh house: wife and marriage / eighth house: death / ninth house: voyages and religion / tenth house: trade and honors / eleventh house: friends / twelfth house: enemies and prison. By kind permission of the Crawford Library, Royal Observatory, Edinburgh.

angles of the chart, the cusps of the houses, the planets and the luminaries, the lunar nodes, and other astrologically significant points like the *pars fortunae*. Within the chart interpretation, the astrologer often paid particular attention to the calculation of "the giver of life" and "the giver of the years" (called, respectively the *hyleg* and the *alchocoden*). The techniques required to find the *hyleg* and the *alchocoden* are too complex to be explained here, but suffice it to say that they included the calculation of a planet's "dignities" in the chart.[132] Planets aquired "dignities," or points, depending on their placement within the celestial sphere. Planets in their "domicile" (i.e., one or more signs of the zodiac believed to be particularly beneficial for a given planet), for instance, acquired five points; in their "exaltation" (the specific degree within a given sign of the zodiac believed to be extremely beneficial for a given planet), four points; in their "triplicity," three points. But planets "in detriment" lost five points, and if they were "in their fall," they lost four. The list of elements that gave planets dignities also included their placement within a certain house so that if, for instance, the planet was in the first or tenth house, it would gain five points, but if it was in the second, fifth, or eleventh, only three.[133] Once the astrologer had established the *hyleg* and the *alchocoden,* he would then proceed with the calculation of the minimum length of a person's life.

A practice related to nativities was the annual revolution of the natal chart, which was based on the examination of the sky around the time of a person's birthday, or, to be more precise, at the exact moment when the Sun returned to the degree of the ecliptic that it occupied in the natal chart. This could occur on the day of one's birthday or on the day before or after, and the interpretation of this chart was valid only for the year ahead, that is to say until the following birthday.

Together with the practice of nativities (also called genethlialogy), three more branches encompassed the practice of Renaissance astrology: elections, interrogations, and conjunctions and annual revolutions.[134] Elections dealt with the proper time to initiate action: this could include a rich variety of undertakings, such as the founding of a building or a city, choosing the best time for a marriage or the conception of a child, waging war or signing peace, or setting off on a journey. Interrogations cast charts at a given time and place related to specific questions the client might have. For instance, if somebody was critically ill, an astrologer could cast a chart for the moment of the interrogation and attempt to determine whether the person in question would die. Ideally, to answer this question, the astrologer would also request the person's time

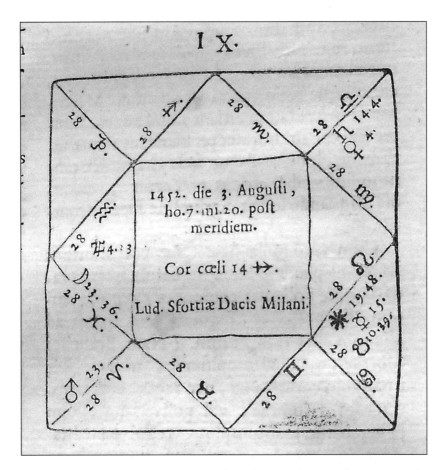

FIGURE 14. Geniture of Ludovico Maria Sforza, from Girolamo Cardano, *De exemplo centum geniturarum,* in *Opera Omnia,* 10 vols. (Lyon: Jean Antoine Huguetan & Marc Antoine Ravaud, 1663), vol. 5, 463. By kind permission of the University of Edinburgh Library, Edinburgh.

and date of birth as well as the moment when he or she fell ill in order to construct two other charts that could be compared with that of the interrogation. Other questions could include whether a person would have a baby boy or a baby girl; whether the newly elected Pope would be favorable to the state of the querent; or whether the querent would win or lose a war.

Finally, astrologers could make predictions based on the revolution of the year. This technique required the erection of a celestial figure for the moment the Sun returned to a precise zodiacal point, which was set as the first degree of Aries. Having cast and examined this chart, the astrologer would go on to express his judgment on a series of classic topics that included the weather, the crops, war and peace, the immediate future of certain classes of people or professions, and political events.[135] Connected to this practice was the analysis of planetary conjunctions of the three superior planets: Mars, Jupiter, and Saturn. Conjunctions of these planets occurred at regular intervals: every twenty years for Saturn and Jupiter, about every two for Saturn and Mars and Mars and Jupiter. Depending on the sign and the triplicity in which they occurred, these celestial occurrences were thought to herald different events on Earth.[136] As noted above, the conjunction of Saturn and Jupiter in Scorpio in 1484 was scrutinized closely by the Pavian student Giovanni Battista Boerio, who could compare the alleged effects of the conjunction with his own experience as a refugee from his university town because of the plague. As suggested earlier, beyond the more personal and local dimension of our Pavian student, this conjunction remains significant for its posthumous association with the date of Luther's birth, which remained an issue of debate among astrologers in the early sixteenth century, and was used by Catholic propagandists to cast him as the Antichrist. Another momentous conjunction of Saturn and Jupiter closely associated with this Catholic-Protestant debate was that of 1524 (this time in Pisces), which many Protestant astrologers interpreted as heralding the end of the world and bringing about the second Great Flood.[137] As noted, these types of interpretations derived their origins from the doctrine of great conjunctions made so popular in Albumasar's *De magnis coniunctionibus,* where the ninth-century Arabic astronomer and astrologer theorized that the conjunction of the superior planets brought with it a change of religions and kingdoms.[138] Thus, the Black Death of 1348 and the Great Schism of 1378 had similarly been interpreted as consequences of the Saturn-Jupiter conjunctions of 1345 and 1365 in the late Middle Ages.[139]

All these techniques, as we shall see, were variously employed at the court of the Sforza and in Italy more broadly. If astrology within late-medieval and Renaissance society did not reach quite the same level of prominence that it held in the caliphal court of Baghdad in the Abbassid period (750–1250 AD), where many of the texts adopted by Renaissance astrology were written and these practices refined, from the thirteenth to the sixteenth century, it still gained considerable favor among a number of European courts.[140] Renaissance Milan was certainly one of them. As this book illustrates, following the legacy of their Visconti ancestors, the Sforza dukes who reigned in the second half of the fifteenth century did not disdain the advice of the astrologers. Rather, they occasionally received unsolicited prognostications and often actively sought the astrologers' advice to solve problems, answer questions, and procur political guidance in times of uncertainty and turmoil. To different degrees and with different levels of enthusiasm and engagement, all of the Sforza dukes surrounded themselves with physicians and astrologers well trained in the art and received from them personal as well as political advice.

Astrologers trained at Italian medieval and Renaissance universities were, on the whole, very competent calculators. They mastered the various techniques outlined above with confidence and drew on a solid textual tradition and the strength of its authorities to provide competent counsel and inform decision making. If at times their predictions might sound to a modern reader conveniently formulated to fit what a lord would have liked to hear, this does not necessarily mean that their art was fraudulent. Court practitioners may have felt particular pressure to produce certain prognostications to please their lords—a factor that, as we shall see in Chapters 3 and 4, did not escape the attention of contemporaries—but on the whole they operated with intellectual honesty and professional integrity. Some court astrologers may have been more inclined to compromise their intellectual integrity to gain the favor of their lord and enrich themselves, but that does not mean that they did not have faith in their art nor know its principles. The majority, therefore, operated with a level of integrity that was common to all university-trained professionals, and that distinguished them from the numerous unlicensed empirics and quacks who populated Italian towns.[141] Even so, elections and interrogations came to be seen as particularly dubious practices even by some Renaissance practitioners.[142] The parallel between astrology and economics made by historians Ann Geneva and Anthony Grafton seems particularly fitting: nowadays there are good and bad economic advisors, and yet millions of people resort to economic advice of one sort or another every year, in spite of its dubious soundness and level of certainty. Even if economics is not a hard science, people prefer to receive advice on what is likely to happen than to navigate blindly. Moreover, they tend to trust the advice they received, even if, in some cases, they consult more than one financial advisor. As we shall see, court astrology fulfilled a similar human need and operated in a very similar way.

2

✶

The Making of a Dynasty

*Astrology under
Bianca Maria Visconti and Francesco Sforza*

In his early days as lord of Milan, the *condottiere* Francesco Sforza received a series of letters from a certain Antonio da Camera, a Mantuan astrologer keen to provide astrological advice to the new duke of one of the most powerful principalities of Northern Italy. In these letters we learn that da Camera had worked (and continued to work, it seems) in the service of two of Francesco's important political allies, Sigismondo Pandolfo Malatesta, Lord of Rimini, Fano, and Cesena, and Ludovico Gonzaga, Marquis of Mantua. Both were *condottieri* like Francesco.[1] A well-informed astrologer, not only did da Camera cast Francesco's geniture and his annual revolution, but also the chart of his election to the title of Duke of Milan. On this basis, he was able to announce a series of warnings: first of all, Francesco had to beware of poison; secondly, due to the unfavorable position of Saturn and Mars in relation to the ascendant in Francesco's annual revolution, as well as the conjunction of Saturn with the Moon and its opposition to the Sun at the time of his election, Francesco should avoid visiting other people's houses from the day the letter was sent (February 27) until March 7. More dramatically even, da Camera suggested that Francesco pay particular attention on the seventh day and avoid seeing anybody.[2] Documents like this populate the diplomatic archives of many Renaissance principalities, and yet they rarely attract the attention of political historians. They are, however, particularly useful to the historian of science, not only because they reveal how astrological theory was put in practice, but also because they provide unique evidence of the role exerted by astrology in political praxis.

This type of astrological advice belongs to that branch of judiciary astrology that included elections and annual revolutions. Clearly pragmatic, it was meant for immediate application. With its focus on predicting future events, Renaissance astrology clearly vied for attention with other contemporary predictive "sciences" in what I like to call the early modern "predictive market." Such a space comprised astrology; religious or sibylline prophecy; medical prognostication; and the science of prodigies and omens. All these "sciences" responded to the need, felt by most people, to cope with the uncertainty and precariousness of their times and make sense of what could not be easily understood or controlled. To a newly installed lord, therefore, this kind of advice may have appeared particularly timely. But did Francesco perceive it to be useful?

If getting this sort of unsolicited advice may have been relatively common, this does not say much about how it was received. Did Francesco Sforza believe in astrology? Such a question is not easily answered. As we will see in this and in the following chapters, focusing on the issue of "belief" may be reductive when discussing court astrology. We cannot be privy to the most intimate thoughts of long-dead Italian rulers, and for lack of unequivocal statements, all we can do is to assess how people like Francesco reacted to unsolicited advice, or, indeed, if they sought such advice themselves. It is clear that the attitude of Francesco and his contemporaries toward astrology was varied: there were some who were keen consumers of astrological advice and others who resorted to it only rarely or never. Nobody, however, ignored it completely. Their level of buy-in, moreover, may have differed depending on the kind of astrological techniques and theories used. For example, elections, which included choosing the best moment for traveling or waging war, found less enthusiastic consensus than meteorology or natal astrology, even among the astrological practitioners themselves.

Unfortunately, we do not know whether Francesco followed Antonio da Camera's advice. We do know, however, that, to the utmost joy of the Mantuan astrologer, the Duke of Milan replied to his letter. This in itself is significant: if Francesco was completely disinterested in da Camera's advice, he would have avoided replying. For his part, da Camera answered back, thanking the duke for his letter, reassuring him of his loyalty (which, he added, was equally divided between him and the Marquis of Mantua), and providing him with further astrological advice.

Retracing the history of Milan from the end of the Visconti dynasty to the rise and fall of the Ambrosian Republic and Francesco's elevation to the duchy, da Camera connected the rise of powerful historical figures to God's will and celestial dispositions. In hyperbolic fashion, Francesco was compared to Moses, Christ, Mahomet, and all the great political leaders of the past. Francesco's nativity, he said, was ruled principally by Mars, and then by the Sun and Jupiter "all signifying rule, richness and victory against the enemy."[3] Comparing Francesco's nativity with his election to the duchy, moreover, da Camera was able to establish that his rise to power was particularly momentous for the history of Italy, since at the time of his election the fixed sign of Leo was ascending, and its lord (the Sun) was with Jupiter in Jupiter's mansion (Sagittarius). Thanks to the favorable placement of the Sun, Jupiter, Mars, and Venus, the astrologer predicted that the Duke of Milan would be successful against his enemies. The enemy in question was certainly Venice, against which Francesco had rallied his allies, Mantua and Florence. According to da Camera, the alliance among these three powers was sealed in the heavens: Aries, Sagittarius, and Leo, respectively the ascendants of Florence, Milan, and Mantua—in this case, represented by their leaders' ascendants—were all in perfect harmony (*perfecta amicitia*) and of the same fiery "complexion."[4] A last warning was necessary, however. Da Camera stressed again that Francesco had to beware of poison, especially from women. This cautionary note was followed by a list of favorable and unfavorable times and days: midday on March 30 was particularly favorable to launch a military enterprise; April 7 and 8 were generally good, but April 12 was more problematic, for Francesco could risk displeasing his wife. April 28 was favorable to forge alliances and friendships with all but religious people; and on May 2 he had to make sure that no council was held in Florence with the enemy or else they could come to an agreement. Finally, May 23 was a very favorable day for any enterprise and would bring Francesco much honor, status, and utility.[5] In the same letter, da Camera announced his departure for Florence and then Rome, and indeed there are fleeting traces of his presence in both cities.[6] It was indeed from Rome that da Camera sent Francesco a brief *iudicum* for the year 1453/54, in which he provided astronomical data followed by political forecasts for the year to come. These were divided into a series of paragraphs specifically devoted to Francesco, Ludovico Gonzaga, the Republic of Florence, and the city of Bologna.[7]

We do not know if Francesco replied to Antonio da Camera expressing appreciation for his services. But we know that da Camera wrote again to Francesco a few years later, in the summer of 1457, once again offering political advice by means of astrology. In a first letter dated June 14, 1457, da Camera warned the duke of a nefarious constellation due to appear at the end of July, heralding difficulties: a treaty against his state, he specified, could be signed by his enemies.[8] In the same letter, he also recommended that the duke pay attention to his health and govern his body appropriately so that, by following the right regimen, he could counteract the negative influence of the stars and avoid the plague. Looking into September, da Camera commented further on the eclipse that would take place on September 3 and on its evil influx, saying that this would cause the death of a great man and that the Duke of Milan could either benefit or suffer from this, depending on his ability to make the most of the situation.

A much shorter letter a few months later offers further evidence of da Camera's relationship with the Sforza. In this letter, he limited himself to advising the duke to be particularly careful around November 18 because of a dangerous celestial configuration. By paying particular attention during the two days both preceding and following the date indicated, however, the duke could easily escape its malignant influx.[9] Antonio da Camera also sent a long and detailed *iudicium* for the year 1458/59. This was dedicated expressly to Francesco Sforza and more or less followed the traditional format discussed in the Introduction. After the astronomical data, da Camera provided information about the weather virtually month by month, followed by a detailed discussion of possible diseases associated with planetary motions, the condition of the harvest, the fate of various professional classes, political leaders (Francesco included), and the citizens of the most important cities. Finally, the *iudicium* closed with a brief note on the Turks.

Admittedly, these documents do not say much about Francesco's attitude toward Antonio da Camera's astrological advice. As far as we know, the advice was unsolicited. What we can say, however, is that as the new prince of a flourishing Italian principality, Francesco received astrological advice from astrologers who had also been advising his allies, the Gonzaga and the Malatesta. Furthermore, had he ignored Antonio da Camera's initial letter, or indeed any of those that followed, no more letters would have probably come his way. Instead, he decided to reply, thus tacitly inviting da Camera to write again. We can safely say, therefore, that Francesco was not adverse to astrology and indeed that he may have cautiously fallen victim to its powers.

In fact, further evidence supports the hypothesis that Francesco was not indifferent to "astrological intelligence." He certainly paid attention to exceptional celestial events and to astrological prognostications, at least at the level of "news." Sometime in 1456, for instance, Francesco obtained a copy of Antonio da Camera's 1456 *iudicium* on the appearance of a comet (known to us now as Halley's comet), which had originally been addressed to Ludovico Gonzaga, a sign that Sforza deemed this type of "astrological intelligence" of some value.[10] This hypothesis is reinforced by the presence among the same documentation of two lengthy annual prognostications, one by the Cremonese astrologer Battista Piasio and the other by the Bolognese astrologer-astronomer Giovanni de Fundis; neither was addressed directly to Francesco. Piasio's *iudicium,* which survives in two copies, is addressed to Francesco's contemporary and patron of astrology, the Ferrarese duke Borso d'Este, while de Fundis's *iudicium* is addressed to the city of Bologna.[11] Their presence among Francesco's papers allows us to speculate that these two prognostications, like that of Antonio da Camera, were copied in Ferrara, Bologna, and Mantua and then sent to Francesco. As we shall see in Chapter 3, the practice of requesting copies of *iudicia* became common under the reign of Galeazzo Maria Sforza, and one cannot help but wonder if the more scant documentation during Francesco's reign is not due simply to the dispersion of this type of ephemeral material among the Sforza diplomatic correspondence.

By all accounts, Milan must have appeared to be an attractive site of astrological patronage, if, much like da Camera, other astrologers wrote to Francesco offering their services. The noble Paduan mathematician and astrologer Nicolò de' Conti (better known by his Latin name Nicolaus de Comitibus), for example, penned a long letter offering his astrological skills to Francesco. A student of astronomy and astrology at Padua, early in his career Nicolò dedicated his treatise *De motu octavae sphaerae* to Malatesta Malatesta of Pesaro.[12] While now virtually forgotten, de' Conti must have held respectable credentials among his peers if the famous German astronomer Johannes Regiomontanus praised him in an oration delivered at Padua in 1464.[13]

De' Conti's knowledge of astrology was extensive, and he was particularly interested in astrological meteorology and the interpretation of natural events, such as earthquakes, volcanic eruptions, and plague epidemics.[14] Although he believed the latter to be related to celestial movements, de' Conti explained that they could spread via water or air.[15] Once again, de' Conti's letter is best read against the background of Italian diplomatic history: the Paduan astrologer prefaced his astrological advice with an apology for not having written to Francesco earlier, "but as God, with His infinite Goodness, has sent this quiet peace and union, which was indicated in the skies and I myself had predicted in my judgment (*iudicio*) of last year," he explained, "it is now time to redress the errors of the past, if such they can be called, and share with you my astrological studies."[16] De' Conti was clearly referring to the recent Peace of Lodi of April 1454, which ended the war between Milan and Venice that had occupied Francesco in the early years of his reign. As a citizen of the Serenissima, de' Conti was clearly not in a position to write to Francesco before 1454, but, with the new alliance signed, he could now offer his services to the Duke of Milan.

Wrapped in the rhetoric of gift-giving, de' Conti's letter accompanied his judgment for the following year based on the revolution of the duke's nativity and that of other Italian lords. Unfortunately, I have not been able to locate the *iudicium* that de' Conti sent, but already in the letter he anticipated some of his predictions for the forthcoming months, saying that, as a consequence of the solar eclipse of 1453 in the sign of Sagittarius (which was, as noted, Francesco's ascendant), his life was in danger. In his *iudicium,* however, de' Conti had volunteered further precious information as to how the duke, "with the help of his many court astrologers," could avoid danger.[17]

Not surprisingly, de' Conti's gift of a free astrological consultation was not completely disinterested. As in many other letters written at the time, it contained a supplication for himself and his sons (all of appropriate age to serve the duke, he specified) and particularly for his half-brother (*fratello uterino*) Francesco Ariosto, uncle of the more famous Ludovico, who later found employment at another astrological court, that of the Este in Ferrara. Once again, what is missing in this context is Francesco's reaction to this kind of letter. We know for sure that he did not believe it appropriate to employ either Francesco Ariosto or de' Conti as requested in the letter, and yet it would be much harder to argue that he did not pay any attention to de' Conti's astrological advice.

Francesco Sforza may not have been an active supporter of astrology. We may even conjecture that he was mildly skeptical of this art, and yet

the presence of copies of annual *iudicia,* together with the variety of astrological prognostications that reached him, suggest that he did not dismiss this
type of information entirely. Such astrological intelligence was obviously
considered relevant to Renaissance politics, and some of the information contained in these prognostications may have been quite sensitive and potentially
destabilizing. If used appropriately, however, astrological intelligence could
likewise help a lord navigate the rough seas of Renaissance diplomacy. Many
astrologers were certainly convinced of this; they were aware of the impact
that these predictions could have if made public. In Antonio da Camera's
words, "many astrologers describe in public every prince's business, and either
out of love or fear of their lords, they either neglect to mention or say many
things; and having seen this, the wise princes prevent these things from
happening."[18]

For the same reasons, the virtually unknown Brescian physician-astrologer
Giovanni Boioni made a *iudicium* based on Francesco's nativity that possibly
never reached the duke. He wrote to Francesco explaining that he had had
problems delivering the *iudicium* to him because citizens from Brescia were
denied access to Francesco's territories (an indication that the letter dates to
before the Peace of Lodi), adding the following curious remark: "do not be
amazed, My most Humane Excellency, if in this *iudicium* the name of your
illustrious lord does not appear, because in this way, if it falls into the wrong
hands, it will not be possible to establish who it was about."[19] In time of war
at least, astrological prognostications that discussed political matters were
clearly seen by both parties, namely the producers and consumers, as extremely
sensitive. For this reason, they needed to remain private and had to be kept
out of the hands of potential enemies. This, as we shall see, was still valid
during the following decade or more but started to become increasingly problematic with the advent of printing and the wide circulation of cheap copies
of university professors' annual prognostications.[20] Within the predictive market
of Renaissance Italy, astrological prognostications possessed an intrinsic political value: they could provide advice for a lord, but also sensitive information
to his enemies. As such, they were often shrouded in secrecy.

Even though Francesco does not emerge from the documentation as a
particularly active patron of political astrology, it seems that he maintained
an active interest in political prognostications, even at the simpler level of
potentially useful information. Considering the perils of archival documentation, it cannot be excluded that more letters of this type reached Francesco

over the years. Even still, what is left is enough to indicate that astrologers saw in him a potential patron and felt encouraged to write. This may well be because of his acquired relationship with the Visconti, who were notoriously keen on this predictive art. If Francesco was presumably a more reluctant consumer, this was certainly not the case for his wife, and indeed for his sons Galeazzo and Ludovico. Indeed, the latter was possibly one of the most avid consumers of astrological prognostications of the Italian Renaissance. It seems therefore useful to contextualize further Francesco's figure by comparing his interest with those of his wife, Bianca Maria, and other personalities of the time. In what follows, therefore, I shall first look at a short but intriguing work of astrology by the Milanese humanist Pier Candido Decembrio written under Filippo Visconti's rule, then at Bianca Maria's own personal interest in some specific forms of astrological prediction, and then at the uses of astrological advice in another sumptuous court of the Renaissance, the Gonzaga of Mantua, who, in those years, had a very close relationship with the Sforza.

Filippo Maria and Pier Candido Decembrio: Astrology under the Last Visconti

In his *Life of Filippo Maria Visconti*, written around 1447, the humanist-secretary Pier Candido Decembrio (1399–1477) recounts how Gian Galeazzo Visconti, the first Duke of Milan and father of Filippo Maria, preferred Filippo to his older brother, Giovanni Maria, and believed he would make a better leader of his territories. His preference was based not only on his observation of his two sons, but also on the opinion of his court astrologers who had often asserted that, if Filippo Maria lived to reach mature age, he would bring much glory to the family name.[21] According to the astrologers, the qualities of a lord were largely inborn and determined by the position of the planets at the time of his birth. In other words, Filippo's rule over Milan was written in the stars. As we shall see in the course of this book, this was only one instance among many where astrologers formulated predictions of various kinds about the progeny of the dukes of Milan.

Decembrio wrote his *Life of Filippo Maria Visconti* with hindsight, when the astrologers' forecast had come true and not only was Filippo Duke of Milan, but also one of the most feared and respected leaders of the Italian peninsula. Bianca Maria Visconti, the woman destined to marry Francesco Sforza and give birth to Galeazzo and Ludovico—respectively, the fifth and the seventh Dukes of Milan—was born from Filippo Maria's mistress, Agnese del Maino, in 1425. It was largely through this materlineal Visconti line, as we shall see shortly, that astrology passed down to the Sforza. Astrology featured prominently in Filippo Maria's court as much as it had in that of his father. Although Decembrio's work had a eulogistic hue, it also offered a vivid and sometimes harsh portrait of its main character. In this work, Decembrio did not hesitate to portray what he considered the weaknesses of his lord, and among those he included his superstitious nature. In a series of brief chapters describing Filippo's character, Decembrio recounted his lord's superstitious beliefs and fears of (among others) darkness, the singing of birds, and lightning.[22] Decembrio dedicated a separate chapter to Filippo's belief in astrology:

> He gave so much credit to astrologers and astrology as a *scientia* that he attracted the most experienced practitioners of this discipline, and he almost never took any initiative without first consulting them: among those who were held in the highest esteem were Pietro from Siena and Stefano from Faenza, both very experienced in the art, while toward the end of his time as lord of Milan he drew actively from the advice of Antonio Bernareggi, sometimes Luigi Terzaghi, and often Lanfranco from Parma. Among his physicians he counted also Elia, the Jew, a famous soothsayer.[23]

Under their guidance, Decembrio reported, Filippo decided what were the best days to wage war and sign for peace, which were best for traveling and which for rest. It seems, however, that Decembrio did not share his lord's enthusiasm for astrology (or possibly, more simply for astrological elections), as he quipped: "I do not know for what reason he was induced to believe such credulous things."[24] Together with the movements of the planets, which Filippo seems to have read from the clock devised by Giovanni Dondi he kept in his library in Pavia, Filippo also paid particular attention to bad omens.[25] Decembrio reports how before the death of Filippo's father, a comet had been seen in the skies and had remained visible for three months, until, on the day of his death, it disappeared. Other strange omens, apparently, accompanied the three years preceding Filippo's own death. Filippo, Decembrio stressed, gave great importance to these celestial signs.[26]

From the *Life* we may conclude that Decembrio was skeptical of astrology and inclined to interpret it as mere superstition. Other elements, however,

suggest that he may have had a scholarly and even personal interest in the
subject, and thus that his dislike may have been only for certain astrological
techniques and applications, and not for astrology *tout cour*. Probably to
engage with the interests and beliefs of members of Filippo's court, Decembrio
penned a brief treatise on embryology and gynecology entitled *De genitura
hominis et de signis conceptionis, et de impedimentis circa conceptionem* that
contains much astrological embryology. Little is known of the circumstances
of composition of this astro-medical work, but it seems certain that the text
was conceived and written in the 1430s, while Decembrio was still residing in
Milan, employed as Filippo Maria's secretary.[27]

The text, which may have been inspired by the medical interests of
Decembrio's maternal grandfather, a well-known physician from Pavia,[28] was
later collected in his *Historia peregrina,* which he dedicated to the Milanese
jurist Nicolò Arcimboldi.[29] This *Historia* comprised two other seemingly
unrelated works: the *Cosmographia,* and the *De muneribus romanae rei publicae.* It is not clear why Decembrio chose Arcimboldi as his dedicatee, but we
know that the two were friends and members of the same humanistic circle.
Arcimboldi belonged to a distinguished Parmense family of jurists and eccle-
siastics, became *consigliere ducale* under Filippo Maria, and was often
employed by him in diplomatic missions. Decembrio and other Lombard
humanists regularly met at Arcimboldi's house to discuss classical texts, and,
for this reason, we may presume that Decembrio's early work was relatively
well known among Milanese intellectuals and physicians. The text could not
have been easily ignored by the Milanese medical and intellectual community in which Decembrio had taken his first steps: by the 1470s the *De genitura hominis* circulated widely across the Italian peninsula, often appearing
separately from the other two works which had accompanied it originally in
the *Historia peregrina,* and seeing at least nine printed editions between 1474–
1500, a clear indication of its great success.[30]

Although he was not an ardent republican, at the fall of the Visconti regime,
Decembrio took an active part in establishing the Ambrosian Republic. Thus,
when Francesco Sforza declared himself Duke of Milan, Decembrio preferred
to seek permanent employment at the papal court in Rome rather than try to
gain a courtly position with the Sforza; probably for this reason the *De genitura
hominis* was first published in Rome. This decision was not taken in order
to flee the Milanese court at the arrival of the new duke, however. Rather,
Decembrio's desire to leave Milan had arisen already in the final years of

Visconti rule, as the humanist became increasingly unsatisfied with his salary and tried to improve his career prospects by seeking employment within other courts.[31] Even if he did not reside within the territory of the Duchy of Milan, however, Decembrio remained actively involved in Milanese politics and courtly life. Ample correspondence between him and Francesco's trusted secretary, Cicco Simonetta, together with other letters to Bianca Maria and Francesco, clearly indicate that over time Decembrio maintained a strong and consistent relationship with the Duke and Duchess of Milan. Indeed, while in Rome he became an important intermediary between Francesco and the Holy See; so much so that in 1458 he pleaded for Francesco's imperial investiture with the Pope, and continued to act as Sforza cultural and political broker in the Eternal City in the following years.[32] Admittedly, with hindsight, Decembrio's relationship with Francesco seems rather one-sided and his efforts ill placed, but there is no doubt that Francesco saw in Decembrio a useful and valuable informant and promoter of Milanese interests.

While his relationship with Francesco was never particularly warm, Decembrio seems to have had a more intimate correspondence with the pious Bianca Maria.[33] A letter that he wrote directly to the duchess in the summer of 1460 after being informed by Francesco Visconti of her illness strongly suggests that his contacts with Bianca Maria and other members of the Milanese elite remained lively and his standing within the court good.[34] In this letter Decembrio—who was in Milan at the time—explained how, at the suggestion of Pope Nicholas V, he had translated the stories of Job and Tobias in the Bible into the vernacular because he had himself experienced that those stories, so inspiring and full of hope, had a therapeutic effect on people afflicted by illness or melancholy.[35] He had mentioned these stories to Bianca Maria's court physician, Guido Parati da Crema, but unfortunately they were not yet written down properly, and for this reason Parati did not mention them to her. In the letter, however, Decembrio asked Bianca Maria to advise what he should do. To please her and help her recover, he was more than happy to send the stories to her even if they were not as perfect as he would have wished. In the Middle Ages and the Renaissance, medical care was conceived as being as much physical as spiritual, and the reading of the Scriptures was clearly one way patients could remain positive and find some hope of recovery.[36] In this respect, therefore, Decembrio was simply following principles of contemporary medicine by offering a remedy for Bianca Maria's condition, thus fulfilling his duty as a citizen of the duchy and former member of the Visconti court.

As noted, possibly because of his grandfather's profession, Decembrio seems to have been in contact with Bianca Maria and Francesco's court physicians, and his *De genitura hominis* is likely to have circulated at court. Indeed, a copy of the *Historia peregrina* was probably housed in the ducal library together with other works by Decembrio.[37] The *De genitura hominis* is a brief vademecum about human conception, a theme that was particularly popular among late thirteenth- and fourteenth-century university physicians, who penned numerous treatises on human embryology.[38] It was, however, much shorter and more accessible than the medical treatises of the time, and this added to its circulation. Decembrio's text still awaits thorough study, but its format and content are relatively traditional: after discussing the way in which conception happens, it carries on to discuss the signs of conception, providing also recipes for conceiving a male instead of a female child. The rest of the work, which is divided into a series of brief chapters, treats things as varied as the nourishment of the fetus in the womb, how to determine if a woman is still a virgin, the length of gestation, and monstrous births.

Particularly interesting, however, is the section in which Decembrio analyzes the influence of the planets on the fetus.[39] In this section, Decembrio draws heavily on the medieval tradition of the *De secretis mulierum* attributed to Albertus Magnus to explain how man is made of matter and form (i.e., body and soul) and how certain virtues of the soul derive from planetary influence. According to this tradition, the baby would derive the faculties of reason and understanding from Saturn, magnanimity and other virtues of this kind from Jupiter, courage from Mars, concupiscence and desire from Venus, the faculty of knowing and remembering from the Sun, pleasure and enjoyment from Mercury, and the vegetative faculty, which is said to be the core of the organic faculty of the soul, from the Moon.[40]

In the subsequent chapter, Decembrio proceeds to illustrate those planetary virtues that are infused into the fetus before birth: Saturn governs the first month of gestation and impresses the vegetative and natural faculties upon the fetus,[41] Jupiter, with its warmer nature, predisposes the matter to take its form, while in the third month, Mars's dry heat separates the matter into well-formed members: the legs, the arms, the neck and the head. Then comes the Sun, which creates the heart and gives the fetus its sensitive soul,[42] and then Venus, which shapes the various members by adding the mouth, the nostrils, the ears, the genital organs, the feet and the hands (including toes and fingers). This happens in the fifth month. In the sixth month Mercury forms the instrument of the voice, the eyebrows, the eyes and the eyeballs, and, finally, it makes hair and nails grow. In the seventh month, the Moon intervenes to perfect and complete the structure of the fetus. The Moon, together with Venus and Mercury, provides moisture and nutrients for the baby; and if the baby is born that month, Decembrio explains, it is complete and will survive. The eighth month, however, is governed once again by Saturn, which has a dry and cold nature, and for this reason a baby born in this month is destined to die.[43] That this theory of planetary influence on the fetus was relatively well known among astrologers and university teachers is suggested by its inclusion in the commentary to the *Sphaera* of Sacrobosco written by the Bolognese professor of astrology Giovanni de Fundis, who, as we shall see, was one of the many astrologers to offer his services to the Sforza court.[44]

The phases of the Moon are also said to influence further the development of the fetus, but what seems to influence in greatest measure both the appearance and the character of the baby are, once again, the various planets. Slow and dark Saturn gives the baby (and the adult that will form subsequently) a dark complexion, a thick beard, and a small chest. He will be evil and mean of spirit. Jupiter, instead, gives the baby a beautiful face, clear eyes, a round beard, and big front teeth. The baby will have a nice complexion and long hair, will be good-natured, law-abiding, and generous. Mars gives the baby a dark-reddish complexion as if he were sunburnt; the baby will grow to have little hair, crooked eyes, and a chubby body; he will be immodest, haughty, and feisty. The Sun, king of the planets, will make the baby plump, his face well proportioned, his eyes big, his beard full, and his hair long. Some will say he will be a hypocrite, others that he will be of royal or religious disposition, wise, and joyful. Venus, a benevolent star, will make the baby beautiful, of medium build and height; he will be good with words, take pleasure in the care of the body, and be interested in music. Babies born under Mercury will be of graceful constitution, interested in philosophy, good-natured, immune to evil and betrayal. Those born under the Moon, the fastest planet, are born wanderers; they will have a gay expression, an average height, and one eye bigger than the other.[45] Each of the planets, in short, endowed the baby with particular physical and intellectual traits that were related to the planet's own characteristics.[46]

While Decembrio may have approached astrological interrogations and elections with some skepticism in his *Life of Filippo Maria Visconti,* this was clearly not the case for the phenomenon of celestial influence in general and for the stars' influence on human character; otherwise it would be difficult to explain why Decembrio wrote the *De genitura hominis*. Within a worldview that saw an intricate web of correspondences between the macrocosm (the celestial spheres) and the microcosm (the world below), celestial influence held a place of honor in people's understanding of the universe. The principle of celestial influence was firmly present in the works of Aristotle, both in his *Metaphysics* and in his natural philosophical works (*Physics, On the Heavens, Meteorology, On Generation and Corruption,* and *On the Generation of Animals*), where Aristotle formulated a nexus of causality between the movement of the heavenly spheres and generation and corruption in the world below.[47] People may have questioned the astrologer's ability to interpret the movements of the stars correctly, but the idea of planetary influence remained largely unchallenged, at least until the publication of Pico della Mirandola's *Disputationes adversus astrologiam divinatricem* in 1496, when Pico forcibly argued that the heavens acted upon the sublunary world only through movement, heat, and light, and any other occult influence was unproven.[48] Even assuming that Decembrio was not a staunch believer in all types of astrology, he shared with members of the court the basic belief that terrestrial events were influenced by celestial movements, and that a child was marked to an extent by the configuration of the skies at birth.

After having been a long-standing member of the Visconti court, moreover, Decembrio continued to remain in contact with the Sforza court, and particularly with Bianca Maria. In addition, he was on friendly terms with some of the court physicians, and he himself penned a book on astrological embryology, as we just saw. Furthermore, while skeptical about the idea of following astrologers' advice to the letter, Decembrio clearly listened to their advice in 1457 when he travelled to Naples, possibly on a ducal mission for Francesco. On the basis of celestial observations, the trusted court astrologer Antonio Bernareggi had advised Pier Candido to delay his departure for a few days. Decembrio must have followed his advice, thus escaping the terrible earthquake that hit the city a few days before his arrival. Having reached Naples safely, Decembrio wrote as follows to the duke's secretary, Cicco Simonetta:

> Magnificent and Highly Honorable Lord, when I left Milan to come to this part of the world, I did not imagine I would find so many impediments: I will spare you an account of the conditions of my travel, which

was never harder because of the mud, rains and constant floods of rivers that kept breaking their banks. It may be enough to write of the terrible news of the earthquake, which has devastated this region as never before. Thanks to the kind advice of Magister Antonio Bernareggi (whom I have now proved to be a sound astrologer), however, I left Milan a bit later than initially planned. Had I left earlier, I would have been caught in the earthquake, but I was still in Rome when the earthquake stroke Naples, and I barely perceived it. Those who were in Naples saw such a dangerous scene that it scares me even just hearing about it. On the thirtieth of December, a Thursday, there was another earthquake, but very different from the first, because much of the matter had settled, and it was one of those that lasts the time of two Paternosters, and, had it lasted a third, I doubt much would have remained of Naples.[49]

The rest of the letter describes in great detail the level of destruction that met Decembrio's eyes when he looked at the city: one could only distinguish heaps of stones among the trees, houses were being destroyed daily as too damaged to remain standing, Castel Sant'Elmo was in ruin, and the Certosa of San Martino was full of cracks. It was believed that up to 30,000, possibly 40,000, people had perished, and this reminded Decembrio, he said, of the necessity to live mercifully and according to the Scriptures.[50]

While the letter opened with a praise of Bernareggi's skills as an astrologer, it also closed with other remarks strongly suggesting that, despite some possible skepticism towards some astrological practices, Decembrio was no disparager of this predictive art: learned astronomers, he concluded, "are awaiting a conjunction of the three [superior] planets similar to that which brought the great plague all those years ago."[51] For more information about this phenomenon, however, Decembrio suggested that Cicco consult Master Antonio Bernareggi and Master Lanfranco Bardone (Lanfranco da Parma), who were much more knowledgeable than him about these things. It is possible, of course, that Decembrio expressed his appreciation for Bernareggi's predictive skills and his faith in astrology only to please his addressee, but why then suggest to Cicco to consult the two Milanese astrologers to gain more information about the forthcoming conjunction? It seems obvious that Decembrio had confidence in natural astrology and believed that the movements of the planets had an influence on Earth and on its people. He had some reservations, however, on the practice of one specific astrological technique belonging to judicial astrology, that of elections. It is also worth noting that Decembrio praised Bernareggi and expressed his gratitude to him confident that he would find in Cicco a sympathetic listener, thus indicating that the astrologer occupied a prominent position within Francesco's household.

Bianca Maria Visconti and Astrology

As noted, astrology was clearly part of courtly life under both the Visconti and the Sforza. Filippo Maria Visconti was an avid consumer of astrological advice; it is thus not totally surprising to discover that this interest was shared by his only daughter, Bianca Maria. Of the various astrologers employed by Filippo who were mentioned by Decembrio in his *Life of Filippo Maria Visconti,* only one, however, seems to have maintained a prominent position at the court of Francesco Sforza and Bianca Maria Visconti: the Milanese physician and astrologer Antonio Bernareggi just mentioned.[52] Son of a physician himself and member of the Milanese College of Physicians since 1422, Bernareggi became ducal physician sometime before 1440, with the specific duty of caring for the health and well-being of Filippo's mistress, Agnese del Maino, and their daughter, Bianca Maria.[53]

At the death of Filippo Maria, Antonio and other family members took an active part in establishing the Ambrosian Republic, and yet this did not prevent him from being employed as ducal physician by Francesco Sforza at his assumption of the duchy of Milan. As indicated by extant documentation, this was undoubtedly due to the privileged relationship that Bernareggi had forged with the female members of Filippo's household, and particularly with Bianca Maria. Indeed, it was Bianca Maria herself who took pains to provide Antonio Bernareggi with a flourishing career at the Sforza court after Francesco's ascension to power. This is clearly illustrated by a letter that the Duchess of Milan wrote to her husband in the summer of 1453, where she pleaded Bernareggi's case and asked her husband to reinstate his membership in the College of Physicians of Pavia. For his part, Francesco replied to her with words that clearly indicated his admiration for Bernareggi's service as a physician, and particularly for treating Galeazzo's illness successfully on a former occasion:

> My dearest wife, having been informed of what you write in your letter of the 24th of the past month of your order to the prior of the College of Physicians of our city [of Pavia] that they accept and reinstate Magister

Antonio Bernareggi's membership in the College as he was at the time of our Illustrious Lord, our father-in-law [Filippo Maria], and of the request you make that I write to them, we say that we want to do much more than this for Magister Antonio, for his merits and virtues towards us and our people, and particularly in this illness of our son Galeazzo, and we know how much he is attached to us and to our State. But as we fear that we will have opposition, given the agreement (*capituli*) we have with that community of Pavia, we do not think, in all honesty, that we will be able to satisfy your request; rather, we think you need to withdraw that order, because we can compensate Magister Antonio with other things in our control.[54]

Having just instated himself as Duke of Milan, Francesco was obviously wary of doing something that might compromise his position with the neighboring community, and especially members of its elite. We do not know in what circumstances Bernareggi lost his membership in the College of Physicians of Pavia, but it is clear that Francesco did not feel it was worth damaging his relationship with the College in order to fulfill the request of Bianca Maria's protégé. In this instance, therefore, Bianca Maria's plea on behalf of Bernareggi was not completely successful. Nevertheless, it is obvious that while Bernareggi did not regain his membership to the College of Physicians of Pavia, he maintained a role of prominence within the court.

Despite the change of power, therefore, we can trace a certain level of continuity between the intellectual and cultural environment of Visconti Milan and that of the Sforza period that followed. This is not surprising. Francesco Sforza's claim to the duchy of Milan was based on his marriage to Bianca Maria, with whom he built a successful partnership based on trust and respect. It also hinged on the difficult balance between his own power and that of the Milanese nobility and people of the old regime.[55] With more pressing issues awaiting him on the national and international scene, it was not in his interest to abruptly change more than was absolutely necessary. He clearly saw no reason why he should change physicians and astrologers when his father-in-law and his wife had retained some of the finest minds trained at the local university and were satisfied with their services. Bernareggi, therefore, represents a useful case study of how members of the Visconti court fared, some more smoothly than others, in the transition from the Visconti to the Sforza entourage. As the letter above suggests, Visconti women were key to this process.

After looking after Bianca Maria's health as a child, Antonio Bernareggi continued to care for her, her mother, and her children, thus benefitting from her direct patronage.[56] In September of 1453, we find Bernareggi caring for Agnese and her grandson Galeazzo;[57] in 1456, together with other court physicians, he attended to Elisabetta Sforza's wet nurse and, a few days later, he provided medical advice on the quality of some rooms that had recently been built above a well in a court of the castle.[58] In 1457 he wrote to Bianca Maria, together with the court physician Cristoforo da Soncino, to report about the health of the noblewoman Margherita da Santoseverino and the wife of a certain Messer Gentile. He added news about the city, the health of her own mother Agnese, that of Francesco and their children, and recounted how that day, with the favor of God, they had "purged" Galeazzo and Ippolita (presumably by bloodletting). Bernareggi was able to reassure her that the two children had complied with the operation without complaining and that this had worked well. Filippo and Ludovico, instead, had simply received some *cassia* (a purgative administered orally), which had also worked very well.[59] Confirming that Bernareggi was part of the team of doctors looking after the Sforza progeny, in December 1458 we find him and Soncino caring for another of the Sforza children, Elisabetta, who had a temperature.[60] This is further proven by a series of letters from the year 1459 that indicate Bernareggi was in the habit of visiting Francesco and Bianca Maria's children up to twice a day.[61]

Caring for the ducal children was certainly burdened with responsibilities, but it obviously came with a number of privileges and must have been an extremely desirable job. But while looking after Bianca Maria's children, Bernareggi also remained at the service of his most important client, Bianca Maria herself. In the summer of 1460, we find him at Bianca Maria's bedside; she was unwell for over a month.[62] Receiving contrasting reports about her health, Francesco wrote directly to Bernareggi and Soncino, the two physicians caring for her, asking for clarifications and recommending that they take care of his wife and not go overboard with those syrups and medications that they usually administered.[63] The tone, halfway between ironic and annoyed, finds a nice counterpart in another letter written by Francesco, this one to Bianca Maria herself, where he expressed his delight at the news of her improved health, but also added that presumably she was now well enough not to have her physicians around and that at the first occasion she should impose on them the very same restrictions that they regularly imposed on her:

> After this improvement, you will develop an appetite for food and drink, and the doctors will forbid you from eating and drinking. I would warmly

encourage you to do the same, and forbid them from eating those things they like the most, such as melons, fruit and other similar things. And make them stay an entire day without drinking wine, like I did in the past few days with Magister Gaspare when he was sick, as we forbade him from drinking wine and eating those things that he likes, so that he now knows what he does to us when I am not well. And you can ask Magister Antonio Bernareggi, who knows about this: and in this way they will relent and allow you to eat the things you have an appetite for.[64]

It is difficult to make sense of Francesco's words in relation to the treatment of his physicians: Was he serious or joking? The rest of the correspondence does not help us answer this question, but we may assume that, like many patients, he felt a healthy skepticism toward the numerous remedies administered by the doctors and the long list of things they forbade their patients from doing. Despite the optimistic tone of Francesco's letter, however, that very night the duke received news that Bianca Maria was not well once again, and in a letter dated July 26, he recommended that she not make the effort to write by herself, but that she rest and take care of herself until she felt better.[65]

In his letters Francesco expressed regret for the poor health of his wife and requested that he be updated hourly on her status. During this time, letters from Bianca Maria's physicians reached Francesco daily, and Francesco himself sent his courtiers to visit Bianca Maria regularly.[66] Francesco's skepticism about Bianca Maria's doctors, however, re-emerges strongly in another letter, this one dated July 27, where he reproaches the two physicians for not having acted swiftly to stop the progression of Bianca Maria's illness. After all, the doctors had all the means at their disposal to help Bianca Maria, who was not only young, but of good complexion, and intelligent enough to understand her condition very well. Bianca Maria—Francesco indicated—was very obedient to doctor's orders, and they could spend any amount they needed to on her medications. Doctors—Francesco added bitterly—can easily blame the patients for not listening to their orders, but Bianca Maria's physicians knew all too well that this was not true; Bianca Maria followed their orders to the letter, and they were very privileged to be paid to stay with her constantly, with the only task of looking after her.[67] All this caused him distress, Francesco lamented, as he heard that she was fine one day and unwell another. The doctors, he stressed once again, should have stopped the condition from deteriorating further, and this was causing him the greatest distress.[68] Francesco emerges from these letters as a harsh, sometimes reluctant patron of Bianca

Maria's physicians, and yet this may be simply a symptom of his profound concern for her health.

In the years 1460–1463, the care of Francesco and Bianca Maria's children was in the hands of a team of physicians: Guido Parati, a Magister Tebaldo, Benedetto Reguardati, and Giovanni Matteo Ferrari da Grado, but hardly ever our Bernareggi.[69] It is not clear why for a time Antonio abandoned his role of chief physician in care of the Sforza children, but his services may have been required elsewhere. Given his excellent reputation (despite Francesco's occasional negative remarks), Antonio was sent to care for numerous other courtiers and political individuals. For instance, in 1453 Giovanni Matteo Ferrari da Grado and Bernareggi took care of Giovanni Malatesta, who had fallen ill in Pavia; in 1459 Francesco dispached Bernareggi to Borgoforte, near Mantua, to care for the sick Marquis of Mantua, Ludovico Gonzaga; he looked after him again when the Marquis resided in Milan in the spring of 1461; in the summer of 1462, instead, he was sent by Francesco to the bedside of one of his relatives, Count Bolognino degli Attendoli, the *castellano* at Pavia; and in the autumn of 1464 he visited Maria of Savoy, Filippo Maria Visconti's widow, in Vercelli.[70] It is possible that Bernareggi was getting too old to be put in charge of the Sforza progeny. After all, he was the longest serving physician at court: his fame may have encouraged Francesco to send him to the care of relatives and allies, but he may have been otherwise too old to care daily for small, and often sick, children.

Another possible reason for Antonio's absence from the Sforza children's care during these years, however, may be related to Antonio's lecturing in Pavia, as suggested by some correspondence on this theme preserved in the Milanese archives. A letter dated 1458 documents how Bianca Maria wrote to Francesco about some new university lectures and about the salaries of some of the doctors teaching at the *Studium*. "Because of the death of some doctors who read in Pavia," she explained, "there are about one thousand florins left over." As a consequence, she wrote to the Privy Council, which was in charge of the *Studium,* asking its members to set aside five hundred florins until her return to Milan and a consultation with her husband. She expressed her desire that a hundred florins be given to Magister Guido [Parati], others to Magister Antonio Bernareggi, who—she added—wanted to give a cycle of lectures at Pavia, and a hundred or a hundred and fifty more to Giovanni Matteo Ferrari da Grado since they had been promised to him more than once for his lectures, should the money become available.[71] With the deference of any

Renaissance wife, Bianca Maria said that ultimately the decision rested with her husband, and yet from her letter it was clear that she had taken the matter to heart. In this letter we see once more how actively Bianca Maria was involved in promoting the careers and interests of her favorite physicians, going so far as to obtain confidential information regarding the budget for university teaching, and not hesitating to interfere as much as possible in the distribution of university places and salaries. If this did not meet with full success, it was certainly not for lack of effort on Bianca Maria's part.

Francesco's answer was sent to her only on November 3, and, once again, must have been disappointing: the members of the Council had written to him saying that they had kept the university appointments on hold while they waited to hear from him after receiving Bianca Maria's letters. Francesco asserted, however, that the Council could not satisfy Bianca Maria's request to raise the salaries of the university lecturers she had recommended. "Should we listen to all the requests of those who ask something and consent to them," Francesco quipped, "the thing would go on ad infinitum, and we would end up paying more in salaries than what we gain from Pavia."[72] For this reason he had decided to leave every decision to the Council, which had, he explained, a better sense of the competence, value, and condition of those who lecture in Pavia, thus clearly leaving Bianca Maria no hope that he could satisfy her request. Francesco explained further that Bianca Maria should reiterate the same to those who pleaded with her, and she should write to the Council straightaway to inform its members that they should go ahead and appoint the lecturers.[73]

As noted, these documents offer a rare glimpse of Bianca Maria's acts of patronage toward her own favorite courtiers and of her attempts (in some of these instances at least, rather unsuccessful) to influence the decisions of the Privy Council regarding university appointments.[74] We do not know if Bernareggi succeeded in returning to teach at Pavia and offer the cycle of lectures, as he so greatly desired. The survival of an astrological *taquinum* for 1461 by him could be an element in favor of this possibility, but firmer evidence is lacking.[75] We do know, however, that Bernareggi had had a flourishing career at the Pavian *Studium* under Filippo Maria, holding his first chair in logic from 1414 to 1418 and then in natural philosophy from 1418 to 1425. He is listed under the teaching of astrology from 1425 to 1429, and he imparted evening lectures both in medicine and astrology from 1429 to 1432, followed by practical medicine and astrology in 1432–1433, medicine and astrology in 1439–1440, and theoretical medicine alone from 1443 to 1447. In short, Bernareggi pursued a successful academic career as professor of medicine and astrology for over twenty years before losing his membership in the College of Physicians of Pavia and seeing his employment restricted to the caring for the duke's family.

By all accounts, Bernareggi's academic career should be considered extremely successful: by the time of his last recorded appointment under the Visconti in 1447, he received the healthy salary of four hundred florins for his teaching in medicine. These records, unfortunately for us, stop in 1448, and we have no other mention of him in the *rotuli*.[76] Bernareggi's name re-emerges briefly in the autumn of 1464, when we find him in charge of Francesco and Bianca Maria's son, Ottaviano, and again in the winter of 1464, when he was involved in the physical examination of Dorotea Gonzaga, Galeazzo's promised bride, about which more will be said shortly.[77] After this time, however, we lose track of him. We know for certain that he had made his last will in 1463, and that he was dead by 1467, when Bianca Maria issued a *privilegium* for his brother Protasio and his widow Caterina. At his death, his patrimony was quite solid: three houses in Milan, two in Pavia, and some land in the Lombard countryside, a clear sign of increased prestige and wealth acquired due to his profession as ducal physician and astrologer. In all of his wills, including his last one, Bernareggi made special mention of his library. At his death, part of the library was to go to his heirs, and part was to be given to the convent library of Saint Thomas of Pavia and thus be accessible to poor students who could not otherwise purchase those texts.[78]

As indicated in Chapter 1, the University of Pavia did not have its own library, and for this reason on occasion professors borrowed from the ducal library at the Castle of Pavia, while students seem to have resorted to college or conventual libraries in the city or borrowed directly from their professors.[79] I have already mentioned how the books donated by Bernareggi to Saint Thomas responded to the need for instruction in judicial astrology of students at an advanced stage in their four-year degree in arts and medicine when specializing in astronomy and astrology. More importantly, Bernareggi's donation of Guido Bonatti's *Liber astronomiae,* Haly Abenragel's *Liber de iudiciis astrorum,* and Alfraganus's *De aggregationibus scientiae stellarum* reflected the increasing popularity of Arabic authors, who came under ferocious attack in the early decades of the following century, when Arabic astrology, and especially conjunctionist astrology, became the object of increasing scrutiny after Pico della Mirandola's scathing critique of the discipline, and some astrologers started to pursue a reform of astrology along its classical (i.e., Ptolemaic) roots.[80] But a critique of Arabic astrology had yet to come, and at least for the period with which this book is concerned, there is little evidence that astrologers made a clear distinction between classic Ptolemaic astrology and its Arabic equivalent. Rather, what we generally find in fifteenth-century texts is the constant juxtaposition of these two intellectual traditions as if no real hiatus separated the authority of Ptolemy from that of Albumasar or Alcabitius.[81] By reflecting the tastes of fifteenth-century Italian astrologers, therefore, Bernareggi's booklist is particularly important in helping historians flesh out the history of the teaching of judicial astrology within Italian universities.

My discussion so far has been limited mostly to Bernareggi's practice as a physician. Unfortunately for us, the letters that Bernareggi wrote over the years are scanty on details concerning his use of astrological medicine in the treatment of patients. We should resist interpreting this fact, however, as evidence that astrological medicine was not part of his intellectual baggage and practice. As Marilyn Nicoud and Chiara Crisciani have noted more generally in relation to the treatments to which court patients were subjected by Milanese physicians, there is a significant epistemological gap between the formal training that these doctors received, the knowledge they acquired at university and the theoretical writings they authored, on the one hand and the way in which the illnesses of their patients were described in their correspondence with various members of the court on the other. The fact that fifteenth-century therapy abided by the theory of the six non-naturals and was grounded in humoral theory, for instance, barely emerges explicitly from fifteenth-century courtly letters written by these physicians. Yet we know for certain that these doctors treated their patients according to these theories.[82] While it is essential to avoid generalizations, we can thus speculate with some justification that a physician like Bernareggi, trained in medicine and astrology from an early age and a teacher of those subjects himself for almost twenty years, did employ principles of astrological medicine in his daily practice.

Why these astrologico-medical principles do not emerge in the correspondence is better explained by considering the specific context in which courtly medical discourse developed (in the form of courtly letters that were mostly short and to the point) and the audience for which these documents were written (largely non-professional physicians, but more simply courtiers of family members). This hypothesis seems to be supported by the fact that, in a later period, physicians wrote similar letters regarding the health of the then

Duke of Milan, Gian Galeazzo Sforza, to his uncles, Ludovico and Ascanio, but this time referring more often to principles of astrological medicine.[83] In this case, however, the audience to which the Milanese physicians addressed themselves—almost always comprising Ludovico and Ascanio Sforza—was reasonably well versed in astrological lingo, and thus recourse to principles of astrological medicine not only was comprehensible to their addressees, but would also provide authority to their practice. Moreover, by the 1480s the leading figure at court in terms of medical expertise was not a physician renowned for his *Regimen sanitatis* like Benedetto Reguardati, but the physician-astrologer Antonio Varesi da Rosate, who fashioned a brief but stellar career from his expertise in the science of the stars. We can even speculate that Ludovico's privileging of a physician-astrologer within his entourage may have led other physicians to embrace astrological medicine more boldly than in the former period.

Beyond Bernareggi's medical correspondence as one among many Milanese physicians caring for the Sforza family, other documents exist that are more revealing of his profession as court astrologer. Quite interestingly, these revolve around two interlinked issues that seem central to Renaissance astrological practice: the influence of the heavens in informing a person's character and physical disposition and the choice of one's bride or groom. Both issues are clearly interrelated with the kind of astrological embryology discussed in connection with Pier Candido Decembrio's work. As noted, these theories had common currency in university instruction and beyond. It therefore seems eminently plausible that Filippo Maria Visconti—who, after all, was a staunch believer in astrology—found in astrological embryology a perfectly reasonable set of principles to apply when attempting to produce an heir and that Decembrio's work provided a crisp and readable summary of well-known theories circulating at court. It is not difficult to imagine that Renaissance elites may have planned their marriages carefully to make sure not only that they were favored by the stars, but also that the conception of their progeny happened under the most auspicious sky. But were these only theories, or do we have evidence of their application at the Visconti-Sforza court?

Unfortunately, much of the material pertaining to the Visconti period is now lost.[84] The situation is more fortunate, however, in the following period. Perhaps unsurprisingly, the documentation available for the period of Francesco's rule is once again related to the maternal side of the family, and particularly to Agnese del Maino and Bianca Maria Visconti. The evidence I shall discuss reveals how astrology was employed to explore the future of the Sforza children and to secure successful dynastic marriages. As will be discussed in Chapter 5, this early example finds a nice echo in Ludovico Maria Sforza's similar actions half a century later concerning his own marriage and that of his nieces, Anna and Bianca Maria, thus suggesting that the practice of planning marriages astrologically was relatively common among the ruling dynasties of Northern Italian Renaissance courts.[85]

Marriage Strategies:
Dynastic Aspirations, Astrology, and Politics

Francesco Sforza's assumption of power in 1450 was not without problems. As a *condottiere* married to an illegitimate daughter of Filippo Maria Visconti, Francesco lacked real legitimation of blood. As the Holy Roman Emperor, Frederick III, refused to recognize him as imperial vicar in Lombardy, and to give legitimacy to his rule, his chancery was forced to forge a document of donation from Duke Filippo Maria.[86] One aspect of his foreign policy to which Francesco devoted particular attention from the very beginning, therefore, was that of arranging favorable dynastic marriages for all his progeny, and particularly for the heir to the throne.[87] In the early days of his assumption to power, Francesco had sought to marry his eldest son, Galeazzo, to Susanna Gonzaga from Mantua's ruling family. The reasons prompting Francesco to seek an ally in the Gonzaga were dictated by the threat of a Venetian war after Francesco's assumption of the duchy.[88] In the early years of Francesco's reign, such a marriage seemed reasonably favorable, so a formal marriage agreement between Galeazzo and Susanna was signed in 1450 and renewed in 1454.[89] A few years later, however, it became evident that Susanna was developing a deformity of the spine that would have made her unsuitable to bear healthy children. For this reason, Ludovico Gonzaga (1412–1478) himself decided to ask for that marriage contract to be annulled, and for Dorotea, his other daughter, to marry Galeazzo instead.[90] In addition, the new contract included a clause stating that this marriage, too, would be cancelled if Dorotea also developed a hunchback like her sister.[91]

Between 1458 and 1462, the two children spent much time together, with either Galeazzo visiting the Gonzaga court or Dorotea visiting that of the Sforza: Galeazzo was then a teenager (in 1458 he was fourteen years old) and Dorotea still a child (she was only nine at the time). It is said that in the early days of his acquaintance with Dorotea, Galeazzo expressed much affection and attachment toward the young girl. These tender feelings may have vanished, however, once Galeazzo became a boisterous adolescent. Indeed, doubts were raised about the excessive level of intimacy of the young couple.[92] By the end of 1458, Galeazzo had already fathered his first illegitimate male child, Carlo, and was clearly living out the exuberance of his youth to the full. As a new more powerful alliance with the French king, Louis XI, started to become a concrete possibility, concerns began to be expressed about the health of the new bride. Possibly in response to the tremendous levels of scrutiny to which the Gonzaga daughters had been subjected, Barbara Hohenzollern Gonzaga wrote to Agnese del Maino asking for reassurance about Galeazzo's health, character, and behavior. Unfortunately, we do not possess Barbara's original letter,[93] but we still have Agnese's reply, dated December 27, 1460, in which she tried to reassure the Marchioness with the following words:

> Most Illustrious and Noble Lady, my dear honorable sister, it is not that at present I did not believe (or do not believe) that you are somewhat concerned about the Illustrious Count Galeazzo. Yet, at present, you should not have concerns as everything has been put in place and provisions have been made, so that it is almost impossible that any mistakes or negligence will happen. So, My Lady, you should live happy and serene, also because, more for my own peace of mind than out of necessity, *I have asked for his nativity to be scrutinized by many astrologers, and particularly Magister Antonio Bernareggi,* who tells me that I should have no apprehension, as our dear count is not inclined by any planet to any vice; rather, he is strongly inclined to live well and virtuously, being particularly clement, pious, and magnanimous, without fear of any of his neighbors.[94]

Agnese's reply could not be more revealing of the way astrology shaped people's perceptions of character in the Renaissance: to reassure Barbara of Galeazzo's inclinations (which by this point may have been far from noble, at least toward young women), Agnese had consulted a number of astrologers, including Antonio Bernareggi, and was thus able to provide reassuring information to her correspondent. This quite clearly implies a common discourse whereby such an astrological explanation was deemed acceptable by the standards of the time. In Agnese's mind, astrology must have presented a reliable form of reassurance as to the character qualities of his grandson, and she must have thought that this view was shared by Barbara.

Galeazzo's horoscope, however, was not the only one to have been cast at the time of the proposed marriage. In a period of increasing uncertainty and difficulty, Dorotea's own nativity was scrutinized as much, and possibly more, than that of her fiancée. Barbara's attitude toward astrology was possibly not as enthusiastic as that of Agnese and Barbara's own husband, Ludovico, who, as we shall see, seem to have resorted to astrology regularly. And yet this did not seem to have deterred her from requesting (and receiving) astrological advice from her husband's favorite astrologer, the Mantuan Bartolomeo Manfredi, on various occasions.[95]

In 1463, probably aware of Barbara's mild skepticism towards his art, Manfredi enthusiastically reassured the Marchioness about the genuine nature of his prognostications, be they positive or negative. After having looked at her nativity, he also assured her that there was no danger to her state, but that, concerning her body, she should pay particular attention to her health from mid-December onward as she could be prone to an overabundance of choleric humor. Similarly, he added, she should remain alert about contagion and thus avoid conversations with people as much as possible. If she followed his prophylactic measures, Manfredi argued, she would run no danger of contracting the plague. Furthermore, he added joyfully, in December there would be a great conjunction in the heavens, and this would be extremely favorable to her husband's nativity.[96] Only a few days after receiving this letter, Barbara sent another letter to the astrologer, this time requesting Dorotea's geniture. The astrologer replied promptly to this request, first saying that he did not have the chart with him (but providing the details of Dorotea's birth necessary to construct one) and later sending her the chart with his interpretation.[97] Manfredi must have possessed a personal database (either written or mnemonic) of Gonzaga birth dates from which to draw if needed, and he clearly owned copy of Dorotea's horoscope, which he himself may have cast on a prior occasion, possibly at the time of her birth.

Once again, Barbara's inquiry was linked to the planned Gonzaga-Sforza marriage: she demanded reassurance about Dorotea's newly fading chances of marrying Galeazzo, and particularly about her health. To this request, Manfredi responded with as much precision as he could muster:

I wish to point out that her nativity [Dorotea's] is without fault (*difetto*), but that if anything inconvenient happens, it should be attributed to something else. In addition, I looked at the day when she was promised [to

Galeazzo]. It was on Wednesday, September 7, 1457, around the 19th hour and 30 minutes, when the ascendant was at 23° Sagittarius, the placement of the Sun in the radix [i.e., her nativity], and here in the house of Jupiter, at 19° 26' Leo, in the eighth house, strong and oriental, in its triplicity, and fortunate. This does not indicate imperfection—she was promised on the day and in the hour of Mercury, at 10° 58' Virgo, at an angle (*in li forci*) in the ninth house, that of religion, and the ascendant of her nativity. This denotes perfection in religious matters and support from the Holy See. And as by means of directions and revolutions I saw ahead, I can finally reassure my illustrious lord to have good hope to this end, and that this change is explained, as it must be attributed to something else.[98]

Not only did Manfredi look at Dorotea's chart to establish if she was predisposed to physical defects—a question for which Manfredi provided a negative answer—but he also examined closely the celestial configuration at the time when Dorotea was promised to the young Sforza.[99] According to Manfredi, this second aspect, too, did not give reason for concern, and he suggested, therefore, that other reasons were behind Francesco's sudden change.

Barbara's inquiry was prompted by the arrival at the Mantuan court of Francesco Sforza's trusted physicians Benedetto Reguardati da Norcia and Antonio Bernareggi, who were sent there to examine Dorotea's body in search of signs of deformity before the marriage could go ahead. Understandably, Francesco Sforza's request that Dorotea be examined was met with much opposition and resentment. Rumors of Francesco's diplomatic negotiations with the King of France to marry Galeazzo with Bona of Savoy had already reached Ludovico Gonzaga's ears, and he feared that the aim of such an examination was to annul the contract they had stipulated years before. Francesco Sforza, on his part, was aware of the problematic nature of such an annulment, which could undermine his relationship with a neighboring principality. To remedy the situation, Francesco gave instructions to his ambassador to reassure the Marquis that he should not think that the marriage was simply one of convenience, stipulated in 1450 only because at the time Francesco needed Mantua to be his ally against Venice. If this were the case, he argued with Ludovico, he would have abandoned the idea of a Gonzaga-Sforza marriage at the time when Susanna had been found unfit for marriage. Francesco, however, had remained firm in his intent to marry Galeazzo to one of Ludovico's daughters and had thus accepted the offer of marrying his son to Dorotea.[100] With these words Francesco was obviously trying to alleviate the seriousness of the situation, and, at the same time, to make a different marriage alliance palatable to a reluctant ally. For this reason he insisted that Dorotea be examined properly to determine her state of health. Ludovico, however, argued that examining Dorotea naked as requested by the Milanese physicians would have compromised her honor, and he requested that she be examined dressed in a light robe instead. On their part, the Milanese doctors did not give in and explained that to be sure that Dorotea was fine they needed to examine her spine from the lower neck to the tailbone, and they wanted to see her chest too. Comprehensibly, this stalled the negotiations.[101]

We do not know how Barbara planned to use Manfredi's response about Dorotea's chart, or indeed if she used it at all. But her astrological consultation constitutes an intriguing parallel to Agnese's letter, and it cannot be excluded that Barbara may have attempted to use Manfredi's prediction to reassure the Sforza of Dorotea's suitability for the marriage, much like Agnese had done only a few years earlier with Galeazzo.

Be that as it may, the two parties did not reach an agreement and the Milanese doctors were forced to return home. In late January 1464 the Mantuan courtier Giacomo da Palazzo reported to his lord a conversation he had with Benedetto Reguardati regarding the "Dorotea affair." The physician, he said, gave him his reassurance that both the Duke of Milan and he were operating in good faith, and there was no intention to cancel the marriage unless necessary.[102] This was not enough to convince Ludovico to let Dorotea undergo a physical examination. By August 1464, the case remained unresolved and rumors circulated that Bianca Maria was so frustrated with the situation that she had planned a meeting with Barbara Gonzaga during which Dorotea would be secretly examined by Antonio Bernareggi, who would have participated in it "dressed like an old lady."[103] The "Dorotea affair" split the court: some of its members were favorable to the marriage while others were contrary to it.[104] When Bianca Maria finally met Barbara in February 1465, however, all chances to save the marriage had evaporated for good.[105] Despite the astrologer's assurances to the contrary, Dorotea did, indeed, develop a malformation of the spine. Sent to a convent, Dorotea died a few years later at the age of nineteen.[106]

While Barbara had requested Manfredi to examine her daughter's chart, her husband Ludovico had consulted him to know more about his relationship with the Duke of Milan, Francesco Sforza, and the King of France, Louis XI. A busy Manfredi was asked to cast the nativity of both the King of France and of Francesco on at least two occasions: first on November 14, 1461, and, again, on February 11, 1464.[107] By 1461, it was well known that Francesco was trying to reach an alternative marriage arrangement with the King of France, and the situation needed close monitoring.[108] Astrology, therefore, became one of the ways in which Ludovico Gonzaga tried to keep abreast of the situation as it developed. In Manfredi's first letter to Ludovico, the astrologer divided his interpretation into three parts or stages: first, he examined the relationship between Louis XI and the papacy, then he turned to a comparison between Louis's chart and that of Duke Francesco Sforza, and finally he did the same with that of Ludovico Gonzaga himself. Mars, he explained, was lord of the ascendant in Louis's chart and, at the time of his birth, the planet was in Leo and in the house of religion. As Leo is said to be the ascendant of the Roman Pontiff, Bartolomeo argued, this suggested problems with the Holy See (*impedimentum*).[109] Moreover, Saturn, which was in Libra, was the lord of the year in which the king was born, and thus ruler and significator of his majesty. Because Saturn in Libra was the significator of the Christians and Louis was born king, it followed—he explained in elliptical fashion—that the king would seek after the Christian empire, not because of piety and justice, but rather to dominate, and this "because Mars is by nature impious and injust."[110]

The second part of his prognostication was not too different from the first: Francesco Sforza's Sun was in Leo, while Mars was in Leo and in the house of religion in Louis's geniture, and this apparently indicated troubles and Louis's interference in Francesco's alliances in the area of religion (and thus, presumably, the papacy). But, Bartolomeo added, as the Duke of Milan's geniture was very positive, this diminished the number of obstacles. Finally, he reassured Ludovico that Francesco's geniture was more in line with his own than with that of the King of France, with the possible exception of Saturn in Libra, which was said to be the indicator of the Christian religion and was somewhat problematic.[111] The third part of Bartolomeo's forecast compared Louis's nativity with that of Ludovico: in this case, Mars, the lord of the king's ascendant, was in Leo and in the house of religion, as Bartolomeo had indicated before, at about the same degree of the ascendant as Ludovico's. Because his majesty's ascendant was in the house of children in Ludovico's nativity, this indicated that Ludovico would be displeased by the king about his progeny for reasons linked to religion.[112] To complete the picture, however, Bartolomeo had also looked at Ludovico's son Francesco; he could not see any obstacles arising from the comparison of his chart with the king's.

Rather, it was propitious, which Bartolomeo found surprising, as everything else seemed to run contrary to it.

Unfortunately, we do not know how Ludovico Gonzaga interpreted Bartolomeo's words. They expressed a rather promising future for his son Francesco, but they also indicated problems related to religion or the papacy and to his own progeny more broadly. Could this be interpreted as a veiled reference to a likely papal annulment of Dorotea's marriage with Galeazzo? Without any additional documentation at our disposal it is hard to say, but it is certain that Bartolomeo's prognostication for his lord was not entirely positive and heralded difficult times for the Marquis.

Manfredi's failure to predict correctly that Dorotea would develop a hunchback did not diminish the Gonzaga's faith in their astrologer's predictive skills. By 1464 a marriage between Galeazzo and Dorotea seemed very unlikely. Ludovico's position in relation to Francesco Sforza and the King of France, therefore, had become even more precarious and difficult to gauge. For this reason, in February 1464, a few years after asking Manfredi to compare his own chart with those of both Francesco and Louis XI, Ludovico made the same request to his trusted astrologer. This second prognostication (*iudicio*), which made explicit reference to that of a few years earlier, was obviously aimed once again at gaining further insight into local and international politics. Manfredi reminded Ludovico of how Saturn in Libra indicated that soon enough the King of France would claim the crown of the Empire. The celestial events that they were witnessing, Manfredi added, were rare and unique, and their effects would be felt at least for the following twenty years. In recent times, Mars had conjoined Jupiter then Saturn, then the Sun conjoined Jupiter, and the Moon conjoined the Sun and Saturn. At the end of March, Saturn conjoined Jupiter and Venus first, and by April, the Moon also. To be more precise, on April 8 there would be a favorable conjunction of Saturn and Jupiter, and then, on the night of April 21, a total lunar eclipse, while on March 6 there would be a partial solar eclipse, but one too small to notice.

In the context of conjunctionist astrology, such a staggering series of astronomical phenomena (both conjunctions of the superior planets and eclipses) was perforce to have dramatic and lasting effects. The history of great men, therefore, would necessarily be affected by the powerful sequence of celestial phenomena outlined by Manfredi. But our astrologer's prediction dared to be even more precise. On June 25, he explained, there would be a conjunction of the Sun and Mars, and on this occasion, Mars would not be favorably placed.

This, he concluded, would not bode well for the French king, who would not obtain the Empire that year as he wished. In addition, Pope Pius II would want to organize an army against the Turks, but things would be looking up for the Turks from July onward.[113] Of course, with God's will, the Pope would succeed, and yet the signs Manfredi read indicated the papacy's weakness at the time. In a cryptic note, he also added that he who had encouraged the King of France to aspire to the imperial title would regret it, as it would set Christianity on fire.[114] Was this once again a subtle reference to Sforza's dealings with the King of France? It could have been certainly interpreted in this way.

Truly prophetic in light of the events that unfolded with the descent of Charles VIII into Italy in 1494 (of which more in Chapter 5), the note clearly indicates how astrology was actively used to read into the changing fortunes of European and Italian political actors. Astrology was not, therefore, simply part of a passive system of beliefs that has often been classified as superstitious, but, more importantly, a heuristic tool of investigation put at the service of Renaissance politics. The letters just discussed indicate unmistakably that Ludovico Gonzaga had the greatest anxieties regarding his place within Italian politics, and the fact that the planned Sforza marriage had fallen through made him particularly suspicious of Francesco Sforza and his dealings with the French king. Finding it extremely difficult to gather any information about the dealings undertaken between the two houses, Ludovico thought it appropriate to resort to astrological analysis.

In the summer of 1464 Francesco was still trying to quell rumors regarding his son's marriage with Bona of Savoy, but with mixed results.[115] Manfredi, therefore, was consulted once again to express his opinion on any development regarding a potential Sforza-Savoy marriage. The astrologer had been informed that Galeazzo had left Milan for France the previous Friday, August 2. Based on this information, Manfredi cast an interrogation chart, from which he was able to deduce Francesco Sforza's intentions. Unsurprisingly, he confidently stated that the Duke of Milan wanted to strengthen his realm by forging family links with France. Discussing another forthcoming Sforza marriage, Manfredi speculated further that Francesco would rather not send his daughter Ippolita to Naples to marry Alfonso of Aragon as planned, but as this would anger Ferrante he probably had to send her as previously agreed and would try to smooth things out as much as possible once the deal was sealed. Francesco Sforza's relationship with Naples, however, was not genuine, Manfredi indicated, and soon their real feelings would emerge, as he could evince from Mars being direct and in his exaltation, which would bring about war.[116] Once again, this may sound genuinely prophetic of what was yet to come. Only a few years later, Milan's relationship with the Aragonese would collapse under the strain of Galeazzo Maria Sforza's unreliable behavior and his delusions of grandeur. Once again Manfredi was prophetically right.

Beyond the accuracy of any of these predictions, however, what this epistolary exchange reveals unmistakably is that court astrologers like Bartolomeo Manfredi were significantly clued in to the ins and outs of Italian politics, and that they put this knowledge to use in their astrological prognostications, thus conceiving likely and unlikely political scenarios in the fleeting panorama of Renaissance diplomacy. It is obvious that no astrological data would ever provide our astrologer with as much insight into Italian politics as emerges from the Gonzaga correspondence. The court astrologer's job was therefore to interpret the feelings of his lord, garner as much political information as he could, and then provide as plausible a scenario as possible to help guide his lord's decisions. When speaking about the Sforza-Aragon marriage, for instance, Bartolomeo most likely discerned the thoughts of his lord or drew on second-hand information that circulated at court. Be that as it may, he provided an accurate prediction of what was to come. The marriage between Alfonso and Ippolita was not a happy one, and yet it forged a lasting (if brief) peace between Naples and Milan. As we shall see in the following chapters, the situation became more precarious after Francesco's death and the ascent of his temperamental son Galeazzo, only to deteriorate further in the late 1480s when Ludovico Maria Sforza usurped the title of Duke of Milan in all but name at the expense of his own nephew, the legitimate duke, Gian Galeazzo Sforza, who was by then married to Ippolita and Alfonso's daughter, Isabella of Aragon. In the precarious and ever-changing scenario of Italian diplomacy, therefore, astrology served the function of providing a viable interpretation of what was to come. If this was based on reading the stars, or more simply on reading between the lines of courtly correspondence and conversations, it is hard to say. Probably both.

The evidence presented so far indicates that astrology was actively employed as a political tool at both the Sforza and Gonzaga courts, but while Ludovico Gonzaga had a privileged relationship with one particular astrologer, the Mantuan Bartolomeo Manfredi, this does not seem to have been the case for Francesco. No astrologer seems to have become the close counselor of the fourth Duke of Milan. Indeed, possibly because of his military background and his more modest origins, Francesco does not emerge from the sources as a political leader avidly interested in resorting to astrology to obtain political intelligence in the way, for example, Ludovico Gonzaga did. This is particularly striking because, as we shall see, his sons, Galeazzo, Ludovico, and Ascanio, all demonstrated in their different ways a robust interest in this predictive art, resorting frequently to astrology in their political praxis. So far, instead, no evidence has emerged of Francesco's active involvement in requesting astrological advice. It seems, instead, that the active patronage of astrology, and particularly of Antonio Bernareggi, passed through his mother-in-law, Agnese, and his wife, Bianca Maria. Unlike them, Francesco may have maintained a modest distance from astrology and astrologers, and yet astrologers thought of him, quite naturally, as a potential patron. For this reason many documents exist that highlight the kind of astrological advice that was offered to him. Such advice, not surprisingly maybe, was largely political in nature. One possible exception to this trend was natal astrology, where Francesco could have been influenced by his wife. It is true that there is no document proving his requests for horoscopes to be cast, but we know that Francesco's horoscope circulated widely, and thus that somebody had bothered to find out his day and time of birth. This information must have been known to Bianca Maria and to Bernareggi as well, and it is likely that Francesco was kept privy to its astrological significance. He was considered, moreover, the apt recipient of a lengthy *iudicium astrologicum,* a horoscope, for his son Galeazzo. Significantly, however, while the figure on the illumination is that of Duke Francesco, during which reign the work was completed, the work is not addressed or dedicated directly to him, but to his son, Count Galeazzo Maria instead (Figure 15).[117] Was our astrologer attempting to catch two birds with one stone? As we shall see, this hypothesis is likely, as that very astrologer would later find employment in Galeazzo's court. It is therefore to Galeazzo that we shall now turn, this time to examine how the Visconti's penchant for astrology was transmitted to the Sforza children.

ON imerito uidlem sere
niſſime comes ad operis
tanti quod ego q̇.q̇.'indi
gnus confitiendum acce
pi laudem explicandam

FIGURE 15. Raffaele Vimercati donating his *iudicium* to Francesco Sforza, fourth Duke of Milan and father of Galeazzo Maria Sforza, from Biblioteca Trivulziana, MS Triv. 1329, *Liber iudiciorum in nativitate Comitis GaleazMarie Vicecomitis Lugurum futuri ducis* (1461), fol. 2r. By kind permission of the Archivio Storico Civico, Biblioteca Trivulziana, Milan. © Comune di Milano.

3

✦

Astrology Is Destiny

Galeazzo Maria Sforza
and the Political Uses of Astrology

In his *Historia di Milano* Bernardino Corio memorably recorded: "This Galeazzo was much subject to Venus and to filthy lust, in such manner that his subjects were greatly disturbed by this."[1] Galeazzo Maria Sforza could be certainly considered amongst the most eloquent representatives of the world of excess that characterized the lives of Renaissance princes. This was not for lack of effort on his parents' part, however. Both his father and his mother keenly strived to instill in their son the moral and military virtues appropriate to his recently acquired status of future Duke of Milan and invested limitless resources in providing him with the exemplary education befitting a Renaissance prince.[2] Yet the recommendations of his father, his tutors, and his physicians on the importance of measure and moderation seem to have fallen on deaf ears.[3] Quite simply, Galeazzo was not the self-controlled person that his parents wished for.

From a very young age, Galeazzo became notorious for his exuberant sexual behavior and, worse still, was hardly embarrassed by it, reportedly remarking scandalously that he possessed lust in full perfection, having employed it in all the fashions and forms possible.[4] His passion for the young noblewoman Lucia Marliani, and the eccentric behavior that accompanied it, became notorious at court and within diplomatic circles. Galeazzo himself actively encouraged rumors about this affair to circulate.[5] Falling madly in love with the young woman, he bought her from her husband and made her his mistress, lavishly providing her with expensive jewelry and clothes, purchasing for her a beautiful palazzo (whose furnishings alone would cost him over a thousand ducats), bestowing upon her the title of "Countess of Melzo," and giving her and her future progeny the name of the Visconti—all of this while being married to Bona of Savoy.[6]

While there was nothing exceptional in a Renaissance prince having a mistress (or even more than one), these affairs needed to be handled tactfully so as not to offend the sensibilities of the prince's legitimate wife and her family.[7] Galeazzo's own sister, Ippolita, for example, greatly and openly resented the numerous love escapades of her husband, Alfonso of Aragon, with both men and women, causing no end of trouble for her own family and her brother Galeazzo.[8] As if unrestrained love was not bad enough, the duke's lust often took remarkably more dissolute forms. Galeazzo was notorious for sexually abusing young prostitutes and sodomizing his own employees.[9] It seems clear why, in Corio's eyes, Venus, as the planet governing sexual desire and carnal pleasure, aptly epitomized Galeazzo's lustful nature and his sexual exploits.

Corio's reference to Venus and to the concept of planetary influence, however, should not be read simply as a metaphor: in the Renaissance a person's physical appearance and moral tracts were believed to be influenced by the position of the stars at the time of birth, and Corio would have thought no differently of Galeazzo. Education, nonetheless, could have helped the future Duke of Milan correct some of the less pleasant traits of his character. This is clearly what his parents had hoped. Doubts about Galeazzo's disposition, however, had already been expressed when he was an adolescent. As illustrated in the previous chapter, his grandmother had asked the astrologers to scrutinize his horoscope in search of reassurance about his moral qualities and his suitability for a Gonzaga marriage, while his mother patronized the very astrologer who was asked to express his judgment on Galeazzo's qualities. Astrology, it seems, was never far away from the court. When circumstances required it, astrologers could be called in to express a judgment on the character of the ducal progeny and of their prospective spouses.

If Galeazzo took little heed of his parents' concerns about his behavior, this was not so for astrology. Within his family, Galeazzo's penchant for this discipline was second only to that of his brother Ludovico. Unlike him, however, Galeazzo did not develop a privileged relationship with one single astrologer, but preferred to draw on the expertise of a series of them, to be consulted often at the same time. Galeazzo's interest in astrology was not exceptional in the context of Renaissance court culture. As already noted, Ludovico Gonzaga resorted with some frequency to astrology in times of crisis, and Borso and Ercole I d'Este of Ferrara were keen patrons of astrologers as well.[10] Yet this aspect of his life has passed largely unnoticed: even Gregory Lubkin's meticulous monograph on his reign barely makes any reference to it.[11]

As this chapter illustrates, however, Galeazzo's existence was fashioned by the stars in more ways than one: not only did he patronize astrologers at his court, but he was also victim of astrologers' speculations and negative prognostications, to the point that, especially in the years preceding his death, he tried to exert as much control as possible over the circulation of this type of information for fear that it would undermine his authority and create social unrest. Far from being a passive agent seemingly at the mercy of astrological determinism, however, Galeazzo always strove to grasp and dominate the influence of the stars: he avidly collected astrological information about himself and those who surrounded him and actively endeavored to control and manipulate the heavens to his own benefit. He did so not only by surrounding himself with astrologers who could offer him their professional expertise, but also by asking his agents and ambassadors to canvas information about other astrologers' predictions and send them back to him. He was thus, by all accounts, a patron and consumer of astrology.

Following the example of his grandparents and his own mother, during his life Galeazzo demonstrated a vivid interest in astrology's diverse personal and political applications, establishing astrology as one of the politically valuable domains of intellectual inquiry at the Milanese court. The present chapter illustrates several ways in which astrology was inextricably linked to his reign and highlights how Galeazzo's interest in astrological predictions was part of a wider political context that valued such information as one important form of political "intelligence." Such intelligence, as we shall see, could be used for personal, diplomatic, and military ends, depending on the need and the occasion. This chapter serves thus three functions: first, it documents Galeazzo's relationship with contemporary astrologers and their services; then, it provides examples of the ways in which astrological prognostication informed Galeazzo's political considerations, influencing his foreign policy and his relationship with other Italian states; finally, and more broadly, it illustrates the ways in which astrological prognostication, political propaganda, and diplomacy interwove with one another to shape Renaissance political praxis.

Destiny Is Astrology: Galeazzo's Natal Horoscope

Like many other prominent rulers of his time, Galeazzo Maria Sforza had his own natal horoscope cast. Among elites, this was hardly unusual. In general, it was believed that the interpretation of a person's geniture could provide useful information regarding his or her own future, thus giving a person the chance to actively dominate the stars, counteracting the evil influx of the malignant planets or certain unfavorable aspects, while maximizing one's chances to make the most of the positive influx of the benevolent planets and of any positive aspects in the chart. As the stars could only incline and not necessitate—or, in a phrase often attributed to Ptolemy, "the wise man rules the stars"—man's free will was, at least theoretically, safe-guarded. Seen in this light, astrology could be commended as the discipline that would allow men to take control of their destinies.[12]

Although we do not know for sure if Galeazzo's horoscope was analyzed at the time of his birth, it is certain that his geniture was cast and analyzed numerous times in the course of his lifetime. We have already seen in Chapter 2 how his own grandmother had requested the interpretation of his chart from numerous astrologers and how she had used it to reassure Barbara Gonzaga of her grandson's flawless character. We know also that his horoscope, that of his father, Francesco, and of his first prospective wife, Dorotea, were scrutinized intensively by the Mantuan astrologer Bartolomeo Manfredi in the months preceding and following the annulment of the marriage contract between Galeazzo and Dorotea. Furthermore, the astronomical data necessary to cast Galeazzo's horoscope are appended to the verso of the first folio of a manuscript now at the Ambrosiana Library containing Alcabitius's *Introductorius*, an unidentified excerpt from a work by Zael, and a number of astronomical tables, suggesting that the time and day of Galeazzo's birth were known among Lombard astrologers.[13]

When casting horoscopes, however, some astrologers went to greater lengths than others. The little-known astrologer Raffaele Vimercati, for instance, did not save any efforts to impress Duke Francesco and his son Galeazzo with a lengthy interpretation of Galeazzo's horoscope. The lavishly decorated manuscript of Vimercati's *iudicium* now in the Trivulziana Library, we are told, was duly presented to Francesco in the summer of 1461.[14] With its beautifully illuminated frontispiece, this small booklet containing a detailed reading of Galeazzo's nativity is clearly revealing of the delicate relationship between patron and client. In this image Vimercati kneels respectfully in front of Francesco Sforza and extends his hand graciously to offer him his precious booklet (the leather cover painted in the manuscript's illumination looks exactly the same as the one preserved to this day). In the other hand he holds Francesco's hat while Francesco receives the blessing of God from above (in the shape of a crown). The illumination celebrates Francesco's power and magnificence: not only does the Duke of Milan wear his distinctive red and white *calze* (stockings)—the Sforza colors—but he is also clothed in a finely embroidered golden robe. To complete the picture, the Visconti coat-of-arms stands firmly at the bottom of the page surrounded by Galeazzo's initials.

While the iconography is not particularly original, there is clearly nothing in this image that the illuminator and the astrologer left to chance.[15] Every detail was studied with care to ensure that the product would attract the full appreciation of the two dedicatees, Francesco and Galeazzo Maria (see Figure 15 in Chapter 2). The strategy of paying tribute to both Francesco and his son was hardly accidental: in this way Vimercati ingratiated himself both to the current Duke of Milan and to his successor, for whom, as we shall see, he would eventually cast other horoscopes.

MS Trivulzianus 1329 is a beautifully illuminated presentation copy that reveals distinctly both the ambition of its commissioner and that of its dedicatees. With its sixty-three folios, this parchment manuscript is possibly the single longest astrological interpretation of a fifteenth-century horoscope still preserved.[16] It is certainly one of the most remarkable examples of the fine intricacies and sheer complexity of Renaissance astrology. The text contains a wealth of information deemed relevant to the life of Galeazzo. Ad-hoc astrological counsel, it was believed, could help him foresee and prevent future difficulties and dangers, both personal and political. As extensive commentaries on natal charts are relatively rare for the fifteenth century, this text provides much insight into the life and works of a Renaissance astrologer, revealing at least as much about Vimercati's astrological competence and personal aspirations as about the political and personal ambitions of his patron.[17] Unfortunately, little is known about the author of this work, who is very likely to have

studied medicine and astrology at the local university of Pavia and was probably seeking more permanent employment at court. Vimercati did not publish
anything during his life, and his service as an astrologer, as far as we know, was
limited to the court of Milan.[18] In the next few pages, therefore, I shall provide
a brief analysis of the text and discuss some of the motivations that led Renaissance astrologers to produce such artifacts for their prospective patrons.

Writing such a lengthy *iudicium* was no trivial matter. It required a high
level of competence in astronomical computation and astrological interpretation, and such a lengthy *iudicium* would probably take a few days to be produced. The first step in casting a geniture was to produce an accurate chart
(Figure 16). Vimercati did this admirably, going through a series of intricate
calculations to rectify the chart and obtain the most accurate celestial figure.
Scholarly sources generally report that Galeazzo Maria Sforza was born in
Fermo (Italy) on January 14, 1444.[19] The examination of the geniture in
Vimercati's manuscript, however, allows us to be more precise than this. The
astrological data in the geniture records the position of the stars in the sky at
about 2 a.m. on the following morning, January 15 (Table 2, col. 1).[20] The
same celestial figure tells us also that Vimercati adopted astrological tables for
45° latitude. We can presume that these may have been the very popular
Alphonsine tables, which were widely used in Northern Italy at the time, or
that our astrologer may have used an astrolabe.[21] To obtain the most reliable
celestial chart, Vimercati rectified the chart by applying the Arabic theory of
animodar, a complex series of calculations aimed at gaining greater precision
as to the time of birth.[22] Then he divided the chart into the customary twelve

Table 2: Astrological data for Galeazzo's natal chart.

Planet	Vimercati	Alphonsine Tables	Cardano	Modern
Moon	5°54' Sagittarius	6°42'	6°36'	5°52'
Mercury	10°3' Capricorn	10°22'	11°	8°75'
Venus	1°10' Aquarius	1°36'	0°52'	0°59'
Sun	3°22' Aquarius	3°47'	3°47'	3°21'
Mars	13°22' Virgo R	13°26' R	13°20' R	15°57' R
Jupiter	11°43' Taurus R	11°43' R	11°48' R	10°52' Gemini R
Saturn	27°24' Gemini R	27°22' R	27°15' R	26°01' R

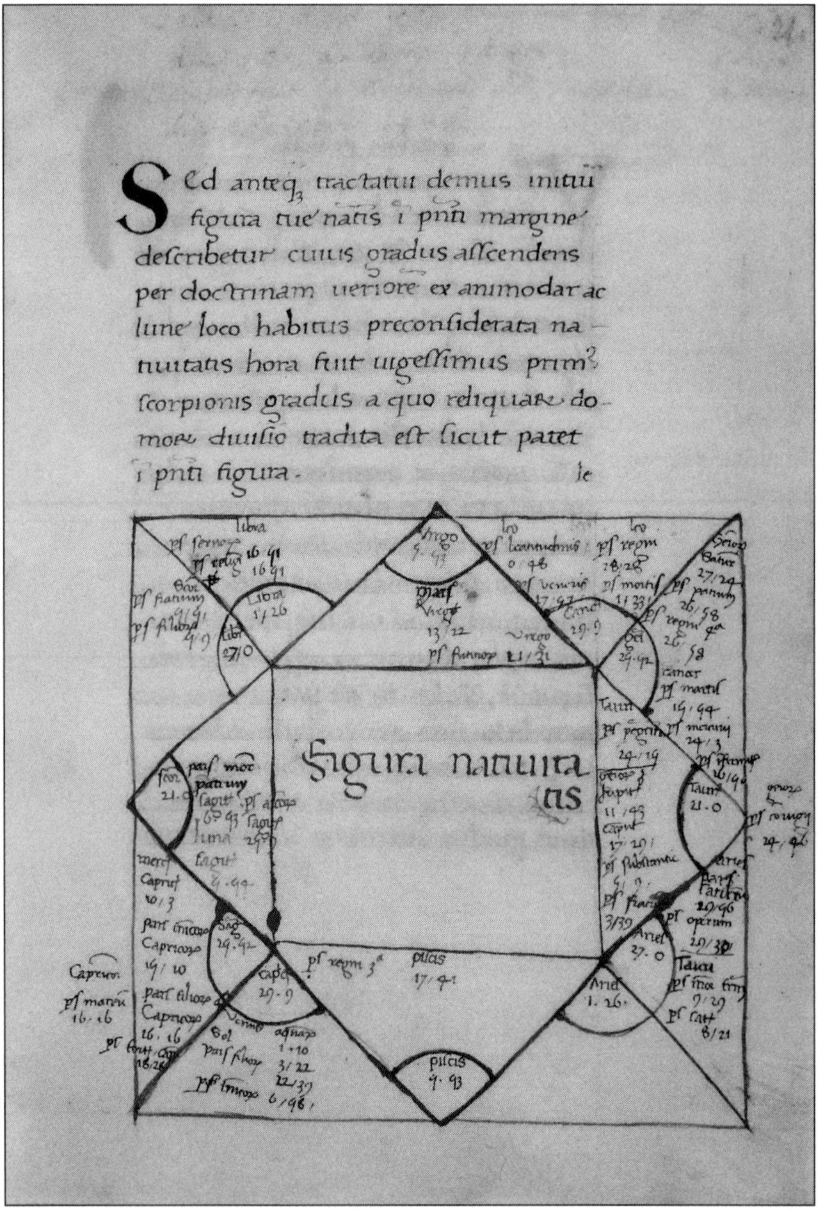

FIGURE 16. Geniture of Galeazzo Maria Sforza, from Biblioteca Trivulziana, MS Triv. 1329, *Liber iudiciorum in nativitate Comitis GaleazMarie Vicecomitis Lugurum futuri ducis* (1461), fol. 21r. By kind permission of the Archivio Storico Civico, Biblioteca Trivulziana, Milan. © Comune di Milano.

houses, adopting the standard house division system usually attributed to Alcabitius. Once again, Vimercati's calculations demonstrate a high degree of precision (Table 2).

While casting a celestial figure demanded precise computational skills, its interpretation allowed the astrologer much more room to maneuver. What elements to consider and how to interpret them were things that astrologers would learn from books and by practicing, but no one astrologer would interpret the same chart in the very same way, and thus subjectivity, personal taste, and familiarity with different scientific authorities would lead to different readings. In other words, mathematical certainty and precision gave way to conjectural elucidation.

Vimercati offered a lengthy and detailed interpretation of Galeazzo's geniture: his commentary addressed an impressive array of topics, and it would be impossible to do justice to the sheer complexity of the text in the space of this chapter. What is significant for our purpose, however, is that Vimercati indulged in extended calculations of one aspect of Galeazzo's chart that had both enormous personal and political import: the duration of the duke's life. In the same *iudicium* Vimercati also explored and analyzed the times and causes of his illnesses, and he predicted how Galeazzo would die.[23] Thus, while the mathematical aspect of the chart gave apparent objectivity to the process, the interpretation that accompanied it touched on the most personal and intimate aspects of a person's life, but also—in the case of the leader of a principality—on aspects closely related to his public *persona*. Aspects that could be considered of political import.

Attempting to address these delicate topics while the client was still alive was no small task for any astrologer, let alone one whose patronage so directly depended on the good grace of his patron, as in Vimercati's case. The calculation of the length of life was a particularly delicate matter. For this reason it required care and sound judgment on the astrologer's part.[24] The length of life could be calculated from the natal horoscope following a complex series of calculations and manipulations. This practice conformed to a specialized technique called *prorogation,* which had been codified by Ptolemy in classical antiquity and further developed by Arabic astrologers in the Middle Ages.[25] In *Tetrabiblos* III.10, Ptolemy had likened the lifespan of man to an arc of the celestial ecliptic. The arc would start at a particular point on the ecliptic, the *hyleg* (the planet or point considered to be the "giver of life" in the chart). From there the life would be cast forward with a greater or lesser force, depending on the strength of the *alchocoden* (the planet that was deemed to be "the giver of the years"). This trajectory could be arrested, or reduced in length, by encountering one or more destructive points or planets met in its trajectory (the *anaeretae*). All these factors, considered together, would determine the actual lifespan of a person.[26]

To calculate the duration of Galeazzo's life, Vimercati needed, therefore, to establish two things: the *hyleg* and the *alchocoden* of his client's chart. Not surprisingly, the calculations required to determine the *hyleg* and *alchocoden* form one of the most esoteric and complex sections of Vimercati's treatise. Such obscurity may not have been wholly intentional: after all, Vimercati was dealing with one of the most challenging aspects of Renaissance astrology. The process is hardly intelligible in Ptolemy, and it did not become clearer as a series of Arabic authors tackled it in their writings. There was, in fact, little consensus among Arabic medieval sources as to how the *hyleg* was to be determined, and this issue was often debated by astrologers.[27] Yet there were other reasons to be obscure. Vimercati was certainly aware of the serious risks involved in arriving at an unfavorable result and predicting a short life for the duke. If the astronomical data were unfavorable, could he really openly declare that the duke would die young? Aware of the personal consequences involved, any prudent astrologer would have been very wary of offering such a prognostication.

While astronomical data bound the astrologer to the chart, freedom of interpretation and the use of competing authorities laid the chart open to extensive manipulation based on personal, professional, and political interests, regardless of accuracy. Vimercati's obscurity, therefore, may have not been motivated simply by complexity. There were, in fact, more pressing reasons to account for why he may have tried to obscure the *arcana* of his calculations, which, as we shall see, led him to make a remarkably optimistic prediction. In his history of Milan, Bernardino Corio, a near contemporary of Vimercati, recounted how Galeazzo severely punished those who dared to make negative predictions on his life, something that, apparently, was not uncommon. Bernardino recalled clearly how Galeazzo:

> was cruel, to the point that, when the duke asked a priest how long he would reign, he answered that he would not reach the eleventh year, and for this reason he put him in jail, sent him a small piece of bread, a glass of wine and a wing of capon, and let him know that he would not receive

anything else. The man survived on these things, even eating his own excrement, for twelve days. Then he died.[28]

There is little doubt that Vimercati would have been well aware of the pitfalls of astrological prediction as well as of his duke's bad temper. It is even possible, as we shall see shortly, that he knew the story of the poor priest that was let to die in prison.

We should now return briefly to the chart and its interpretation. After elaborate considerations and having performed some simple arithmetical sums, Vimercati established that the *alchocoden* of the chart was "Jupiter, who looks at the position of the Moon, where its house, triplicity, and term exert their power." He then proceeded to calculate the duration of Galeazzo's life as follows:

> [. . .] the giver of your years will be Jupiter himself who, since he is in aspect, will bring you your greater years, which, according to the opinion of the learned, are said to be seventy-nine. However, because it recedes and was at the end of its retrograde motion (as Albumasar agrees) the fifth part of the decades of your years must be subtracted; with this taken away sixty-three years, seventy-two days will remain. In these years, the aspect of the Moon toward the said *alchocoden,* which it receives, will add its lesser years, that are twenty-five, but from these a fourth part is subtracted because of the vicinity of the *cauda draconis* to the Moon of less than twelve degrees, and thus eighteen years and nine months will remain. These, once added to the years of the *alchocoden,* will constitute eighty-one years, eleven months in total, and these are the years that, it seems, your life will last, unless the misfortunes of the *hyleg* make it shorter.[29]

While Vimercati did not explicitly indicate which planet was the *hyleg* of the chart, some quick calculations allow us to determine that the only way to obtain Jupiter as the *alchocoden* is by establishing the Moon as *hyleg.*[30] Having established both the "giver of life" (*hyleg*) and "the giver of the years" (*alcochoden*), Vimercati proceeded to determine the duration of Galeazzo's life based on the maximum, median, and minimum value given to each of the seven planets (including the luminaries) depending on their placement in the chart (Table 3).[31] Vimercati thus reached an enviably positive result, skillfully predicting that Galeazzo would live to reach almost eighty-two years. But even in what is seemingly a straightforward set of arithmetical additions and subtractions, we can detect a level of flexibility and personal interpretation. Another look at the geniture makes this clear.

Table 3:   Lesser, middle, and greater years of the planets, depending on their position
            of strength or weakness in the chart.

Planet	Greater	Middle	Lesser
Moon	108 years	66 years	25 years
Mercury	76 years	48 years	20 years
Venus	82 years	45 years	8 years
Sun	120 years	69 years	19 years
Mars	66 years	40 years	15 years
Jupiter	79 years	45 years	12 years
Saturn	57 years	43 years	30 years

Galeazzo's nativity shows Jupiter at 11°43' Taurus, in a trine aspect with Mars (which, Vimercati had established earlier, was the *almuten,* or "lord of the geniture").[32] To give Jupiter its maximum years—which, according to traditional sources, are seventy-nine—Vimercati greatly emphasized the positive aspect of Jupiter and Mars and the fact that Jupiter is "received" by the Moon.[33] To this value he subtracted one-fifth (roughly fifteen years) because in the geniture Jupiter was retrograde (a detrimental factor), and he added twenty-five years because the Moon "received" Jupiter (this allowed him to add to the calculations the Moon's minimum years, which are twenty-five). To the number obtained, he further subtracted one-fourth because of the vicinity of the Moon to the *cauda draconis* (also deemed a detrimental factor). This allowed him to conclude that, "unless the misfortunes of the *hyleg* make it shorter," Galeazzo would live a very long life. By arguing that Jupiter's position was extremely favorable—a conclusion that, as we shall see, was not shared by other astrologers—Vimercati was thus able to provide an audaciously optimistic prognostication.

So far we have followed Vimercati in his calculation of Galeazzo's length of life. Later in the text, however, Vimercati addressed another delicate question: how Galeazzo would die. Once again he was treading on difficult and dangerous ground; once again he reached an optimistic conclusion. His duke would have wanted reassurance that his death would not be violent and, ideally, as painless as possible. Not surprisingly, Vimercati firmly reassured Galeazzo that he would die in old age of natural causes "by the extinction of inborn heat because of the failure of the radical moisture" and most likely from an illness of the chest and of the intestines.[34] A rather unimaginative (and yet extremely safe) prognostication about anybody's death! Perhaps even more significantly, however, Vimercati hastened to reject the possibility that Galeazzo might die a violent death.[35] No duke, we may presume, would have been happy with such a prediction.

Vimercati's decision to address these aspects of his duke's chart in writing was certainly daring. Most of the extant medieval material seems to indicate that this kind of prediction was often drawn retrospectively, and it is evident why it was safer to do so.[36] So how should we interpret Vimercati's prognostication? Should we see it as a purely self-serving exercise aimed at gaining the favor of his duke? Did he really follow what he believed were proper procedures to achieve the correct results? The answer is both yes and no. The first hypothesis would not explain why Vimercati went to such lengths to produce a truly accurate chart: astrological calculations, after all, required hours of labor and were intellectually challenging; and rectifying the chart according to the theory of *animodar* would have added considerable time to an already lengthy process. Vimercati did not cut corners.

The second hypothesis, however, seems disingenuous for reasons that will become apparent shortly. It seems more plausible to assume that a different cultural and intellectual process took place: astrology, as a conjectural art, lent itself by definition to uncertainty. This was due both to the complexity and number of the elements to be considered—the position of the planets in relation to the zodiacal signs, the planetary aspects, the planets' placing within the twelve houses—and to the human factors involved in the interpretative process. More to the point, the stars could only incline and not necessitate. This gave room to the astrologer to interpret the apparently objective astronomical data in a personal way. Thus, while the politically calculated nature of Vimercati's interpretation cannot escape the modern historian, it may be unwise to interpret it cynically as nothing but personal expedience. It was more like a balancing act between the rigor of professional expertise and the necessity to please and exalt one's client and patron.

In the Renaissance calculating the life expectation of a client was not unusual. The risks involved, however, were self-evident. It is telling that as horoscopes entered the public sphere with the help of the printing press, the interest in prorogations grew consistently through the fifteenth and sixteenth centuries. Steven Vanden Broecke has recently suggested that the burgeoning fortune of prorogations in the early sixteenth century may be due to the early modern appetite for personalized predictions.[37] Prorogations, for example, loom large in the works of both Cardano and Gaurico, two early modern astrologers whose success owes much to the revolutionary power of the printing press and their self-aggrandizing aspirations. By the sixteenth century it became more common to divulge genitures whose interpretation included details as to the modalities of death of the person in question. As early modern astrologers operated in an increasingly competitive predictive market, they were often forced to offer bolder, more detailed interpretations to outdo their rivals.[38] Cardano, for instance, knew well that daring predictions on the life of famous rulers, when correct, could bring much honor, patronage, and money. When they were incorrect, however, they were embarrassingly difficult to justify. Among the examples published in his *Liber duodecim geniturarum,* Cardano himself pointed out how seven were of people still living, four of people who were already dead, and one was drawn while the client was still alive but had since died.[39] This last case was none other than the *cause célèbre* of Cardano's incorrect prediction of the length of Edward VI's life, in which occasion Cardano had initially envisaged a long life for the young king, only to discover a few months later that Edward had died without reaching his sixteenth birthday.[40]

Put on the spot and in need of a justification, Cardano went over the geniture a second time—a process that by his own account took him about a hundred hours—and paid renewed attention to the direction of the *apheta,* namely the "giver of life" (what the Arabic tradition called the *hyleg*) and the destructive planets (*anaeretae*) that may have crossed his path. He concluded his remarks with the following cautionary note:

> [. . .] it is evident that we should not predict the duration of life in weak genitures without having first examined all the directions of the *apheta* [i.e., "the giver of life" or *hyleg*], its processions, and ingressions. And if what I had predicted in the prognostication about them did not happen, I would expect that they would justly complain about me.[41]

Had he originally calculated the directions of the Sun and the Moon, he explained, he could have seen the danger signs emerge. If he had gone on to look at processions and ingressions, moreover, he would have discovered that the young king was doomed to die.[42] Cardano described at great length the difficulties and dilemmas of astrologers who predict the lives and deaths of kings. In his writings, he vividly underscored the personal danger involved in such astrological activities, dwelling on two cautionary examples of people who had predicted violent deaths for their princes and had paid for such daring prognostications with their lives.

The first example, which he drew from his reading of the classics, was that of the Roman astrologer Ascletarion, who was said to have predicted the death of the Emperor Domitian. The second, which he may have heard recounted or read in contemporary chronicles, was that of Galeazzo Maria Sforza's priest, who had unwisely prophesized the death of his duke. The first example drew from Suetonius's account of how Domitian, unhappy about Ascletarion's prediction of his death, had asked the astrologer to predict the way his own life would end. The astrologer's somber reply was that he would soon be mauled by dogs. Desiring to prove the fallibility of Ascletarion's art, Domitian ordered his assassination. To make sure that his own prediction would not be fulfilled, he further ordered that his body be burnt on a pyre. The rest of the story is predictable: Suetonius tells of how the funeral ceremony was interrupted by a sudden storm during which a pack of dogs attacked and ripped apart Ascletarion's body. In Cardano's view this cautionary tale proved both the personal risks faced by truthful (but ultimately unwise) astrologers and the validity of their art. Cardano solemnly concluded that the unfortunate Ascletarion "received, as prize for his own true prediction, his own death."[43] The second story was different but, at least for our purposes, even more revealing. The story of the unfortunate priest recounted already in Corio must have been well known in Milanese circles if the Milanese Cardano chose to use it as a telling tale of the extreme caution to be exercised when predicting the impending death of a powerful lord.[44] As Cardano's stories sum up all too well, predicting a short life to a despotic man was a risky enterprise.

There is little doubt that Vimercati would have been aware of the dangers involved. His interpretation that Jupiter, "the giver of the years," was favorably placed in relation to the Moon allowed him to predict that his duke would live a long life. This interpretation, however, was not the only possible one, as we can see by looking at Cardano's own analysis of Galeazzo's geniture. In his own admittedly posthumous geniture of Galeazzo, Cardano offered his readers a very different interpretation of Galeazzo's nativity (see Figure 16 and Table 2). He confidently asserted that the Moon in opposition to Jupiter and in square with Mars was inauspicious, while its conjunction with the *cauda draconis* manifestly indicated that Galeazzo would die a violent death.

According to his calculations—which, at least on this occasion, Cardano chose not to make his reader privy to—Galeazzo's length of life was a neat thirty-two years.[45]

As he interpreted Galeazzo's nativity while his lord was still alive, Vimercati must have thought it prudent to present every element in the most positive light. When Cardano looked at the same geniture after Galeazzo's own death, he certainly had the benefit of hindsight. Yet he still sought to explain and verify the events in light of what he knew to prove the soundness of his art. While Vimercati decided to accentuate the trine aspect between Mars and Jupiter and Jupiter's reception by the Moon, Cardano focused on the square aspect between the Moon and Mars and the Moon's opposition to Jupiter, thus coming to very different results. One could speculate that Galeazzo would have seen through Vimercati's flattering prediction, but one could also equally imagine that the optimistic nature of his prediction would not have disturbed the flamboyant and ever-optimistic duke.

Galeazzo's Patronage of Astrologers

As we shall see, over the course of his brief reign Galeazzo Maria Sforza employed astrologers in a variety of fashions. Vimercati's beautifully illuminated *iudicium* must have struck the right chord if, as documented, Galeazzo returned to the astrologer's services in a number of instances. Over the following years, Vimercati continued to seek employment at court, offering his professional services directly to Galeazzo. By the late 1460s and early 1470s he was a well-established physician and astrologer operating both at court and in the city of Milan.[46] His *iudicium* was thus only one of a series of astrological consultations—admittedly the most elaborate and splendid one—that Vimercati produced for the court.

In later years Galeazzo resorted to Vimercati's expertise on more than one occasion, asking him to produce judgments on demand, or receiving from him genitures and revolutions as gifts. His astrological consultations were of the most varied kind. In 1472, for instance, Galeazzo's ducal secretary, Cicco Simonetta, was instructed to approach Vimercati and another astrologer—the university professor Francesco Medici da Busto[47]—regarding the apparition of a comet that had engendered great speculation among prophets and astrologers alike. Rumors had been circulating that the Franciscan Marco da Bologna had preached to the Milanese population that astrologers had foretold the plague.[48] In this case the preacher did not use the gift of prophecy to predict God's wrath as others may have done: he instead more simply reported the prediction of the astrologers. Cicco, however, was asked to verify the "news." Raffaele Vimercati and Francesco Medici da Busto denied the validity of these predictions and proposed an alternative interpretation of the celestial phenomenon: they asserted that, in their expert opinions, the comet would bring war in distant lands and troubles for Christianity.[49] It is telling that, with Simonetta as the middleman, Vimercati also took the opportunity to remind the duke of his famous horoscope of 1461, in which he had predicted a long life for him.[50] Astrology, patronage, and personal social advancement were never too far apart.

While it is unfortunate that the 1472 *iudicia* of Vimercati and Medici da Busto are not extant, the Milanese archives still hold copies of similar prognostications on the same comet by other Ferrarese and Bolognese astrologers, further documenting the vivid interest of the duke and his court in these celestial occurrences. These prognostications have been copied neatly in a single leaflet and were penned by the astrologers Pietro Buono Avogario, Battista Piasio, and Giovanni de Bossis Polonio.[51] In total, therefore, Galeazzo would have collected at least five different predictions on the same celestial event. We should notice, furthermore, that Galeazzo's interest was far from unique in the Italian and European panorama of princely courts: Ferrarese astrologers must have briefed their own duke, Ercole d'Este of Ferrara, about the same celestial phenomenon if Galeazzo was to have a copy of such prognostications. Similarly, the Polish court astrologer Martin Bylica, in Hungary, had provided detailed advice to his patron, Matthias Corvinus, about this and previous comets and their impact on Hungary's political situation. As Darin Hayton has rightly highlighted, these celestial events were always put in close connection with the possible political dangers affecting a patron's reign, and Martin Bylica's intellectual career and his success in attracting patronage depended heavily on this type of prognostication, as vague as these may read to twenty-first-century eyes.[52] There can be little doubt, therefore, that this type of astrological prognostication constituted a predictive genre of significant intellectual and political importance in Renaissance Europe.

When it came to practicing astrology, Vimercati, like many contemporary astrologers (Martin Bylica included), presented his patron with a vast array of possibilities. On September 2, 1475, for instance, he wrote directly to Galeazzo to inform him of some prognostications (*iudicii*) he had drawn for him and for his first-born son, Gian Galeazzo. In this letter he recounted how he had brought them to the castle the previous Christmas, but he had not had the courage to deliver them personally to the duke for fear of being considered "daring and imprudent."[53] Besides revealing something of the delicate relationship between patron and client, this letter is particularly significant because it specifically mentions the types of astrological services that Vimercati offered: two genitures—one for Galeazzo and one for his son—and an additional chart that Vimercati drew for a political event, namely, Galeazzo's ascension to the duchy (*intronizatione*), the revolution of which, he boldly claimed, could yield precious information regarding the military challenges that might await him.[54] Once again Vimercati spontaneously offered his services to the duke in the hope of maintaining his favor. Once again the gifts offered were similar to those that other astrologers offered to their lords: Matthias Corvinus had his own geniture, as well as the horoscope of his coronation, painted on the ceiling of his palace.[55]

The relationship seems to have been one of mutual necessity if a few months later Cicco approached Vimercati again, this time to request an annual prognostication (a *iudicio de qualitate temporum et singularitate dierum anni futuri*). Once again Galeazzo insisted on sounding out the opinion of more than one astrologer—the other one, this time, was Nicolò da Arsago[56]—and specified that neither of them should be informed that the other had been consulted so as to guarantee the impartiality of their prognostications.[57] The very same year, once again through Cicco, Galeazzo requested another *iudicium de qualitate temporum et singulorum dierum anni futuri* from the Dominican astrologer Annius of Viterbo, from whom he had already received an annual prognostication in 1473, and a more personal one, on parchment, in the same year.[58]

These documented instances of astrological consultation suggest that Galeazzo resorted to astrologers fairly regularly. While more examples will be provided shortly, it is clear from this evidence that Galeazzo approached astrological intelligence as a wise statesman would, namely with caution. Aware that astrology was not a certain science and alert to the potential tendentiousness of some prognostications, Galeazzo applied "quality control" over his information (and by extension, over his informants) by requesting more than one prognostication to be cast. We know little of how Galeazzo actually used annual prognostications of the kind illustrated so far. The simple fact that he collected them, however, suggests that he believed in their potential value as sources of information. We can imagine them to be the equivalent of an economist's annual projection of how the global markets will behave, a sort of blueprint that brokers and governments would consider prudently when planning future actions. Even admitting that these predictions were tendentious and speculative, they had wide circulation and could affect both public opinion and the diplomatic policies of other states. They could, in fact, contribute to creating a general state of anxiety or expectation and even prepare the ground for political action against a certain state or leader. As we shall see, this was exactly the use to which astrological prognostications were put in the period preceding Galeazzo's assassination. It seems obvious that, had Galeazzo not given credit to this type of information, he would not have patronized astrologers and requested their services.

But how formal and institutionalized was Galeazzo's relationship with astrologers such as Vimercati? Was Vimercati paid for his services, for instance? So far no evidence has emerged of Galeazzo's direct payments for these kinds of services. Monetary remuneration, however, was often not the way Renaissance lords provided for their courtiers, and there were many other forms in which the dukes of Milan could reward their astrologers.[59] Many of them held other positions at court, most commonly that of physician, or were rewarded with other posts over which the dukes had direct control. A common solution was that of providing the astrologer with a post at the local university of Pavia, as in the cases of Francesco Medici da Busto and Gabriele Pirovano, two other astrologers who operated between court and university.[60] Another way was to bestow upon them and their families other privileges and honors. In the case of Ludovico Maria Sforza's favorite astrologer, Ambrogio Varesi da Rosate, these could take the form of a fief and a *vitalizio* (an honorary, life-long stipend).[61]

In the case of our astrologer Raffaele Vimercati, the benefits were similarly of a non-monetary nature. One particularly significant example is documented in a letter dated November 23, 1474, where Vimercati addressed himself directly to the duke urging him to procure a suitable position for his son, Pietro Paolo. The letter reads:

My Illustrious Lord, not only do I desire to serve your highness with worthy reverence and I do my very best to please you as much as I can, but I would like my sons to have a way to demonstrate to you equally the faith and devotion that we all have to serve your highness and your state. As my son Pietro Paolo has been writing for the court for quite some time now, and as he desires to have some firm place in your court, so that you could make the most of his virtues when you so desire, I wish and pray that you make him co-helper (*coadiutore*) in your Privy Council with the honors that come with that office, until he will be qualified for a higher place which we hope to be able to obtain once again from your highness thanks to your great magnificence, and for the exceptional devotion that we feel toward you.[62]

It is not particularly surprising to discover that Pietro Paolo Vimercati was elected as *coadiutor rationatorum ad cartam* on January 1, 1475, only a few months after his father's request, while Gaspare Vimercati, Pietro Paolo's brother, was elected to the same job on November 10, 1479, apparently in his brother's place (*ad hodierna die ad beneplacidum loco eius fratris Petri Pauli*).[63] As with other forms of non-artistic patronage, that of astrologers is difficult to document—no contracts existed to formalize such relationships—but, as this example indicates, this does not mean that scientific patronage did not exist.

I began this chapter with the concept of celestial influence used to justify Galeazzo Maria Sforza's excessive sexual drive. I then looked at the way in which Galeazzo's geniture was read and manipulated to offer the most favorable interpretation of his horoscope. I also illustrated how other astrologers, by emphasizing different aspects of the geniture, were able to reach contrasting conclusions, and what could have been the risks involved had Vimercati read the chart, say, like Cardano did. From there I went on to consider the nature of Galeazzo's relationship with Vimercati, highlighting the material aspects of his patronage of astrology. As I have pointed out, Galeazzo was a keen consumer of astrological prognostications. I shall now turn to other examples of the ways in which Galeazzo used astrology, this time for overtly political ends.

Predicting the Death of Enemies: The Case of Ferrante of Aragon, King of Naples

Together with annual predictions, genitures, and their revolutions, Galeazzo seems to have resorted to other types of astrological prognostications. Sometime between 1 and 2 p.m. on November 24, 1475, the prophet-astrologer Giovanni Nanni da Viterbo—better known as Annius of Viterbo—was asked to cast an astrological interrogation posing the following question: is King Ferrante dead, will he die, or will he escape his illness? This question—or,

once again, to use its appropriate technical astrological term, this interrogation—was posed by the Podestà of Genoa at the request of none other than
Galeazzo Maria Sforza, and it had a manifest political nature. Galeazzo's
request was prompted by the illness of his relative and long-term adversary,
King Ferrante of Aragon, who had fallen sick in Naples that month.

By their very nature, astrological interrogations had a marked pragmatic
aim. Their scope was to find out almost immediately the outcome of a particular event, or at least the likelihood for something to happen. The response
of the astrologer was almost instantaneous, and the client could verify its
accuracy shortly after the question was posed. As in the case of elections, the
client might turn to the astrologer in search of confirmation as to how to act.
What was, therefore, the function of astrological interrogations in the economy
of information that characterized Renaissance diplomacy? What political
meaning could we ascribe to this type of astrological prediction, if any? Why
did Galeazzo feel the need to resort to this type of astrological prediction?
Seeking an answer to these questions requires us to place this interrogation
within the political-diplomatic context in which it was formulated.

As Vincent Ilardi convincingly argued, the relationship between the
Kingdom of Naples and the Duchy of Milan holds the key to a deeper understanding of the entire political and diplomatic scene of Renaissance Italy.
More importantly, political historians have seen in Ferrante's troubled relations with the two sons of Francesco Sforza, Galeazzo and Ludovico, "the fatal
linear progression leading to the French invasion of Italy in 1494."[64] It is therefore important to understand what led Galeazzo to interrogate the stars about
Ferrante's fate and indeed whether or not Galeazzo wanted him to die.

Famously depicted by Jacob Burckhardt as the most ferocious prince of his
time (Burckhardt attributed to him the establishment of a museum of his
enemies' mummies), Ferrante was in fact an astute diplomat of wide geopolitical acumen.[65] As historians have noted, his appalling reputation may
have been generated by the violent controversies that characterized his reign.[66]
Ferrante's ascent to power was troubled both by the contested legitimacy of
the Aragonese rule over the Kingdom of Naples (on which the French cadet
branch of the Anjou had a genuine claim) and by his own illegitimacy. His
position in the early years of his reign was thus particularly problematic.[67]
Pope Calixtus III had been reluctant to legitimate Ferrante's claims to the
throne, and this constituted a serious blow to his credibility as a ruler. In his
correspondence with the other Italian states, and Milan in particular, Ferrante

showed awareness of the problematic nature of his relationship with the papacy.[68] For this reason he went to great lengths to reassure all the other major Italian states that he did not have any aspirations to expand his dominions in Italy beyond what he had inherited from his father, Alfonso the Magnificent.

In the delicate equilibrium that followed the Peace of Lodi (1454) and the establishment of the Italic League the following year, Milan represented a key factor of stability.[69] Despite Francesco's protestations of faith with the Angevins, the Duke of Milan had as much to lose from a French presence in the peninsula as Ferrante did. We should not forget that Francesco Sforza's ascent to the duchy was itself problematic: he had married the illegitimate daughter of the last Duke of Milan, Filippo Visconti, but this was not enough to secure his rights as legitimate regent of the state. The situation was complicated further by the fact that the French house of Orléans could make legitimate dynastic claims over the Lombard lands in the Visconti name. The marriage of Valentina Visconti, the daughter of Giangaleazzo Visconti, to Louis I of Valois, Duke of Orléans (and brother of King Charles VI of France) in 1387 meant that Charles VIII's cousin, Louis of Orléans, could exert his rights over the Lombard lands. The interests of Milan and Naples, therefore, to a large extent coincided.

In typical Renaissance fashion, the early political alliance between the two states of Naples and Milan was sealed through ties of *parentela*.[70] In 1455 Francesco and Ferrante planned a double marriage: Ippolita Sforza, daughter of Francesco, was promised to Ferrante's son, Alfonso II, Duke of Calabria, while Francesco's eldest son, Sforza Maria, was to marry Ferrante's daughter Eleonora of Aragon.[71] The Franco-Milanese alliance that followed at the end of 1463, conversely, neutralized the French threat and sanctioned a new diplomatic era for both Milan and Naples, representing the long hoped for realization of a peaceful alliance among these major European states.[72] Ferrante repeatedly professed his debt of gratitude to the Sforza for having defused the potential threat of a French intervention into Italian affairs.

The relationship between Ferrante and Francesco's son, Galeazzo, however, was much less satisfactory.[73] As Ilardi aptly put it, "Galeazzo Maria had only to stay the course abroad and moderate his appetites at home, and in all likelihood he would have basked in the glory of being a great statesman and would have escaped the assassins' daggers years later."[74] Instead, he demonstrated little willingness to follow the policy of peace and equilibrium so actively pursued by his father, and this, together with other sorts of excesses, made him extremely unpopular among the Milanese nobility. His relationship with his mother and brothers, moreover, did not fare much better. Instead of cherishing the expert advice of his mother, Bianca Maria, who had often assisted and counseled her husband in governing Milan, he forcibly excluded her from political leadership, inducing her to leave Milan for her dower city of Cremona, where she took refuge. The relationship became so strained that Bianca Maria was forced to make appeals both to the King of France and to Ferrante for support and help against her ungrateful son.[75]

Ferrante, however, had not forgotten how important Milan was in his own foreign policy, and he did all he could to advise Galeazzo in political diplomacy and to contain his youthful exuberance. The dispatches of the resident ambassador in Naples, Antonio da Trezzo, reveal as much.[76] The young duke's ambition, however, proved hard to contain. Unwisely, Galeazzo soon formulated expansionist plans to regain the formerly Visconti cities of Bergamo, Brescia, and Crema, which Francesco had ceded to Venice with the Treaty of Lodi of 1454. Mostly for this reason, the years following Francesco's death witnessed a disconcertingly frequent series of diplomatic moves and countermoves (too complex to be summed up effectively here) that pushed apart the two houses of Naples and Milan.[77] Galeazzo's heedless attempt to use the fall of the Venetian outpost of Negroponte to the Turks in December 1470 as a fortuitous opportunity to rekindle his interests in Bergamo, Brescia, and Crema struck Ferrante as exceptionally selfish and politically irresponsible. Aware of the growing danger posed by the Ottoman Turks to his own kingdom, on January 1, 1471 Ferrante signed a secret alliance with Venice that also provided for the defense of their respective states from internal attacks.[78] Their diverging diplomatic ties with Venice made the rift between the two houses grow wider. Possibly in a desperate attempt to mend the situation, on July 23, 1472, Galeazzo and Ferrante sealed the marriage of Galeazzo's son, Gian Galeazzo (then only three years old), to Gian Galeazzo's eighteen-month-old cousin, Isabella, the daughter of Ferrante's son, Alfonso II, and Galeazzo's own sister, Ippolita Sforza.[79] This, however, was not enough. Complete rupture was reached in the autumn of 1474 when, to Ferrante's greatest surprise, Milan allied with Florence and Venice against the Pope and Naples in the battle for the control of Città di Castello.[80]

In November 1475, at the time of the astrological interrogation, however, things took a particularly bad turn for the Aragonese. Both Ferrante and his son, Alfonso of Calabria, fell ill with tertian fever, probably of malaric nature. Both the king and his son seem to have been, at least briefly, in danger of dying. Alfonso's wife, Ippolita Sforza, wrote daily to her brother, Galeazzo, informing him of the situation and providing him with daily medical bulletins regarding the health of her father-in-law and her husband.[81] We should not imagine, however, that in writing these letters Ippolita was acting as a loving wife concerned about her spouse and his family. Her marriage was not one of love. The relationship was marred from the start by Alfonso's frequent infidelities and Ippolita's own obsessive jealousy. As suggested earlier, from the beginning Ippolita refused to accept both her passive female role and her husband's love affairs, and this caused considerable diplomatic embarrassment at both courts. Indeed, she never warmed to the Neapolitan court or to her husband, often actively operating against Neapolitan interests in favor of her own natural family in Milan.[82] She realized, however, that the illness of Ferrante and Alfonso could seriously threaten the safety of herself, her children, and the entire Neapolitan kingdom. For this reason Ippolita wrote to her brother pleading for assistance in securing her husband's succession.[83] Galeazzo was equally concerned. While he had often tested the limits of this delicate relationship with Ferrante, he knew that if Naples found itself without a king, the Angevins could certainly claim the realm. A French descent in 1475 would have threatened not only his sister's reign, but also, quite possibly, his own. For this reason Galeazzo reassured Ippolita, her husband, and Ferrante himself that he was ready to intervene should the circumstances require it.[84] At this point two questions arise: first, did Galeazzo hope that Ferrante would die? Second, why did he commission an interrogation from Annius? Having examined the troubled nature of the duke's relationship with the Neapolitan king, it seems clear that Galeazzo's feelings for Ferrante were neither of admiration nor of love. We could speculate that Galeazzo would have been pleased to see one of his most hostile enemies exit the political scene. He was aware, however, that Ferrante's sudden death constituted a political problem: the events required careful monitoring.

What made the situation particularly difficult in this instance was the fact that Galeazzo could no longer count on the trusted intelligence of his resident ambassador. By April 1475 the tension between the two houses had escalated to the point that Ferrante had decided to recall his ambassador in Milan, Antonio Cicinello. To retaliate, Galeazzo unwisely decided to do the same with his own ambassador in Naples, Francesco Maletta, who, by August of the same year, had left Naples.[85] This meant that in November, when the king and his son fell ill, Galeazzo had limited access to information and little way to find out what the situation was like on the ground. Ippolita's letters were helpful, but Galeazzo did not find them sufficient. Thus, as he was preparing himself for military intervention—the Milanese troops stationed in Romagna and Lombardy had been alerted already—he became increasingly unsure about Ferrante's fate. To remedy this situation, Galeazzo had sent the newly elected bishop of Piacenza, Sacramoro da Rimini, to Naples, but Sacramoro did not reach Naples until November 29, when Ferrante was seemingly already out of danger.[86] Ferrante's royal secretary, Antonello Petrucci, moreover, had not written to Galeazzo with more positive news until November 27, and we can presume this letter took a few days to reach its addressee.[87] It seems reasonable to assume that it was precisely because of this great state of uncertainty in the days preceding Sacramoro's arrival that Galeazzo finally resorted to Annius's services. From a fifteenth-century perspective, this was a legitimate way to dissipate doubts and gather potentially useful information, at least if, as in this case, the person requiring such services was a ruler like Galeazzo, who had frequently resorted to the predictive skills of astrologers in the past.

We may now wonder how Annius responded to Galeazzo's interrogation. Would Ferrante die? If so, when? Annius sent Galeazzo a two-page interpretation of the figure he cast in Genoa at the moment of the interrogation (Figure 17). He answered with confidence that Ferrante would not escape his death and would soon die of his illness. Annius's response to the question was divided into three conclusions based on the chart. The first was that the king would not recover from his illness; the second, that he would either die that week or remain ill; the third, that indeed the illness would be prolonged but that he would not recover from it.[88] In practice, Annius concluded that Ferrante would not die that week, but sometime soon after, and that he would die of the disease he suffered at the time of the interrogation. Such a precise prediction was certainly audacious, but given the fact that the question was posed when Ferrante was believed to be seriously ill (something that Annius was likely to have known), it is probably not very surprising. Our astrologer, in any case, gave himself some room to maneuver. To avoid being blamed (or worse) should his response prove inaccurate, he added that, had he known the day of the decumbiture (*decubitum*)—namely the day and time of the onset of the illness—he would have been able to offer a more conclusive response to the question.[89]

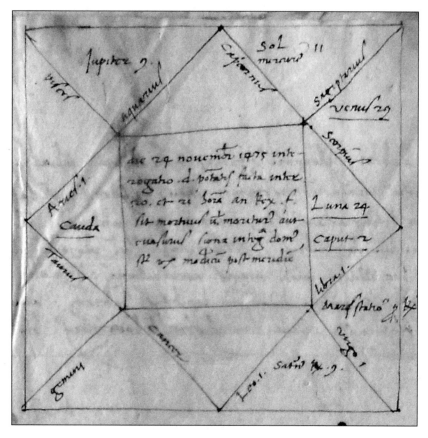

FIGURE 17. Annius of Viterbo's interrogation about the possible death of the King of Naples, Ferrante of Aragon (chart), from ASMi, *Sforzesco,* Miscellanea 1569 Annius of Viterbo to Galeazzo Maria Sforza, Genoa, November 24, 1475. By kind permission of Archivio di Stato, Milan.

We know that Annius's prediction proved incorrect. Not only did Ferrante escape death in 1475, but he also lived for another nineteen years, dying only in 1494 at the age of sixty-three (as feared already in 1475, his death effectively opened the way to the French descent into Italy).[90] No political crisis ensued that required Sforza intervention. How did Galeazzo take the fact that Annius's prediction was inaccurate? Was he infuriated? Did he feel misled? If this was the case, we have not been made privy to his reaction in the correspondence. It is more likely that Galeazzo simply accepted that such a prognostication could only be tentative. As noted in Chapter 1, much like modern-day economics, astrology was considered a conjectural science. While operating according to a set of rules and norms, it dealt with the elements of the sublunary world, and as such it was contingent upon its variations. Additionally, as Annius stated in his conclusions, he had not been given all the information necessary for precision.

Much like his contemporaries, Galeazzo was not a blind believer in astrology. He was, more simply, an avid consumer. He knew that astrology could provide some advice and guidance as to the future, but he did not believe one's fate was inevitable. Ferrante's certainly was not. Once placed in its political context, therefore, Annius's interrogation may be considered simply as one way in which Galeazzo tried to cope with growing uncertainty and turmoil, and, by extension, to predict what the future had in store for the house of Aragon and the Neapolitan kingdom. Together with Ippolita's medical bulletins and other diplomatic correspondence, therefore, Annius's astrological interrogation represents an additional way in which Galeazzo attempted to assess the political situation and maintain control over events in times of political turmoil. In this way, he was not different from Ludovico and Barbara Gonzaga, who years before scrutinized Galeazzo's horoscope and that of his father to gain greater insight into the political situation that was to unfold.

Despite the fact that interrogations were often discarded after the events (and thus may have suffered the perils of time more than other documentation), recourse to this type of "intelligence" may have been less unusual than we believe was the case. We cannot exclude, therefore, that Galeazzo may have resorted to similar means in other circumstances, and especially in times of crisis. As we shall see in due course, his brother Ludovico likewise resorted to astrological interrogations for political ends.[91] Furthermore, this was certainly not the first time that Galeazzo had resorted to Annius's astrological skills. As noted, the astrologer from Genoa had already served the duke in this capacity a few years earlier, first on January 10, 1473, when he sent Galeazzo a prognostication for that year divided into four distinct sections (the first discussed the duke's health; the second, the wars that would be relevant to the duchy; the third, political success; the fourth, the future of his progeny).[92] Later in the same year he also dedicated to Galeazzo a *iudicium de qualitate temporum*.[93] Another *iudicum* of the same type was requested in 1475, thus seemingly proving both Annius's solid reputation as an astrologer and Galeazzo's own inclination to collect and evaluate this type of information as a Renaissance form of political "intelligence."[94]

Annius was thus one of a number of astrologers that Galeazzo trusted and to whom he turned for answers in times of uncertainty. The relationship between Annius and Galeazzo was one of political and pragmatic necessity. It seems natural, therefore, that it ended only when Galeazzo was murdered. Adapting to the changing political scene and his vacillating fortunes, Annius shifted political alliances. A few years later we find him championing Ferrante as the leader of a new crusade against the Turks and defending the Genoese cause against the Sforza.[95] His activities after leaving Genoa in 1480 have been in large part reconstructed.[96] What seems clear is that Annius's career as a Sforza court astrologer ended with the life of his client.

As I hope to have amply illustrated, in using astrological "intelligence" Galeazzo Maria Sforza was not a fool. He did not use astrology as a form of courtly entertainment, but rather as a form of political expediency. In short, the picture described so far clearly indicates that he took these types of predictions seriously. No other event illustrates this more pointedly, however, than the prediction of his own death, to which we shall now turn.

Predicting the Duke's Death: Chronicle of a Death Foretold

Galeazzo did not live the long life that Vimercati had predicted. On December 26, 1476, the fifth Duke of Milan was brutally assassinated while attending mass in the Church of Santo Stefano in Brolio. The three culprits were three young Milanese patricians, Giovanni Andrea Lampugnani, Carlo Visconti, and Gerolamo Olgiati. The basic events that preceded and followed the assassination are fairly well known: Galeazzo's tyrannical style and rather volatile temper had not endeared him to the Milanese nobility (See Figure 1 in the Introduction). Allegedly, the three young men planned and executed the murder after being inspired by the reading of Sallust's *The Conspiracy of Cati-line* under the guidance of their humanist teacher Cola Montano.[97] As we shall see in the next two chapters, Galeazzo's death ushered in a period of political instability: while the title of Duke of Milan nominally passed to his young son Gian Galeazzo (who was only eight at the time of the murder), the duchy came to be controlled in all but name by his uncle, tutor, and guardian, Ludovico Maria Sforza. When Gian Galeazzo himself died under suspicious circumstances at the age of twenty-five, the way was paved for the rise to power of his uncle Ludovico.

Despite the fact that the political events leading to the murder have been recounted numerous times, political historians have never told a conspicuous part of the story surrounding Galeazzo's murder, namely, the extraordinary flourishing of astrological prognostications that preceded and followed the event. In the concluding section of this chapter, therefore, I wish to revisit Galeazzo's infamous murder in the light of this remarkable documentation. On a micro-level, the following analysis reveals the importance of this "trafficking" of astrological information first in building up expectations, and then in justifying Galeazzo's death. On a macro-level, however, this case study exemplifies the political uses of astrology within the information economy of Renaissance Europe.[98] This exchange network of astrological information was wide enough to encompass both the court and the city of Milan and to extend beyond the regional boundaries of the duchy to include the court of Ferrara and the universities of Ferrara and Bologna. More importantly, the ripple effect of the murder was considerable, arguably opening the way to a political crisis that culminated in the French descent into Italy and led to the first phase of the Italian Wars.[99]

As noted, Galeazzo was certainly less cautious than his father in evading resentment and hatred. His approach to local politics seems to have alienated many of the Milanese noble families that had previously supported his father's ascent to power. His relationship with his mother and his brothers, in addition, was less than idyllic. At his father's death in 1466, Galeazzo was reluctantly chosen to succeed his father. Despite being twenty-two years old, he was nonetheless placed under the tutelage of his mother, Bianca Maria. He finally assumed full command of the duchy two years later, in 1468. His succession to the head of the duchy, however, was not uncontested. His relationship with his mother quickly deteriorated, and both his brothers and the Milanese nobility resented his despotic personality and his increasing tendency to exclude them from political power.[100] In June 1476 this dissatisfaction materialized in a first alleged assassination attempt orchestrated by his younger brothers, Sforza Maria and Ludovico. The two were later dispatched to France, officially at their request, to "see the world."[101] Historians concur in saying, however, that their visit to France was in fact a temporary exile from the duchy.[102] This climate of suspicion and conspiracy was well known both within and outside the court. At the time of the second, successful attempt on his life, Galeazzo was, therefore, on guard. As Corio recalls, he had taken the habit of wearing a *corrazzina* (a protective garment) when attending public events, and he was going to wear it also on his way to mass on the day of the murder. A few minutes before leaving the Castle of Porta Giovia, however, he had decided to take it off, claiming that it made him "look fat."[103] Such an act of vanity cost him his life.

If we are to believe Galeazzo's contemporaries, this was a classic case of a death foretold: a number of "signs" had announced what was to come. Corio recounts how, just before Christmas, while the duke was stationed in Abbiategrasso on his way to Milan, a falling star was seen in the skies. Comets, it was well known, could portend the news of death.[104] Around the same time, moreover, his room in the Castle of Porta Giovia caught fire. These events alarmed the duke enough to consider not going to Milan at all for Christmas. He had finally decided to depart when he saw three black crows fly crying over his head. Corio further recounts how he asked to be given a weapon (a *stambichina*) to kill the three birds.[105] This fact, like those that had just preceded it, the duke read as bad omens. Before leaving the Castle of Porta Giovia on the day of the murder—Corio recalls—the duke had been acting strangely, almost knowing that his day had come, and he had asked to see first his daughters, Bianca Maria and Anna, and then his sons, Gian Galeazzo and Hermes, seemingly not wanting to part from them.[106] Galeazzo was evidently disturbed by the events that had preceded his arrival in Milan.

An anonymous chronicle narrates how, in fact, Galeazzo's death had been predicted by several astrologers and pious men (one was no doubt the priest who had been left to starve to death in prison). For this reason he had alerted his army to be ready to intervene.[107] Galeazzo, therefore, had taken these "rumors" extremely seriously. Already in December 1469 he had written to his then ambassador in Naples, Antonio da Trezzo, insisting that he press the king to forbid astrologers from naming him in their annual prognostications for the year.[108] Similar concerns were expressed again a few years later, thus showing that Galeazzo was particularly sensitive to the circulation of this type of astrological information. In the summer of 1474, for instance, Galeazzo became particularly concerned about certain rumors about him and his state that had started to circulate. Addressing himself to his trusted ambassador in Ferrara, Sacramoro da Rimini, on July 14, 1474, he lamented:

> You know how serious things become when some evil opinion spreads among the populace about some looming calamity, as it is often done by daring and vain astrologers who, in divining freely about occult things which are known only to God, unwisely predict the death of princes, wars and famine, and they even come to identify unambiguously the person who should meet such a terrible fate. Serious and honest people pay little attention to this type of prognostication; the populace, however, listens to them and waits in suspense, often giving rise to ideas that create chaos in those states and principalities. It is our opinion that His Holiness [i.e., the Pope] should excommunicate all those astrologers and mathematicians who will have the presumption, in their prognostications, to name or specify a prince or a lord, either by making explicit or implicit mention of him, but that it is allowed to them only to express universal predictions, because the particular ones can throw up chaos and are dangerous, and religion and the Catholic faith forbid this type of superstition.[109]

In the concluding part of his letter, Galeazzo encouraged Sacramoro, then bishop of Piacenza, to solicit the Pope to censure such practices, adding, "while these prognostications go against the will of God, they are also done in bad faith, to please the will of their lords by flattering and wheedling them." He cited the recent example of a Ferrarese astrologer who, to please his duke, Ercole d'Este, had prognosticated good things about the Duke of Burgundy and bad things about the King of France.[110]

Galeazzo was clearly concerned that some astrologers would spread bad rumors about him and even go so far as to predict his death. This is evident from another letter written by Giovanni Simonetta, brother of Galeazzo's trusted secretary Cicco, to the Milanese ambassador in Bologna, Gerardo Cerruti. In this letter Giovanni asked Gerardo to sound out the opinion of the Bolognese astrologers as to the prognostications about the Milanese duchy for that year. "Gerardo," Simonetta ordered, "we want you to shrewdly find out from those astrologers from whom you sent this year's prognostications if they could foresee something specific about us [i.e., Galeazzo] in their prognostications, and similarly if in their judgments they could perceive something about ourselves and our state."[111] He requested that "everything be written and clearly spelled out" and that he make sure that from then on those astrologers would not mention anything regarding Milan and its duke, or, if they did, they should do so only in positive terms.[112] As Galeazzo knew too well, rumors could encourage enemies to act. Managing the sources of such potentially disruptive rumors was thus extremely important to him.

Galeazzo's concerns about the content of some of these prognostications only grew when his court astrologers examined in detail those prognostications that had been sent to Milan from Ferrara and Bologna. The annual prognostication of the Ferrarese astrologer Pietro Buono Avogario was particularly troublesome: while not mentioning Galeazzo's name directly, it clearly predicted that his own safety and that of his state were in danger.[113] While reporting these facts in a letter, Galeazzo resolutely instructed his *familiare*, Giovanni Battista da Cotignola, to go and speak to the Duke of Modena and Ferrara, Ercole d'Este, and ask him to admonish the astrologer never to write again, either implicitly or explicitly, about him or any of his allies. Galeazzo's request was followed further by two strong admonitions: if the astrologer did not listen to his warning, he would show him his displeasure more "tangibly"; likewise, if Ercole did not intervene, Galeazzo might use his own astrologers to spread similar rumors about him (he added, however, that this was not the kind of behavior suitable to a royal prince).[114] The letter concluded with a transcription of those passages which, according to the Milanese astrologers, predicted that his life and state were under threat. It is significant that it mentioned specifically those "hideous brothers" who, as Galeazzo no doubt knew, were ready to foment sedition among the Milanese nobility.

Avogario's prognostication is still perfectly preserved among the Sforza documents. It was copied in a neat hand in a booklet format and sent from Ferrara to Milan. We can thus compare the "censored" passages of the prognostication indicated in the letter—those deemed to be referring to the duke's death—with those in the actual prognostication: they correspond almost verbatim to those of Avogario's prognostication, thus showing the effectiveness of Galeazzo's ambassadors and courtiers in collecting and attempting to control politically sensitive information.[115] Another letter exists that is addressed directly to Avogario: this warns the astrologer of Galeazzo's reaction to his prognostication and clearly mentions Galeazzo's threat. The sender, one

Lorenzo Belleto, advised the astrologer to watch out for Galeazzo's men (who are described and named):

> Magister Petro Bono, you predict and cast *iudicia* about other people but you cannot foresee or prognosticate your own imminent dangers. The Duke of Milan has sent there a man to have you cut into pieces and he is sending another man right now, so that if one fails, the other will do it; so that you know I am telling the truth, if you check the entries to the city you will find that a certain Giorgio Albanese, a man of short build and with a dark complexion, will come that way. The other is a Giovanni de Lucoli, a larger, ruddy man with long chestnut hair, who has a limp. Be careful as I speak with reason.[116]

One wonders if the letter ever reached Avogario. Regardless, a copy has somehow made its way into the Sforza archives. Be that as it may, its tone is sufficiently menacing, and makes Galeazzo's vengeful intentions abundantly clear. If he received it, there is little doubt that Avogario would have taken such an intimidating letter very seriously indeed.

Rumors about Galeazzo's imminent death were not limited to Ferrara. They had appeared elsewhere. As a consequence, similar threats were extended to two other astrologers in Bologna, Girolamo Manfredi and Marsilio da Bologna. Once again Galeazzo's secretary Cicco quoted almost verbatim from their *iudicia,* jotting down those passages that seemed to suggest that Galeazzo would die that year. Like Avogario, Manfredi had suggested that the threat may come from within, divining that "one king or lord will endure great danger either from brutish animals, or from hidden enemies."[117] This prognostication, too, needed to be censored. Galeazzo sent instructions to Giovanni Battista: he should also go to Bologna and ask the rulers of that city to similarly make sure that the two astrologers not express any judgment about him, either explicitly or implicitly, ever again. "If they do otherwise," Galeazzo warned, "we will show them our displeasure about their prognostications, and they will regret it."[118] Like in Avogario's case, a letter describing the men sent by Galeazzo to harm them was sent to the two Bolognese astrologers.[119]

A third *iudicium,* authored by a Giovanni Artoni, includes a section on the Duke of Milan (in a different hand), which has been glued over the original prediction, possibly suggesting that the piece of paper was added to correct a pre-existing negative prediction by the astrologer.[120] The political response was immediate: Giovanni II Bentivoglio himself wrote reassuring the Duke of Milan, and Giovanni Battista followed suit soon after, penning a letter indicating that the Bolognese government would admonish the two astrologers.[121] It is probably because of the Bolognese government's intervention that the astrologer Marsilio da Bologna felt the need to write a contrite letter to the duke in which he apologized for the misunderstanding, commented on the controversial passages in his prognostication, and attached Galeazzo's geniture complete with a gloriously rosy prediction for himself and his family. Our astrologer did not fail to warn the Duke of Milan that after his forty-second birthday he might encounter perils and difficulties, adding, however, that he could easily avoid them with the help of astrology.[122]

It is obvious that, despite Vimercati's overly positive prediction in 1461, by 1474 Galeazzo was very concerned about his personal safety and took "astrological intelligence" extremely seriously. It also seems clear that the circulation of astrological information predicting his death in itself provided a challenge to his authority as well as a pretext for acting upon it. Astrological intelligence could be manipulated for practical ends, and while controlling this type of information was essential, it was not particularly easy.

Avogario's prognostication about the murderous intentions of the "hideous brothers" and that of Manfredi about "hidden enemies" both proved correct. These prognostications were certainly not neutral documents. Although it is difficult to prove that Manfredi and Avogario had any direct involvement in the two conspiracies (no incriminating evidence has come to light so far), it seems nonetheless crucial to notice that both Avogario and Manfredi had collaborated with Cola Montano—the humanist teacher of the three murderers, Lampugnani, Olgiati and Visconti—on his edition of Ptolemy's *Geographia* (1477).[123] This may have been more than mere coincidence. Equally interesting is the fact that years later, Avogario wrote letters to Galeazzo's brother Ludovico. These letters are, both in tone and content, very different from those that Avogario would have written to Galeazzo. In one of these, Avogario pleaded with the seventh Duke of Milan to exert his influence to resolve some personal matter regarding a good friend of his. Ludovico, on the other hand, resorted to Avogario to determine through an election when would be the most favorable time for a wet nurse to travel.[124] Avogario seems also to have cast Ludovico's own horoscope, as attested by a copy of Ludovico's natal chart preserved in the sixteenth-century collection of genitures by the Paduan astrologer Antonio Gazio.[125]

Galeazzo's dramatic death seems to have confirmed to contemporaries the truly predictive value of astrology.[126] Not only did Cardano see his death in the stars, but later astrologers revisited his chart in search of meaning. When, years later, the Paduan astrologer Antonio Gazio went over Galeazzo's horoscope, he had much to say about the reasons and the manner of his death. In his manuscript collection of horoscopes of famous men, he included four celestial charts for Galeazzo Maria Sforza. The first one was his nativity, the second was a revolution of the year of his death, the third, a profection for the same year, and the fourth, another nativity, this time with a longer and more detailed interpretation.[127] In every case he painstakingly examined the astrological data. Each time, he wrote down meticulous comments. He seems to have concentrated on astrological data that his other colleagues had decided to ignore, for instance, the placement of Saturn in the ninth house of religion. This fact, he said, explained why Galeazzo had died in a church.[128]

On the other hand, the revolution of his death at almost thirty-three years of age demonstrated what Gazio already knew about the duke's demise, namely that he would be killed by his subjects. An inauspicious retrograde Saturn in the twelfth house—the house of enemies—announced instead that he would be killed by foes.[129] The profection, with Mars located in a human sign, indicated that he would be murdered,[130] while the fact that he was wounded in the abdomen could be explained by the position of retrograde Mars in Virgo, which in *melothesia* rules the abdominal area.[131] Gazio, in other words, read Galeazzo Maria's life into his chart in search of confirmations of the truthfulness of his art, and he found that it all fit very well with the astrological data. He also explained Galeazzo's personality astrologically. Like Corio, Gazio mentioned Venus. The fact that Galeazzo was luxurious, he noted, could be easily explained by the position of the planet in the duke's geniture: Venus was in a hot sign (Aquarius) and conjunct the Sun.[132] Astrology thus seemed to fulfill two related purposes: to predict events and thus encourage action and to verify and better understand the events and personalities of the past.

Galeazzo Maria Sforza lived in a dangerous present, aware of the threats to his life. For this reason, he tried actively and persistently to control and limit the circulation of any type of astrological intelligence that would undermine his political authority. If he was ultimately unsuccessful, we have to assume it was in part because of his own hubris.

In this chapter I have tried to show how the circulation of astrological knowledge in fifteenth-century Italy could take many routes and serve multiple purposes. One purpose was certainly to create a heightened sense of expectation about future events. In our case such "astrological news" could prepare the ground for action. Although Galeazzo was on alert, it is hard to believe that he would have escaped the destiny so carefully crafted by his enemies. If he had escaped murder on the night of December 26, 1476, he would likely have encountered it in some other form a few years later.

While Galeazzo did not shun the use of astrology for his own political ends, he objected to the fact that astrological prognostications could deliberately be used as negative propaganda against him. He obviously realized the potentially destabilizing nature of this "news" and was determined to defuse its effects. Galeazzo was rightly apprehensive regarding the value of the astrologers' prognostications. Whether he believed these astrological predictions to be true or not is almost irrelevant. What seems significant is that he took them seriously, was deeply concerned about their value as "news," and was convinced that this type of "rumor" could have concrete political consequences for himself and his role as leader. Indeed, they contributed substantially to the climate of tension and suspicion that led to the murder.

My aim has been to demonstrate how astrological prognostications could either support the duke's persona by predicting a long life for him (as Vimercati had done) or be instrumental in creating a sense of expectation about his death. In both cases astrological prognostications could hardly be interpreted as disinterested pieces of "scientific information." After all, astrological predictions were often written by astrologers who were willing to predict what pleased their patrons or friends (as Galeazzo himself had pointed out), even at the risk of displeasing others. The trafficking of astrological intelligence, thus, served practical political goals; for this reason, it was often manipulated to shape actions and affect events. Machiavellian political expediency took many forms in the Renaissance and spreading rumors was certainly one of them.

4

⋆

The Star-Crossed Duke

Gian Galeazzo Sforza and Medical Astrology

The previous chapters have highlighted the political uses of astrology at the court of Milan, emphasizing the way in which astrology was employed both to consolidate and to undermine political power. However, this was not the only way in which astrology was practiced at court; it was also employed in various branches of Renaissance learned medicine to assist in the treatment of members of the court. As noted in the Introduction and in Chapter 1, medical astrology was an important aspect of the Renaissance physician's approach to illness, and many fifteenth-century university-trained court physicians in Italy were reasonably well versed in the main tenets of the discipline. In the case of Milan, this was certainly evident in the physician's approach to the treatment of fevers, which was often framed within the Galenic theory of the critical days. Such a doctrine, very popular in the Renaissance, established that an illness would reach a crisis a certain number of days after onset, to turn towards recovery or death depending on the "signs" that appeared in the patient's body. According to this theory, these could be profitably interpreted by the physician to express a prognosis.[1]

This chapter focuses on the case of Gian Galeazzo Sforza's illness as paradigmatic of this particular application of astrology to medical practice. The correspondence between Gian Galeazzo, his doctors, and his uncles, Ludovico and Ascanio Sforza, is relatively well preserved among the large body of documents that make up the intimate correspondence of the Sforza family members. It encompasses dozens of letters sent between Pavia, Rome, and Milan in the years 1470–1494 and provides much information about Gian Galeazzo's disease and its treatment. Yet its relevance does not end here: its political import is equally notable, particularly in light of allegations of murder by poisoning directed at Ludovico Maria Sforza by some of his contemporaries. Although a definitive answer to the vexed question of whether the wicked uncle really murdered his helpless nephew may never be found, an examination of Gian Galeazzo's personal correspondence yields precious insights into the ways in which his illness and ultimate death were perceived within and outside the court and how different political powers reacted to the news of his demise.[2]

While most histories of the Italian Renaissance include some discussion of the role of Francesco Sforza and his sons, Galeazzo and Ludovico, in the political landscape of fifteenth-century Italy, Galeazzo's son, Gian Galeazzo Sforza, is mostly dismissed as a sort of "duke that never was." Historians rightly assume that the process of decision-making during his reign was not firmly in his hands but in those of his uncle, guardian, and tutor, Ludovico. While this is certainly true, the fact that Gian Galeazzo (and not Ludovico) was the legitimate Duke of Milan was of great significance for the stability of the Italian peninsula. As the son of Bona of Savoy—the sister of Louis XI's wife, Charlotte—Gian Galeazzo was closely related to the King of France; as the husband of Isabella of Aragon, he was the son-in-law of the King of Naples.[3] These dynastic ties made his position enviably secure within the panorama of Renaissance principalities.[4] His death in the same year as his father-in-law, Ferrante of Aragon (d. 1494), therefore, was extremely momentous and carried severe consequences for Milan and for Italy more broadly. Sixteenth-century European history might have been dramatically different had the young duke retained his health and, consequently, the command of his duchy. Therefore, the nature and causes of Gian Galeazzo's illness to be examined here are directly germane to the history of the duchy and the entire Italian peninsula at the end of the fifteenth century.

Starting from this premise, and attempting to go beyond the question of whether Gian Galeazzo died of natural causes or was murdered, this chapter tries to understand the nature of his illness and to assess the political impact of his death within the increasingly unstable scenario of fifteenth-century

Italy. In the following pages I trace the circumstances of Gian Galeazzo's birth, the early signs of chronic illness, and the final deterioration of his health up to the moment of his death. An examination of Gian Galeazzo's brief reign through his personal letters sheds new light on the role played by medical astrology in the treatment of Renaissance elite patients: not only was the progression of his illness interpreted according to Galen's theory of the critical days, but malevolent celestial configurations and his own natal horoscope were adduced as the most obvious explanations for his death in October 1494. Taking such notions to their logical extreme, moreover, his doctors justified his death as due to his reckless behavior, his poor constitution, and his ill-fated nativity. As we shall see, the posthumous interpretation of his horoscope by the sixteenth-century Paduan astrologer Antonio Gazio (the same person who had analyzed his father's geniture with admirable dedication) emphasized once more his inauspicious geniture, this time to assert—with the confidence of a post-eventum interpretation—that his chart indicated that the young duke would die a violent death. While, rightly or wrongly, we may smile at posthumous conclusions of this sort, it is significant that our astrologer did not fail to add to his astrological comments the poignant remark that Ludovico was responsible for his nephew's murder.

Perpetuating the Lineage:
Gian Galeazzo Sforza as Ducal Heir

As noted in the previous chapter, Gian Galeazzo's father, Galeazzo, was far from the measured, capable ruler that his parents had desired. He seems to have learned little from the advice and guidance of his physicians, preceptors, and caretakers, especially regarding moderating his sexual activities.[5] Despite all the efforts to curb his passions and educate the new prince to a life of moderation, Galeazzo was a man prone to excesses, be it money, food, sex, sport, or other leisure activities. When he married Bona of Savoy, he was not new to fatherhood. Well known for his sexual precocity, by the age of fourteen he already had an illegitimate child. By 1468, when he married Bona at age twenty-four, he had already fathered four acknowledged illegitimate children.[6]

The political significance of Galeazzo's marriage with the French-raised Bona is evident. Galeazzo was certainly aware of the political benefits that such an alliance brought to the Duchy of Milan. He was also aware of the

importance of this alliance for the internal equilibrium of the entire peninsula, proudly announcing that his marriage would guarantee the "universal
peace of Italy in perpetuity."[7] What Galeazzo and Bona (and their respective
families) now desired most, of course, was the birth of a male heir to perpet-
uate the lineage.

Lombard marriage customs required that the marriage be celebrated first
at the bride's home and then at the groom's home, but Galeazzo preferred not
to travel to Paris and sent his brother Tristano instead. Tristano married Bona
by proxy on May 10, 1468, with a ceremony that included the symbolic consummation of the wedding (Bona and Tristano kissed, climbed into a bed,
and touched each other's bare legs).[8] Soon after Bona's arrival from France,
the couple repaired to the Castle of Vigevano, outside Milan, for the consummation of their marriage. Galeazzo's mother asked the physician Guido Parati
and two other courtiers to advise her of "that which ensues."[9] On July 5, 1468,
Galeazzo proudly announced to members of the court that he had consummated the marriage "as required by the regulations of the Holy Church,"[10]
and on June 20, 1469, barely a year after their marriage, Bona gave birth to
the couple's first child—a boy.[11] As expected, the birth of a male heir was
received with great enthusiasm and joy as it guaranteed legitimate continuity
to the Sforza dynasty. Public celebrations were held across the dominion, and
Galeazzo eagerly informed all the major European states, and particularly the
French king, of the birth of a healthy son.[12] On July 25 the young infant was
baptized in the Duomo of Milan with the name Gian Galeazzo Sforza.

Galeazzo Maria Sforza gave his son the same name that Galeazzo II
Visconti had given to his son and successor. This, of course, was also the
name of the first Duke of Milan, Gian Galeazzo Visconti.[13] The choice of the
boy's name, together with the scale of the celebrations, was far from insignificant. As his father, Francesco, had done before him, in calling his first born
after a famous ruler of the Visconti family, Galeazzo himself deliberately
chose to strengthen the idea of a dynastic line descending from the original
ducal family of Milan to the Sforza. This choice can be seen as a conscious
attempt to give legitimacy to a rule that was based on the very tenuous blood
relation between Filippo Maria Visconti and Francesco Sforza's wife, Bianca
Maria, Filippo's only offspring, who was both female and illegitimate.

The importance of stressing the Visconti lineage did not escape Galeazzo's
contemporaries. In a letter written a few days after Gian Galeazzo's birth, the
event was greeted by the city of Tortona with the following words:

Most Illustrious Prince and Excellent Duke, my most caring Lord, having recently read the letters of Your Highness written to me of the fruitful and joyous delivery of the illustrious and excellent Lady the Duchess, your most worthy wife, not so much for having delivered without danger to her person, but for having given birth to a beautiful baby boy, I received immense pleasure and joy; this for the reverence that I feel for Your most sublime Highness, which deserves to find in me a most affectionate and devout servant. So with this letter I wish to demonstrate my loyalty to you and congratulate myself with Your Highness, praying Him who reigns above everything that as your son is born from such a glorious father and mother as Your Highnesses and in such an excellent and glorious State such is that of Your Highness, so by his nature he will deliver those glorious and triumphant successes that Your Highness desires, *to the perpetual glory and happiness of Your Highnesses and the eternal memory and praise of Your name and that of the excellent house of the Visconti.*[14]

It was clear that legitimacy was of prime concern for Galeazzo. The birth of a son by his wife, Bona, was, therefore, a most auspicious event. Galeazzo had high hopes for his young heir. As mentioned in Chapter 3, on July 23, 1472, when Gian Galeazzo was only three years old, his father successfully secured a marriage contract between him and his cousin, Isabella of Aragon, the niece of Ferrante of Naples and the daughter of Galeazzo's sister, Ippolita, and Alfonso II, Duke of Calabria.[15] Politically, this was another very successful dynastic alliance. Like the French House of Orléans, the Aragonese of Naples had raised claims to the Duchy of Milan at the death of the last Visconti. The union of Gian Galeazzo and Isabella clearly aimed at putting to rest the issue of any possible claim to the duchy by the Aragonese.[16]

Despite early signs of the birth of a healthy boy, however, Gian Galeazzo's health soon deteriorated. From his early years, he proved a sickly child in need of constant care. It is hard to establish the reasons for such poor health, and indeed the documentary evidence does not yield any significant clues as to why Gian Galeazzo was so prone to illness from early childhood. Diet, however, may have been a concurring cause. Italian courts were notorious for their high consumption of meats, wines, and other luxury produce (such as sugar, which they imported from the Orient), and the Sforza were no differ-ent.[17] When Galeazzo's sister, Elisabetta, married Guglielmo Paleologo, Marquis of Monferrat (who was five times older than she was), it was remarked that with Galeazzo "she ordinarily ate meat four to five times a day, which totally ruined her complexion." In Monferrat, however, she was "eating more correctly" and was doing very well.[18] This seems to be supported by Machiavelli's account of Galeazzo's famous visit to Florence in 1471. Commenting on the increasingly corrupted customs of the Florentines, Machiavelli remarked how Galeazzo and his court, defiant of the dietary restrictions of Lent, had happily eaten meat in great quantity "without showing respect for the Church or God."[19] One can only speculate that the inclination of the Milanese court to consume large quantities of meat, in addition to other sorts of excesses, may have been a factor contributing to the boy's poor health as he was growing up. We know for certain, however, that on November 8, 1473, when Gian Galeazzo was only four years old, Galeazzo had written from Vigevano to the Milanese courtier Boldrino Crivelli, asking him to exhort all those monasteries of nuns and friars devoted to the duke to pray for the health of his son.[20] Numerous other letters to Galeazzo written between 1473 and 1476 document the ill health of the boy, who often seemed plagued by fevers.[21]

At the time of Galeazzo's dramatic assassination on December 26, 1476, therefore, his wife and son were hardly prepared for it: Gian Galeazzo was only eight years old, and Bona was a young foreign duchess in a hostile and dangerous environment. As noted earlier, Galeazzo's succession was itself problematic, and only a few years earlier his brothers had attempted to gain control of the duchy by conspiring against him.[22] With Galeazzo's death, the way was cleared for a sustained challenge, this time against Bona and her son, the now legitimate sixth Duke of Milan. Historians agree in saying that it was mostly due to the political skills of Galeazzo's secretary, Cicco Simonetta, that Bona maintained the reins of power and the command of the duchy on the death of her husband.[23] A key figure under Galeazzo's leadership, Simonetta assumed a leading role in Bona's government, essentially helping her counter the attack of Galeazzo's brothers, the future cardinal Ascanio, Sforza Maria, Ottaviano, and Ludovico, spurred on by their first cousin, Roberto Sanseverino, and a mercenary soldier who had been close to their father, Donato del Conte.[24] After uncovering a conspiracy against her and her son, Bona temporarily banned Galeazzo's brothers Ascanio, Sforza Maria, and Ludovico from Milan (Ottaviano died crossing the Adda in an attempt to flee, while another brother, Filippo, remained faithful to Bona and Cicco). Sforza Maria was confined to Bari, Ascanio to Perugia, and Ludovico to Pisa. However, fears of possible claims to the duchy by local and foreign powers and the renewed attacks of the Sforza brothers and Roberto Sanseverino must have worn poor

Bona out. From exile, Ludovico wrote insistently asking for forgiveness and to be allowed to return.[25]

Persuasion could take many forms and there were people at court who certainly encouraged the Sforza brothers' return. The historian Bernardino Corio recounts how Bona—artfully convinced by her *cameriere* Antonio Tassino of the brothers' good intentions and their genuine contrition— allowed them to return to the duchy. Her faithful secretary, Cicco Simonetta, allegedly responded to the news of their return with the ominous words: "My Beloved Duchess, I will lose my head and, you, for your part, will soon lose the State."[26] This is, indeed, what happened. By September 1479, Ludovico, his brothers, and Roberto Sanseverino were back in the city.[27] A letter addressed to Roberto Sanseverino documents the changed political atmosphere and foreshadows the political execution of Galeazzo's loyal secretary Cicco Simonetta:

> Magnificent Lord, having discovered in full that the bad fortune and ruin suffered by the state and our citizens and almost the rest of Italy have proceeded from the inadequacies and perversity of Cicco Simonetta, his brother, Giovanni, and Orfeo da Ricavo, we have decided to take care of them in a timely manner, for the safety and quiet of the state and our people. Today we had all three incarcerated. Such a remedy, and many others, will give to all our citizens so much happiness, good will and tranquility that it is impossible to put into words.[28]

With Cicco out of the way, Galeazzo's brothers were finally in a position to gain direct control over the duchy. Ludovico emerged as the fiercest contender. On November 3, 1480, Bona officially relinquished the tutelage of her son—which she had acquired on January 9, 1477, a few weeks after her husband's death—so that she could retire almost completely from the political scene.[29] Gian Galeazzo's tutelage officially passed into Ludovico's hands.[30] While the personal reasons adduced by Corio cannot be excluded, one could speculate that Bona's difficult decision to step down as regent of the duchy was prompted, at least in part, by Gian Galeazzo's poor health. She probably realized that her son would never be fit to rule the duchy and that the control would almost inevitably pass to one of Galeazzo's brothers.

As Pietro Bembo recounted in his *History of Venice,* Ludovico "gradually removed Galeazzo's wife and his other ministers from the governance of the duchy, something she had taken on after her husband's death so that she might hold the state in trust for her son Gian Galeazzo, still at the time very young."[31] Ludovico, then, raised Gian Galeazzo in such a way that "he appeared to have made every effort to see that the boy would never come to anything," neglecting to instruct him in the arts of war and literature, "or any skill or discipline befitting a ruler," and employing people, instead, "to corrupt and deprave his childish nature, so that in their company, Gian Galeazzo might become habituated to every sort of indulgence and idleness."[32] Unsurprisingly, under Ludovico's tutelage Gian Galeazzo's health further deteriorated. This allowed Ludovico to exclude his nephew completely from power, thus entrusting himself with full control over the duchy.

A Web of Correspondence: Gian Galeazzo's Illness and Renaissance Medical Astrology

While Ludovico installed himself in the Castle of Porta Giovia, effectively occupying the seat of power, the fragile Gian Galeazzo was consigned to the countryside, to the Castle of Pavia. The links between the two residences, however, were constant. An intricate web of correspondence developed between Gian Galeazzo himself, his uncles, Ludovico and Ascanio, his attendants and doctors, and, later, Gian Galeazzo's wife, Isabella. Letters from the Castle of Pavia were sent daily to Milan while Gian Galeazzo was ill. Often Gian Galeazzo himself wrote personally to his uncles, and to Ludovico in particular, to inform them of the progress of his illness. More importantly, this correspondence reveals a certain familiarity with the theories that the doctors adduced to explain and treat his fevers. In the fall of 1483, for instance, Gian Galeazzo himself informed his uncle Ascanio of his poor health with the following words:

> As I wrote to you in other letters, today—which is the *fourth day*—a certain alteration (*alteratione*) appeared, which was accompanied by cold and hot fever. All day yesterday and tonight we felt very sick and unwell until the 9th hour, to the point that it was too much for our *complexion* and young age. Nonetheless, around the 9th hour it started to get better and we have been feeling quite well for the remaining part of the day. For this reason we rest our hopes in the divine clemency that we will sail to a safe harbor and be free from this condition. And to reassure you, we wanted to inform you, so that, if you have read otherwise, you can have some peace

of mind, and you can communicate this to the illustrious duke of Calabria [Alfonso of Calabria, father of Gian Galeazzo's future wife Isabella].[33]

Words like *alteratione* and *complessione* (from the Latin *complexio*) were common terms used by physicians to explain a person's state of health or illness. According to Hippocratic-Galenic medicine, *complexio* was the term used to refer to the natural balance of the four qualities of hot, wet, cold, and dry resulting from a mixture of the four elements (fire, earth, water, and air) in the human body.[34] In antiquity, this theory was grafted upon that of the four humors of the body (blood, yellow bile, black bile, and phlegm) to create a rich taxonomy where the predominance of one of these humors gave rise to different individual temperaments (sanguine, choleric, melancholic, phlegmatic) that explained the different psychological and physical dispositions of individuals. These qualities, moreover, were not exclusive to the sublunary world: different planets were made up of different elements, possessed different qualities (and thus "characters"), and exerted their direct influence on the world below.[35] In his *Tetrabiblos*, arguably the most influential text of classic astrology, Ptolemy had authoritatively established which qualities pertained to which planet and connected them with the four elements and the four humors of the body. He had, in other words, closed the cycle that established a direct correspondence between the microcosm (man) and the macrocosm (the universe). This influential taxonomy of the four elements, the four qualities, and the four humors and their relation to the different planets was the basis of all medieval and Renaissance astrological medicine.[36]

According to this humoral world view, if a single individual's overall balance of the four qualities was temporarily altered (hence the term *alteratione*), sickness could ensue. The causes of disease could be multiple and related to issues such as location, the quality of the air, or the different seasons (what we would now call environmental factors; in the Renaissance these would be called *causae ex radice inferiori*), or they could be linked to broader celestial movements that brought about changes in the sublunary world (*causae ex radice superiori*), particularly the corruption of the air (the plague was sometimes explained this way).[37] In both cases, the physician's task was to restore the balance of these qualities (and of the humors themselves) to their healthy state. Given the complex nature of some diseases, the physician's approach was sometimes twofold and had to take both factors into consideration.

The treatment had also to cater for individual differences. Different patients had different natural temperaments. Young people like Gian Galeazzo, it was believed, were naturally warmer and moister, while older people, having lost some of these qualities with time, had drier and colder complexions.[38] In the case of fevers, moreover, humoral theory was complemented and enriched by Galen's theory of the critical days, which provided an elaborate taxonomy of fevers and their periodicity. Such periodicity, it was believed, was closely linked to the motion of the Moon in the sky—its "phases"—and thus could be charted with the use of astrological tables. Fevers, therefore, fell distinctly under the broad remit of astrological medicine.

Linked to astrological and numerological principles, Galen's theory of the critical days was subsequently taken up and elaborated further by various Arabic and Jewish medical authors and further popularized in shorter treatises.[39] In the process, it has been pointed out, the astrological elements that were present in Galen were expanded considerably to provide a fully fledged astrological backbone to Galen's initial theory. While Galen's theory hinged largely on the phases of the Moon and the concept of the "lunar month,"[40] some later Arabic and Jewish authors integrated and expanded their analysis of the Moon's movement to provide more guidance as to how to interpret the position of the Moon in relation to the zodiac and the other planets in the firmament.[41] This expanded theory of the critical days had wide circulation in the Renaissance. As noted in Chapter 1, instruction in astrology at university was foundational to the teaching of medicine; it was believed that one could not exist without the other. As Albumasar argued in his *Introductorium in astronomiam:* "The physician pays attention to the alteration of the elements; the astrologer follows the basic movements of the stars to the cause of the alteration."[42] A good physician, it was argued, could not ignore astrology without damaging medicine itself.[43]

Gian Galeazzo's illness fits the mold of Galen's theory of cyclic fevers. The cycle of a patient's fever would start from the day the fever appeared and continue for twenty days, with the possibility of repeating itself again in subsequent cycles of the same duration.[44] Galeazzo's reference to the "fourth day" clearly abides by one of the key principles of the theory of the critical days, according to which the fourth day was the last day of the first phase of a disease and thus a *dies indicativus,* namely a day in which the patient's body would provide signs (*signa*) that would allow the physician to express a prognosis as to the course of the disease.[45] Gian Galeazzo's physicians, therefore, would have interpreted the symptoms of his *alteratione* in relation to the movement of the Moon. As we shall see shortly, however, the path of the Moon was often read in relation to the other planets with which she was in aspect, especially the Sun and the malefics, Saturn and Mars.

The use of medical terminology in the daily correspondence of the court was not unique to Gian Galeazzo, however. Ludovico similarly demonstrated a certain familiarity with contemporary medical discourse.[46] Their knowledge seemed informed largely (possibly even exclusively) by the words of the physicians themselves, who probably reported directly to Gian Galeazzo, and certainly wrote regular medical bulletins to his distant uncle. Ludovico himself wrote regularly to Ascanio in Rome regarding the health of their nephew, often making direct reference to these medical bulletins. In a letter written on October 28, 1483, the same day as Gian Galeazzo's letter quoted above, Ludovico informed Ascanio (probably on the basis of a medical bulletin) that Gian Galeazzo was still plagued by fever but nature had helped him expel a great quantity of harmful matter (*materia cativa*) from the lower orifices, and he was now feeling better.[47] The following day, Ludovico sent more news to Ascanio regarding the health of their nephew. Gian Galeazzo was still unwell, but the physicians continued to hope for a full recovery. Once again the message contained specific medical terminology associated with the treatment of fevers and the theory of the critical days:

> Around the 17th hour this Illustrious Lord started feeling cold, and then hot, and this condition persists up to now, the 24th hour. The *paroxysm* has been much less significant than that of the day before yesterday and although the corrupted matter expelled (*materia peccante*) is of such a nature that shows the illness to last a few more days, nonetheless the doctors hope to bring His Excellence to a safe port. They have not ordered yet for him to receive the remedy as they are waiting for Nature to take its course, which so far has been very successful, and *to let the present conjunction pass,* because, as the Moon is [in conjunction] with Mars, this would negatively affect the remedy (*faria furere la medicina*).[48]

Much like "complexion" and "alteration," the terms "paroxysm" (*paroxismus* or *parocismo*) and "corrupted matter" (*humor peccans* or *materia peccante*) belonged to the technical vocabulary of the learned physician and were employed to explain the development of cyclical fevers.[49] Once the symptoms reached its peak (*parocismo*), these were followed by the expulsion of putrid humors; this suggested to the physicians—trained readers of the body's signs—that the disease would last for a few more days. The matter, however, was of such a nature that the physicians were hopeful for a recovery. Things were complicated, however, by the fact that the doctors could not administer the necessary medications because the Moon was conjunct with the malevo- lent Mars. As noted, the movement of the Moon in relation to the signs of the zodiac or the other planets was considered particularly important. As the Moon was associated with the Aristotelian qualities of cold and moist, while Mars—a malefic planet—carried the opposing qualities of hot and dry, it was believed that that their conjunction was to be harmful. Mars's excessive dry- ness adversely affected the Moon's otherwise positive purging effect on Gian Galeazzo's humors, thus exacerbating the humoral unbalance that manifested itself through hot and cold fevers.[50]

Plants, herbs, animals, and stones—the chief ingredients of medical recipes—were also influenced by the planets' action. Indeed, medieval astrology explicitly associated various plants and stones with different planets and their natures.[51] For this reason Gian Galeazzo's physicians decided to wait until the conjunction of the Moon with Mars had passed before admin- istering his medication. The reason, once again, was astrological: Renaissance medical practice advised against it because it would prevent the expulsion of noxious matter, as can be evinced from the note on the appropriate time for the administration of medication written by Giovanni Battista Boerio when studying medicine and astrology at the University of Pavia in the 1480s. In his rubric on how to choose the best moment to take a medication, he signaled: "As far as concerns aspects and conjunctions with the Moon, some are to be avoided, others to be elected. Indeed, the conjunction of the Sun, Mars, or Saturn has to be avoided, as well as that of Jupiter, because it retains the humors."[52] Evidently, the principles that were taught at Pavia were applied in the treatment of the young Duke of Milan.

Over the following months, Ludovico continued to inform Ascanio of their nephew's health. Gian Galeazzo, however, remained prone to fevers. In the courtly correspondence, mention of planetary influence recurs often, par- ticularly in relation to the position of the luminaries. Responding to a letter from Ascanio on October 2, 1483, Gian Galeazzo wrote of how his fever had disappeared for a few days but had returned again the day before. The reason for its return, he explained, was the combustion of the Moon:

In reply to what you wrote in your letters of the second-last day of the previous month, expressing your sympathy for our *alteration and fever,* we say that we are absolutely certain that Your Highness and the Illustrious Lord, the Duke of Calabria our father were very upset. In the beginning [this illness] caused us much pain and discomfort, but since Sunday we have felt much better and we are now without fever. Since yesterday we have had a temperature, and we believe the reason is the *combustion of the Moon.*[53] Thank God around the 22nd hour it went away and we hope in God's clemency that we will soon be free from it as it has been much more moderate than in the beginning.[54]

The near-conjunction of the Sun and the Moon was not considered positive, as the two luminaries possessed opposite qualities, and the vicinity with the Sun was believed to weaken the other planets.[55] At the time of writing, however, Gian Galeazzo was without fever (the Moon, incidentally, was no longer "combust") and this was a propitious sign. Once again medical astrology shaped the response of Gian Galeazzo and his doctors.

A subsequent letter from Ludovico to Ascanio on October 5, 1483, joyously announced that Gian Galeazzo had been feeling better, that the medication had been effective, and that Gian Galeazzo had not expelled any matter and was without fever:

> Most Reverend and Illustrious Lord, my dearest brother, yesterday night I wrote to Your most reverend Highness of the improvement of Our Illustrious Duke;[56] *the medicine has worked successfully.* With this letter I inform you how tonight he slept well and *without any expulsion of matter,* and this morning he felt better and has continued to feel this way until now, the 23rd hour. The doctors say that they have found him *clean and purged* since the 16th hour, and that today is the best day he has had so far, being jovial and in good spirits, and enjoying pleasurable things as if he had never been sick, and this gives us hope that he will soon be on the road to recovery.[57]

Counting the passing of the hours and the days, as suggested by these letters, was essential to optimal medical practice, and so was the expulsion of the humors, which was prompted by the use of purgatives or bloodletting until balance was restored. As already indicated, the vocabulary used in the Sforza correspondence reveals some clear familiarity with one of the main tenets of medical astrology, namely the theory of the critical days.

Gian Galeazzo's physicians would have acquired solid familiarity with the theory of the critical days by reading a variety of classical and Arabic astro-medical texts used for teaching in all the major Italian universities: these may have included *verbum* 60 in ps.-Ptolemy's *Centiloquium,* which was dedicated to this specific topic, the ps.-Hippocratic *Astronomia Ypocratis,* and Albumasar's *Introductorium ad astrologiam.* There is evidence, however, that one particular text was used at Pavia for the teaching of Galen's theories: this text is the ps.-Galenic *Aggregationes de crisi et creticis diebus,* a compilation of Galen's teach-ings on the critical days in his *De crisibus* (*De crisi libri tres*), and *De diebus criticis* (*De diebus decretoriis*), but especially Book III of the latter. Cornelius O'Boyle, who has studied the text's circulation in manuscript extensively, argues that this medieval compendium must have been composed sometime after 1260–1280 by an anonymous author who taught either in Bologna or Paris.[58] For all intents and purposes, the *Aggregationes* represents a *summula* of Galen's longer works conceived and arranged for ease of teaching. For this reason, O'Boyle suggests that the text had a large diffusion among practicing physicians both in the late Middle Ages and in the Renaissance. Its teaching at Pavia can be deduced from its inclusion in Gian Battista Boerio's student notebook, which contains an incomplete version of the text.[59]

As noted in the Introduction and in Chapter 1, medical astrology was often closely interconnected to natal astrology. Examining the planets in the sixth house of a client's nativity, for instance, could reveal much relevant information about the patient's proclivity for certain diseases and not others. If to this we add the practice of casting a celestial figure for the onset of the disease (*decumbiture*) that came to accompany the theory of the critical days (itself based on lunar phases), and we allow for a comparison of the two charts, we can easily see how intricate the interpretation of these celestial signs becomes. The two, sometimes three, sets of signs were read simultane-ously to formulate a prognosis on the outcome of the disease. This complex and sophisticated semiotics allowed doctors to read signs, both bodily and celestial, for meaning.[60] It is quite likely that Gian Galeazzo's nativity was examined regularly during the course of his illness, and it is quite possible that the doctors cast also one or more *decumbitures.*

Despite the duke's poor health, the marriage between Gian Galeazzo and Isabella of Aragon went ahead as planned.[61] The couple was first married by proxy in Naples, and then, in person, in Milan's cathedral on February 15, 1489.[62] Then the couple repaired to the Castle of Vigevano to consummate the marriage. As we shall see in the next chapter, the practice of electing the best time to lie with one's bride with the help of astrology was a common feature of all the Sforza marriages organized by Ludovico, who clearly believed that choosing the most propitious time to celebrate and consummate the marriage was essential for maximizing one's chances of a successful marriage that would produce an heir. It is possible, therefore, that also this marriage was planned with the help of astrology. Yet, at least in this case, it is doubtful that Varesi and the other astrologers were entrusted with the task of helping Gian Galeazzo produce a healthy heir. Rather, it is more likely that Ludovico wanted to do just the opposite. Indeed, contemporary historians such as Bernardino Corio, Girolamo Borgia, and Marino Sanuto asserted that Gian Galeazzo was the victim of a *maleficium* that prevented him from consummating the marriage, thus giving some credit to the hypothesis that Ludovico may have entrusted his astrologers with the task of hampering Gian Galeazzo's chances.[63]

Did the *maleficium* operate through astrological talismans or other astromagical formulas? Our sources do not say. Yet such practices are not undocumented in medieval and Renaissance Italy.[64] Indeed, according to Nicolas Weill-Parot the second half of the fifteenth century saw a real "renaissance" of astrological images that culminated in the publication of Marsilio Ficino's *De vita coelitus comparanda* (1489) and Girolamo Torrella's *Opus praeclarum de imaginabus astrologicis* (1496).[65] More to the point, talismans are discussed at some length in the works of Antonio Guainieri, professor at the University of Pavia and one of its most prolific medical authors, and to a lesser extent in Giorgio Anselmi da Parma's *Astronomia* (a work that, as noted, was donated to the poor students of the University of Pavia by the Milanese court astrologer Antonio Bernareggi), thus showing that astrological images were not unknown within Milanese circles.[66] Similar astro-magical formulae (called *experimenti*) to enhance or hinder procreation, moreover, are contained in a fascinating manuscript of Milanese provenance now at the Bibliothéque Nationale in Paris.[67]

While Weill-Parot argues that astrological images were initially more a notion than a reality, this may not hold so true for fifteenth-century Italy. Emerging evidence about the Gonzaga court seems to suggest that at least some Italian Renaissance astrologers and physicians had some practice in constructing talismans and astrological potions for Renaissance ruling elites.[68] While no document has emerged to prove that Varesi cast a *maleficium,* therefore, we cannot exclude that such a practice took place in Renaissance Milan. What seems certain, however, is that Ludovico had complete faith that astrology could facilitate (or hinder) sexual intercourse and procreation and that he organized Sforza marriages accordingly.

In January 1490, eleven months after Isabella had left Naples for Milan and Pavia, the marriage had yet to be consummated. Ferrante and Alfonso were dismayed and put increasing pressure on Ludovico to do something to remedy this embarrassing situation. Ever more frustrated by Gian Galeazzo's inability to fulfill his marital duties, Ferrante eventually resuscitated the prospect of a marriage between Ludovico and Isabella, first secretly proposed by Ludovico himself in 1487 to prevent Gian Galeazzo from marrying.[69] Ferrante offered Isabella to be Ludovico's wife, affirming that her marriage to Gian Galeazzo could be dissolved and she could be remarried to him. If Ludovico was not willing to take her, however, Ferrante intended to remarry her to somebody else who was able to give her children.[70] Convinced that Gian Galeazzo's marriage would not produce offspring, but unwilling to marry Isabella himself after arranging his marriage with Beatrice d'Este, Ludovico politely refused the offer and tried to gain time. He consulted with Bona, his *consiglieri,* and Gian Galeazzo's doctors and asked that they count the year starting from the moment Isabella first slept with Gian Galeazzo, not from the day when she left Naples.

The news spread quickly within the court and was reported by various ambassadors, including the zealous Giacomo Trotti, who wrote daily to Ercole d'Este, often mentioning the embarrassment that the duke's behavior was causing among members of the Neapolitan entourage.[71] The situation must have pleased Ludovico immensely. As the Florentine ambassador in Milan, Pietro Filippo Pandolfini, reported to Lorenzo de' Medici, Ludovico was firmly convinced that Gian Galeazzo "could not perform," and, "even if, with the help of these people [i.e., the Neapolitans], he managed once," Ludovico believed, "it is clear that it would bear little fruit."[72] Pandolfini reported how Ludovico was confident that Gian Galeazzo "was not going to have heirs"[73] and remarked how, on this basis, Ludovico had constructed his plan to undermine the Neapolitans by interfering in their relationship with the papacy.[74] It is clear that Lorenzo the Magnificent was fully aware of Ludovico's machinations against his nephew, but he was not prepared to intervene.

Despairing that Ludovico would come to the aid of Gian Galeazzo, Ferrante sent two expert women (*matrone*) to advise and help the newlyweds. With the diligence of the two women, Pandolfini reported, it was hoped that the duke would finally be able to consummate the marriage.[75] The Florentine ambassador was in no doubt that Ludovico had every intention of seeing the marriage annulled and Isabella sent back to Naples, but he was acting as if he cared about the ducal couple and was trying to help. He cynically noted that Ludovico was so polite and caring toward Isabella that, had one not known the truth (*secreto*), one would have believed that he genuinely cared about her and even wanted to marry her.[76]

Gian Galeazzo's poor health was certainly not helping him perform his conjugal duties, and concerns about his sexual prowess continued to be expressed in Milanese circles. Many more months passed, but, eventually, the situation took a turn for the better. In April 1490, Agostino Calco proudly informed his father, Bartolomeo, that Gian Galeazzo and Isabella had successfully consummated the marriage:

> Magnificent father, the day before yesterday our duke received with much honor and joy his wife, Isabella, who visited him. After she and her princely relatives changed their robes into the French fashion, and Antonio Visconti was dressed as is customary for a messenger, she was conducted into the bedroom, where she shared the bridal bed with her husband, but he was taken by a fever that impeded all sexual pleasure. *During the night, however, as demonstrated by unequivocal actions, they kept busy with great success.* Therefore, Belprato immediately ordered the messenger to go to Naples in haste with the news, and his father, Ludovico, moved by great joy, with great zeal congratulated his nephew and niece.[77]

The fever that had plagued Gian Galeazzo during his first successful night with Isabella accompanied him also in the following weeks. This last illness must have gone on for months, as on July 15, 1490, Agostino Calco reported to his father that the physicians had diagnosed Gian Galeazzo with double tertian fever:

> Our Lord is affected by a heavy fever, which started the day before yesterday, around the second hour before sunset. Yesterday, the same fever, having started earlier, remained with him until the morning, first cold in nature, then hot. It repeated itself today again, but in a lighter form. The

doctors diagnosed it to be a double tertian fever. I attach the letters so that they can be sent to whom it may concern.[78]

As already noted, Gian Galeazzo's fevers were carefully monitored: the physicians' and the patient's attention to the exact time of day at which the symptoms manifested themselves and the specific nature of these symptoms is clearly revealing of the fact that the physicians monitored the progress of the disease according to theory of the critical days. Gian Galeazzo's fever did not disappear, however. A letter written by Ludovico ten days later lamented once again that Gian Galeazzo was still plagued by fever.[79] Thus, it is hardly surprising that the young duke was not in the mood for love. To make things worse, in the summer of 1491, Isabella too fell ill. Since early August she had been plagued by fever, as the letters addressed to Ludovico by her physicians, the *cancelliere* Dionisio Confalonieri, Agostino Calco, and Isabella herself amply document.[80] By September, Gian Galeazzo was sick again. The cold and hot fevers that had plagued him in July had returned and Confalonieri resumed his task of writing daily to Bartolomeo Calco, informing him of the duke's health.[81]

In subsequent years the correspondence between Pavia and Milan remained very frequent: Gian Galeazzo's entourage reported regularly about his health and daily activities. Gian Galeazzo continued to be plagued by fevers. For this reason, the letters of Gian Galeazzo's physicians, various courtiers, and members of the Sforza family, as well as those of the duke himself, represent a unique source of information on the treatment of fevers in the Renaissance more generally. But their importance does not end here. Once set in the political context of a contest for power between Ludovico and Gian Galeazzo (and his father before him), these letters acquire wider political significance. Despite Ludovico's obvious attempts to present himself to the world as Gian Galeazzo's caring uncle, the dispatches of the Florentine ambassador, Pandolfini, leave little doubt that both Ludovico's allies (including Lorenzo de' Medici) and his enemies (especially Ferrante) were very wary of his affected manners. They clearly sensed that his refusal to take over the duchy by force was part of a carefully calculated plan to keep the government of Milan—and thus his role within Italian and international politics—firmly in his own hands. Therefore, these letters need to be read on two levels: as medical sources *and* as political documents.

Nowhere does their double nature emerge more clearly than in the correspondence between the castle of Pavia and Ludovico in Milan in the years just preceding Gian Galeazzo's premature death. During the final years, the correspondence intensified substantially.[82] Ludovico often received two or three letters a day reporting on the young duke's health. These letters were generally written by members of Gian Galeazzo's entourage, and most often by the duke's physicians. Occasionally, Isabella wrote as well. The picture of a sick man defying the advice of his doctors emerges clearly from the correspondence. While it is hard to gauge the reasons for Gian Galeazzo's rather reckless behavior, this later correspondence often mentions his excessive drinking and eating. Worried about Gian Galeazzo's health in the spring before his death, the courtier Dionisio Confalonieri informed Ludovico that, against the advice of his doctors (and while still unwell), the duke had taken up the habit of drinking in his wife's company before going to bed.[83] Significantly, the same comment becomes alarmingly insistent in the correspondence of the months just preceding his death. It is to this last set of letters that I shall now turn.

The Star-Crossed Duke: Gian Galeazzo's Last Year

The fact that patients would resist the advice of their physicians was part of the physician's training. The concluding part of the ps.-Galenic *Aggregationes de crisi et creticis diebus* is insistently cautious on this point, instructing the physician-in-training about the conditions to be imposed on the client and on those assisting him: they have to obey the doctor and follow his advice. Otherwise, exact prognosis is impossible.[84] Nevertheless, the correspondence between the courts of Pavia and Milan seems to suggest that the doctors did not know how to keep the duke's behavior under control. Gian Galeazzo's proclivity for excessive eating and drinking was a cause of grave concern. Unsurprisingly, Gian Galeazzo was particularly disinclined to take purgatives. On July 19, 1494, only a few months before Gian Galeazzo's death, Isabella herself informed Ludovico that Gian Galeazzo had had an upset stomach and that the doctors blamed it on something he had eaten for dinner.[85] This statement would be innocuous were it not the case that this was the first in a long series of letters outlining in detail the progressive deterioration of the duke's health over the following months.[86] Gian Galeazzo continued to be administered purgatives, and this most certainly contributed to making him weak and feeble. By the end of July, Gian Galeazzo was eating little more than bread soup and lamenting that he found it hard to stand on his feet.[87] The physicians, Dionisio dutifully reported, attributed the illness to Gian Galeazzo's lack of sexual exercise.[88] This was, of course, a traditional explanation. Medieval and early modern medicine was traditionally divided into five parts: physiology, aetiology, semiotics, therapeutics, and hygiene. The first three were part of theoretical medicine, while the last two belonged to practical medicine. With respect to hygiene, a number of factors—called the six "non-naturals"—played a prominent role. These were traditionally treated in the *regimen sanitatis* literature and included: air, food and drink, exercise and rest, sleep and waking, repletion and evacuation (including sex), and passions of the soul. The all-important balance of the humors—it was believed—ought to be maintained as well as restored largely through a savvy management of these environmental factors. This may have required the patient to reduce or augment a certain type of food or drink, to exercise more or less, to have more or less sex, or to sleep more or less, depending on their temperament. Effective therapy also quite often included the use of laxatives or bloodletting, two therapeutic procedures that, like sex, were aimed at favoring evacuation. It is not surprising, therefore, that our physicians' reports touch regularly on at least four of these categories: food and drink, rest and motion, sleep and waking, and repletion and evacuation.

Despite not eating much and not having enough sex, Gian Galeazzo kept drinking against the advice of his physicians. In at least one episode, one of his attendants had followed the orders of Gian Galeazzo's physician, Gabriele Pirovano, and had watered down his wine, thus provoking the rage of the young duke.[89] Such behavior, understandably, was seriously compromising his chances of recovery. By early September, Gian Galeazzo wanted to confess himself, suggesting that he feared he would die.[90] Gabriele Pirovano reported regularly to Ludovico, unsure if the duke was suffering of tertian or quartan fever.[91] The background to this last crisis, however, was more ominous than before. The French King, Charles VIII, had already passed the Alps and was on his way to meet Gian Galeazzo's uncle Ludovico before heading to Florence and then Naples.

On September 13, Gian Galeazzo himself sent a letter to Ludovico inquiring on the meeting between him and the King of France in Asti. On that occasion he hastened to reassure his uncle that he was well, and yet a letter sent by Pirovano the same day reiterated Gian Galeazzo's condition of extreme debilitation.[92] In the following days the health of the duke improved slightly. The two letters sent by Dionisio on September 14 and 16 reported that Gian

Galeazzo felt better, he was eating again, and the physicians Gabriele Pirovano and Lazzaro Dactilo da Piacenza had prescribed him some rhubarb pills (these pills had a laxative effect). In this last letter, however, Dionisio also pleaded Ludovico to write to him and dissuade him from travelling across the river Po in defiance of doctor's advice.[93] The correspondence of the following days depicts a duke weakened by a protracted illness, suffering from violent stomach pain, and looking pale and fragile.[94] By September 28, Gian Galeazzo was seriously ill, but the physician Lazzaro Dactilo da Piacenza was still hopeful for a recovery.[95] The constant exchange of letters and medical bulletins continued. Gian Galeazzo, however, did not want to follow doctor's advice and continued to drink excessively. In treating him, his physicians continued to elect the right moment to administer his medications with the help of astrology:

> Most Illustrious and Excellent Lord, Most Honorable, I am writing to keep advising you as to the health of Our Lord the Duke, as commanded in your letters and as it is my duty. *At the elected hour,* we administered him three pills [of rhubarb] and some "manna," a very useful medicine, and they operated as laudably as it was hoped. It resulted in *the expulsion of putrid choleric matter,* which is the cause of the illness, and pink and yellow pus. And what a wondrous thing this way in which different matters hide in the human body! It is not surprising then that fevers and so many other types of diseases exist. His Highness has been obedient in receiving his medication but he has refused to follow our advice for lunch. He insisted to dine at the table with the Duchess and he has exceeded the right proportion of drinking. So we punished him and did not give him anything until dinner, when he took some "pistata" and he drank only a little more than the limit that it was imposed on him. At the 3rd hour he fell asleep and if God helps him he will be better again.[96]

This is one of few letters related to Gian Galeazzo's final days to once again make explicit reference to medical astrology. We have seen earlier, however, how in the early phases of Gian Galeazzo's illness, the administration of medication was sometimes determined astrologically. The physicians' almost obsessive attention to the various hours of the day in which the medications were administered, and the charting of their effects, strongly suggests, therefore, that such principles were never abandoned and that the physicians continued to treat Gian Galeazzo according to the theory of the critical days, electing when to purge and bloodlet him with the help of astrology. The medication that Gian Galeazzo received—once again rhubarb and *manna*—had worked effectively, helping the duke expel the excess of humors that had brought about the fevers. Gian Galeazzo, however, had refused to eat and drink in moderation, appearing tipsy and often cheating on the physicians who were trying to enforce some limitations.[97] The physician Pirovano, Dionisio recounted, had reproached the duke sternly, saying that if he did not abstain from this type of behavior he would be in danger of dying or remaining crippled. This, apparently, did not stop the duke from behaving irrationally, searching for wine in the middle of the night and breaking all the water decanters. Neither the words of the physicians nor those of Ludovico seemed to bring him to his senses.

At this point, the situation was clearly desperate, and Ambrogio Varesi da Rosate, Ludovico's most trusted astrologer and physician, was dispatched to the young duke's bedside.[98] Two days before his death, the ducal physicians wrote to Ludovico, alarmed at the fact that against the physicians' advice and under the pretence of smelling their scent, Gian Galeazzo had eaten plums, pears, and apples in excess. This, they believed, could seriously jeopardize his chances of survival.[99] Gian Galeazzo, meanwhile, had insisted on seeing his favorite horses and his greyhounds, which were brought into his room, and had once again confessed. The confessor was asked to spend the night in the castle, a clear sign that by that point there was little hope that Gian Galeazzo would survive the night, let alone his illness.[100] On the same day, another medical bulletin was sent with urgency to Ludovico. The letter was signed by four physicians: Nicolò Cusano, Gabriele Pirovano, Lazzaro Dactilo da Piacenza, and Pietro Antonio Marliani. The letter reports details of Gian Galeazzo's medical examination: the physicians had taken his pulse and they had examined his urine twice (both were common practice in Renaissance medicine and must have been performed regularly in the course of Gian Galeazzo's illness, even if, as far as I am aware, they were not mentioned in previous medical bulletins). Despite the fact that Gian Galeazzo had been feeling slightly better, the physicians declared that "it seems that the case is of great difficulty and danger," once again leaving little hope of recovery.[101]

The documentation examined so far prompts us to make two types of observations: on the one hand, there is a consistent body of evidence that documents Gian Galeazzo's prolonged illness and his own unwillingness to follow doctor's orders; on the other, reference to astrological medicine, which is more abundant in the letters dated to the early 1480s, is less conspicuous in the correspondence dated to his last year of life. Is this because astrological medicine was no longer employed and thus no consideration was given to the position of the planets, or can this absence be explained differently? The silence of the documents seems to endorse the first hypothesis, but Gabriele Pirovano's letter of October 12 quoted in full above seems to suggest otherwise. Indeed, astrology could have not been too far from the physicians' minds—and, thus, from their practice—if, moments before his death, they turned precisely to medical astrology and horoscopy to explain Gian Galeazzo's quick deterioration and death. On October 20, 1494, only a few hours before Gian Galeazzo died, Gabriele Pirovano and Nicolò Cusano—the two doctors who were constantly at his bedside in the years and months preceding his death—wrote to Ludovico explaining that Gian Galeazzo's chances of survival had been seriously curtailed by the negative influence of the heavens against which little could be done. They singled out two factors for consideration: the concurrent eclipse of the Moon and Gian Galeazzo's own ill-fated nativity:

> Illustrious and Excellent Lord Most Distinguished, after the notice [we sent] at the 17th hour, the Illustrious Duke woke up again at about the 18th hour, and we think he did not look restored as we expected he would after his meal and some sleep. Rather, he looked very weak, and, having given him a bit of broth, soon after he had a violent start and a tremor in his stomach, and he let off some air from his mouth; this was accompanied by some noise of water that one could hear moving. Further, he had a twinge and a sense of suffocation that affected the liver and the spleen with a rather sharp pain in the liver if pressed. But it stopped, and the noise appeared to descend to the intestines, which is a worrying sign in medicine, for we have very few remedies [for it]. But we will carry on [caring for him], especially because of the *terrible influence of the heavens,* both because of the *eclipse and the direction of his nativity* as his Highness knows from past experience because "etiam in medicina solus casus virtutis est per se signum malum." We will continue with the proper and necessary remedies for as long as we can, and we won't fail to inform you of what ensues, and make provisions for what happens and may occur easily in these circumstances. And if this happens, it is not in our power to restore his health, but he will be in clear danger of dying immediately—God forbid.[102]

This was probably the last letter that the doctors sent before Gian Galeazzo's death. His son and heir to the throne, Francesco, was only three. The following day, his uncle Ludovico Maria Sforza proclaimed himself Duke of Milan.[103]

As we can see in this last letter, astrology emerges strongly as decisive in formulating the doctors' final prognosis about Gian Galeazzo's health—the eclipse of the Moon was particularly negative, and the "direction" of his nativity did not promise anything good either.[104] It is unlikely that the physicians would have resorted to such a tightly formulated astrological explanation had they not been following all along the principles of astrological medicine outlined earlier. The fact that these were not mentioned constantly in the correspondence, therefore, cannot be taken as a good indication that medical astrology had been abandoned in favor of other medical theories and practices. It is much more likely, I would argue, that medical astrology remained a tool of prognosis that helped doctors chart the course of the illness and determine the best time to administer medications, but that these medical details made it into the daily correspondence directed to the court only occasionally, remaining instead often circumscribed within oral medical discourse.[105] As disease could ensue both *ex radice inferiori* and *ex radice superiori,* the Sforza doctors operated on both fronts, attempting to impose on the duke an adequate *regimen sanitatis* that could help him regain strength and vigor, while also keeping an eye on the superior causes that may have compromised his already delicate balance. The duke's poor condition and the celestial configuration at the time of writing, however, made them pessimistic about the duke's recovery.

The rich documentation of Gian Galeazzo's protracted illness and its treatment reveal much about the way court physicians operated and what kind of medical theories and remedies they favored. They tell us much less, however, about the nature of the disease that ultimately killed him. In the example of Gian Galeazzo, the doctors followed what was considered standard medical practice: they examined his appearance, his urine, and excreta, took his pulse and localized his pain by listening to him as well as touching his body. Our court physicians did not name a specific disease such as plague or dropsy, and their descriptions make it impossible to identify in modern medical terms the condition that affected Gian Galeazzo. From a fifteenth-century perspective, however, the doctors did, indeed, offer a diagnosis: fevers were considered to be diseases—not, like now, merely symptoms—and, as noted, their treatment occupied an important place in Renaissance medicine, with separate tracts or sections of classical and Arabic medical texts, such as Avicenna's *Canon,* Galen's *On Different Kinds of Fevers,* Rhazes's *Almansor,* and Ysaac Israeli's *Book of Fevers,* dedicated specifically to their nature and symptoms.

The symptoms the doctors reported in Gian Galeazzo's case included cold and hot flashes, a state of general weakness, lack of appetite, and—especially in the weeks preceding his death—vomiting and severe stomach and abdominal pain. We also know how he was treated: through diet (mostly *pistata* and *pannata,* food that is easily digested), purgatives (largely *manna* and rhubarb pills), and by regulating both physical and sexual activity. Gian Galeazzo's recovery, we are told, was hindered by his unwillingness to follow a regimen of health, but also by superior causes. Both aspects were claimed to have caused his death.

The Political Meaning of Gian Galeazzo's Death

It is interesting to notice that, unlike other Sforza horoscopes, Gian Galeazzo's chart is not one of those included in Girolamo Cardano's collection. Given the fact that, for all intents and purposes, he never ruled as Duke of Milan, this may not be entirely surprising. Yet, while missing from Cardano's collection, Gian Galeazzo's horoscope appears in at least one other collection of genitures that was gathered sometime in the late fifteenth and early sixteenth century, that of the Paduan astrologer and astronomer Antonio Gazio encountered in previous chapters. Gazio erroneously recorded Gian Galeazzo's birth as occurring on June 20, 1468 (and not 1469 as was the case), at about 11:30 in the morning.[106] While Gazio is not as eloquent in his interpretation of the chart as he had been with that of Gian Galeazzo's father, he still penciled in the margin the following revealing note: "They say how he, being unlucky, was first rendered a subject, then poisoned by his uncle, having been made foolish and crazy; and he died in the current year 1495 [sic], purportedly of violent death. For his uncle Ludovico took over the duchy, although it was said that Gian Galeazzo had left the realm to his son to whom it should have passed."[107] Like the court astrologers who had taken care of Gian Galeazzo, Gazio also blamed the recent eclipse as an aggravating cause. Unlike them, however, he noticed that Gian Galeazzo's own chart and the concurring eclipse forecast a violent death, thus giving credit to the rumors that pointed to Ludovico's foul play.

Indeed such rumors were frequent in the weeks and months following Gian Galeazzo's death. Soon after the event, the exiled Milanese humanist Giorgio Valla wrote to his friend Gian Giacomo Trivulzio reminding him of how he had predicted Gian Galeazzo's death twelve years before. Aware of the

sensitivity of his assertions, he asked his friend to burn the letter after having
read it:

> For the rest, I believe you have been informed, now that it is a well-known
> fact, how Gian Galeazzo, the Duke of Milan, is dead, and how Ludovico
> Sforza proceeded among the crowds with a golden mantle and the ducal
> hat, and was acclaimed duke. Maybe you do not remember how I had
> foreseen this already twelve years ago. I will repeat it in case you do not
> remember. As Giovanni Marliani treated the disease of Duke Gian
> Galeazzo, I revealed to that doctor that, because of that illness, he would
> have died (certainly poisoned, as it is clearly recounted). This worried the
> physician, and, without me knowing it, he informed Ludovico, who sent
> one of his *camerieri* to summon me. After I was accompanied to the castle,
> and Ludovico—who was in a meeting—was informed, he arrived, and
> even if he sent some messengers, he eliminated any other witnesses. I
> resisted both messengers. He inquired what I thought about the duke; I
> answered that such an illness would have led him to the grave and that the
> future Duke of Milan would have been Ludovico, who answered 'It will be
> as I wish.' I repeated, 'That's obvious.' As I was leaving, he ordered me to
> keep the thing to myself. I discussed this only with you, to be honest. You
> answered, 'Ludovico is a good man, I would never believe, even if I could,
> that he would proclaim himself duke.' Now you know what is happening.
> For this reason, I believe you will give attention to those things that I said
> in other letters. I believe it will not be long before the name of the Sforza
> will be wiped out.[108]

With hindsight we can now say that not only was Valla's quasi-prophetic
prediction about Gian Galeazzo's death largely accurate, but that he was also
right regarding the ultimate demise of the House of Sforza. Twelve years prior
to Gian Galeazzo's death, Valla had already foreseen that the young duke
would have died poisoned; more disturbingly even, Ludovico had unmistakably claimed ownership of his nephew's life in front of him. As we shall see in
the next chapter, Valla was also right in predicting that Gian Galeazzo's death
would bring about the fall of the entire Sforza dynasty: Ludovico's reign as
Duke of Milan was remarkably short lived, lasting less than five years.

Valla was certainly in a good position to reveal the *arcana* of Ludovico's
machinations. At the time of his first prophetic assertion he was still in Milan
and well connected to a number of important courtiers. A long-time friend of
the ducal secretary and humanist Jacopo Antiquario, he held a lectureship in

rhetoric at the University of Pavia. The ducal physician Giovanni Marliani had been Valla's teacher at Pavia, and Valla had dedicated to him his translation of Alexander of Aphrodisia's *Problemata*.[109] His friendship with Trivulzio, furthermore, reveals that Valla had important connections among the upper echelon of Milan's nobility. Gian Giacomo Trivulzio belonged to one of the most powerful noble families of the duchy and possessed major feudal holdings in Lombardy. In Galeazzo's time, Gian Giacomo had featured prominently at court. Having shared much of his childhood with Galeazzo, he became one of his best friends.[110] Predictably, the once amicable relationship between Trivulzio and Galeazzo's brother Ludovico slowly deteriorated into open antagonism after Galeazzo's assassination, the execution of Cicco Simonetta, and Ludovico's rise to power as Gian Galeazzo's guardian. In the mid–1480s Gian Giacomo, who had served as military captain under Galeazzo Maria, first entered the service of Ferrante of Aragon, and eventually that of the King of France. His knowledge of Milan's military capabilities and of the geography of the area surrounding the city was fundamental in securing the French victory following the siege of Milan of 1499.[111]

In a prior letter between Valla and Trivulzio, Valla himself referred to the great injustices inflicted by the Sforza on his former friend, presumably referring to Ludovico's confiscation of Trivulzio's lands and possessions a few years earlier.[112] Valla had opted for exile for similar reasons. After the execution of his close friend Cicco Simonetta, Valla followed the fate of all of those close to Bona and her secretary. A timely call by the Venetian Senate, who employed him to teach in the city's *scuole* from 1484, allowed him to leave the duchy; he remained in exile in Venice until his death in 1500.[113] Allegations of Ludovico's involvement in Gian Galeazzo's death did not remain confined to the personal correspondence of the two friends. Indeed, rumors seem to have circulated widely, soon becoming almost commonplace in the historiography of the French descent. In his biographical work on the lives of the Visconti and the Sforza, Paolo Giovio recounted of Charles VIII's visit to the dying Gian Galeazzo and laconically commented that possibly he was poisoned.[114] In his account of Charles VIII's descent into Italy, the Venetian Pietro Giustiniani similarly told of the King's visit to the young duke, who lay in bed plagued by a grave illness. Giustiniani quipped, however, that "not long after, and not without the suspicion that he was poisoned by his uncle Ludovico, he [Gian Galeazzo] died; and, with the violent death of his nephew, he [Ludovico] tyrannically invaded that duchy."[115]

Other contemporary historians reported similar allegations. Like Giovio and Giustiniani, Francesco Guicciardini narrated the story of Charles VIII's visit, but the Florentine historian was even more explicit in his accusations. After recounting the king's visit to his first cousin Gian Galeazzo (the emphasis on their blood relationship may not be accidental—the fact that Ludovico did not share the same ties of *parentela* with the French king is indeed crucial), Guicciardini added how the king was painfully aware of Ludovico's machinations, convinced that the young duke would not live long.[116] Yet neither this fact nor Isabella's plea for help and her request not to harm her father and brother convinced the king to abandon his military *impresa*. Gian Galeazzo died only a few days after Charles VIII's visit. In his account, Guicciardini tells of how the Ducal Council, convinced that Gian Galeazzo's son was too young to take up the ducal title, asked Ludovico to become duke. He did not fail to comment, however, how "under this pretext— honesty yielding to ambition—although he feigned resistance to some extent, the following morning Lodovico assumed the title and seal of Duke of Milan, having secretly insisted before that he received them as due him by virtue of the investiture of the King of the Romans."[117]

Indeed in the year prior to Gian Galeazzo's death Ludovico had been involved in a series of negotiations with Maximilian I, the future Holy Roman Emperor, regarding his investiture as legitimate Duke of Milan.[118] Within Europe's geopolitical configuration, Ludovico's only hope for protection against any French claims over the duchy rested on the figure of the Holy Roman Emperor. For this reason, Ludovico claimed that as his brother Galeazzo Maria was born before his father's ascension to the duchy, the title of duke went to him (the second born after Sforza Maria, who was then dead) and not to Galeazzo or his son, Gian Galeazzo. To obtain the investiture more easily at Frederick's death, Ludovico offered Bianca Maria, the natural daughter of his brother Galeazzo, in marriage to Maximilian I. The marriage, as we shall see in the next chapter, received the blessing of the stars: as Maximilian shared with Ludovico a keen interest in astrology, the ceremony was cele- brated *per puncto de astrologia*. But while astrology sanctioned the marriage, neither Maximilian nor Ludovico was moved by noble motives. Increasingly concerned about a Turkish threat to his lands and toying with the idea of crusading against the infidels himself, the future Holy Roman Emperor made high demands of his ally. His acceptance of the marriage proposal was accom- panied by the request of an astounding dowry of 400,000 golden ducats, followed by the promise of Ludovico's investiture at his election as Holy Roman Emperor (which occurred in July 1493).[119] This money, it was hoped, would finance Maximilian's fight against the Ottoman Empire.

In the autumn of 1494, Ludovico was evidently concerned that the rumors surrounding Gian Galeazzo's death would seriously compromise his relationship with Maximilian I. His anxiety was nowhere more evident than in a letter addressed to Melchior von Meckau, Bishop of Brixen (Bressanone), in the days following Gian Galeazzo's death. While thanking him for his support against those who had been spreading offensive rumors against him, Ludovico was overly keen to explain away the allegations of poisoning. The impression is that he was probably too keen. Not only did he thank the bishop profusely for his support, but he also insisted on clarifying the circumstances of Gian Galeazzo's death for him.[120] It is telling that he put a very personal spin on the relevant events: his description of Gian Galeazzo's last months and days departs so dramatically from the picture I have just reconstructed through the private correspondence to encourage speculation that Ludovico had something to hide. As he explained:

> Our Illustrious nephew had been sick for a number of months in the city of Pavia. With him was his illustrious wife, and he was treated by three doctors who knew his constitution from the cradle. His illness, they affirmed, was light and curable. For this reason, nothing written was shown to me. While he was sick, we were busy with the forthcoming arrival of the Christian King of France to Italy, and I was sometimes in Alessandria, sometimes in Annone, sometimes somewhere else and always away from Pavia. Nevertheless, I did not neglect at all going to see him, given the opportunity, and by encouraging the doctors to take good care of him, I fulfilled my duty as a father. When the Christian King of France arrived in Pavia, he had access to him, he saw him with his own eyes, and other French noblemen had access to him; if they wish to admit the truth, they certainly cannot deny that he died of natural causes. Furthermore, does your Reverend Lord truly believe that we are of such evil nature that, ruled by cupidity, we would wish to become universally infamous and lose our soul? [. . .] Nothing was more alien from our nature than the thought of plotting the death of our nephew, to whom we always bestowed paternal charity [. . .].[121]

The tone of the letter is exemplary: Ludovico is both defensive and guarded. In the rest of the letter, he continued to explain how Gian Galeazzo's illness had worsened, but the physicians had not despaired for his life until the day before he died. He himself was in Piacenza to meet the King of France, Ludovico added, when the news of the gravity of Gian Galeazzo's illness reached him; he rushed immediately to Pavia hoping his nephew would recover, but while on the way he was informed that Gian Galeazzo had died.[122] The letter concluded with a further plea for support in championing the truth and defending his reputation, asking the bishop to take up his defence also against those who had criticized his decision to support Charles VIII's descent into Italy.[123]

While Ludovico minimized the gravity of Gian Galeazzo's illness, the ducal correspondence provides ample evidence that by the summer of 1494 the duke's health was severely compromised. Ludovico played down the reports of the doctors, making Melchior believe that they were facing a case of sudden death and that Gian Galeazzo's health had deteriorated only the day before. Such statements are hardly genuine. It is certainly untrue that he had not been kept abreast of the developments of his nephew's illness. As we know, he received daily bulletins. Furthermore, that the duke could have died was already clear in early September, when Gian Galeazzo confessed his sins to a priest on more than one occasion.

Ludovico also knew that Gian Galeazzo had started to openly doubt his uncle's love for him. In early October, Dionisio reported to Ludovico how, after having sent everybody away, Gian Galeazzo had approached him in secret to ask if he believed that Ludovico loved him. To this Dionisio replied that he was surprised at such a question since Ludovico's past and present demonstrations clearly testified to his love. Apparently not satisfied with such an answer, Gian Galeazzo asked if Ludovico was distressed about his illness, to which once more Dionisio answered affirmatively to reassure the young duke of his uncle's honorable intentions. Such thoughts, Dionisio added, were probably spurred by "the malevolent comments of some evil tongues."[124]

It is impossible to know from the available documentation if Gian Galeazzo was actually poisoned or died of natural causes. Regardless, it is not difficult to imagine that somebody close to Ludovico could have slowly poisoned him. Ludovico himself had appointed many of the people who cared for Gian Galeazzo, including his physicians. He could have easily instructed one of them to administer the poison. If this was the case, however, Ludovico covered his tracks exceptionally well. He never put anything even mildly compromising in writing, neither he need to do so.

Contradicting Ludovico's claims to Melchior, Guicciardini reported how the royal physician, Teodoro Guarnieri da Pavia, who accompanied Charles VIII on his visit to Gian Galeazzo, had recognized signs of poisoning on Gian Galeazzo's body during his visit.[125] Ludovico had finally received his imperial privileges regarding the investiture just days before his nephew's death. The historian's conclusion was, therefore, that it seemed more likely that Ludovico's decision to poison his nephew in 1494 was premeditated and deliberate.[126]

The most poignant portrait of Ludovico's wickedness, however, is probably that of the Roman historian Girolamo Borgia who depicted Ludovico's ruthless thirst for power in his history of the Italian Wars:

> Here is [Ludovico] Sforza, the Moor, with the viper as his emblem, revealing his evil intentions, looking forward to the ruin prepared for Alfonso, freed from fear, as it were, and having taken over a broader field for sinning, whence he could more freely derive enjoyment from tyranny (as I mentioned above); he attended to the cure of Gian Galeazzo, son of his brother, an innocent youth, and the legitimate prince of Milan, by administering some poison to him. He who ought to care for his health and protect him from being harmed, killed him with a nefarious goblet, whereas, if he carried on taking care of the young duke as he had done, he could have reigned without envy and hatred and governed the duchy in face of such excessive desire. As everybody—citizens and foreigners alike—spoke of it, there was nobody who did not condemn this man's great impiety with whatever kind of malignant accusation, manifestly execrating in the first place such an incredible, atrocious, miserable, and unchristian deed in any man, and particularly in he who was like a father, a ruler, the organizer of all things related to the duchy and the world, and that he similarly ruled as its administrator at his choice and following his conscience. Who did not prophesize then that this violent deed would fall upon the same tyrant and his progeny and bring them down?[127]

Girolamo Borgia's portrait of Ludovico eloquently characterizes his contemporaries' perception of Ludovico Maria Sforza: a man blinded by his thirst for power, ready to commit a crime made more heinous by the fact that Gian Galeazzo was his nephew and under his care and tutelage. Similar portraits were given by Pietro Bembo in his *History of Venice,* Sigismondo Tizio in his *History of Siena,* and by the local historian Simone Del Pozzo.[128] Whether exaggerated or not, this view of Ludovico closely reflects the opinion of many of his contemporaries and as such cannot be easily discarded.[129]

Gian Galeazzo Sforza's death represented a timely occasion for Ludovico to seize power. With his nephew dead and the French army heading toward the Kingdom of Naples, he was confident that his power would be uncontested. The young Gian Galeazzo demonstrated little penchant for astrology and had no chance to apply it to political praxis as his father had done. Although astrology played a small role in his death, however, it played a much greater role in his uncle's life. In the next chapter, I will concentrate more closely on Ludovico's ascent and fall and on his growing reliance on astrology both in personal and in political decision making.

5

.✦.

The Viper and the Eagle

The Rise and Fall of Astrology under Ludovico Sforza

As the previous chapters have illustrated, the political and personal use of astrology (both medical and political) at the Sforza court was extensive, and it seems to have intensified in times of crisis. It is therefore hardly surprising to note that Ludovico Maria Sforza, the seventh Duke of Milan to rise to power, made frequent use of astrology for personal and political purposes alike. Indeed, no period in the history of the Sforza duchy can be characterized as more uncertain and difficult than that of Ludovico. Possibly as a consequence, in no other period of Sforza domination did astrology flourish as much as a political practice as under his rule (tutor of Gian Galeazzo, 1484–1494; duke, 1494–1499). As this chapter will show, Ludovico Maria Sforza took court astrology to new heights, employing it regularly when travelling, arranging a series of dynastic marriages, deciding the best moment to enter a battle, and inquiring into the uncertain events surrounding his investiture to the duchy. In all these instances, his trusted astrologer, Antonio Varesi da Rosate, was never too far away. The documents available offer some rare insight into the delicate relationship between client and practitioner. Unlike in the case of the thirteenth-century astrologer Guido Bonatti, we cannot glance at Varesi's relationship with his client though one of his works as none seems extant. We do not have, in other words, any work by him that contains occasional accounts of his clientele and the kind of services that were requested of him of the kind of Bonatti's.[1] We do have, however, a series of important documents that are very revealing of the kind of astrological advice he provided to Ludovico. An attentive reading of these letters allows us to probe their professional and personal relationship and recover, to a small extent, the voice of our Milanese astrologer.

Ludovico's reliance on astrology was well known. It attracted the scorn of more than one contemporary, as well as the ferocious critique of one of the most famous Milanese astrologers of the sixteenth-century, Girolamo Cardano, who at the time of Ludovico's fall from power was not even born. Undoubtedly, however, Cardano had heard of the dramatic events that led to the French occupation from his father, Fazio, and other contemporaries who had lived through the difficult moments of Ludovico's demise from power.

In his critique of astrological interrogations, an astrological practice concerned with the "particular" and not the "general," Cardano chose to single out Ludovico Maria Sforza and his astrologer, whom he left unnamed, as prime examples of the ruin that could occur if one were to listen to such advice. "Consider, gentle reader," he warned:

> [. . .] how much ridicule has been caused by those who, by selling their client an astrological interrogation on some single point, brought him to disastrous end, while many, in the same situation, enjoy great success without taking the advice of an astrologer. Let me choose one of many examples, with a clear and remarkable result, and that of a man known to me. Ludovico Sforza was the ruler of the province of Milan. He supported a greedy astrologer who was completely ignorant of astrology (for he was one of those whom Ptolemy rightly criticizes), enriching him with a great fortune of a hundred and more big golden talents. In return for this compensation, this gentleman assigned him the time at which he should begin every enterprise, and in such a ridiculous way that the prince, in other respects a man of great wisdom, had to mount horses in the midst of storms, and lead the whole courtly host of his supporters through rainstorms, through the muck and mud, as if he were in hot pursuit or headlong flight of enemies.[2]

The unscrupulous trickster mentioned by Cardano was certainly Ambrogio Varesi da Rosate, the Milanese astrologer who became famous for his privileged relationship with the duke after saving his life during a terrible illness in 1487.[3] We can be certain of this fact thanks to an official document signed by Gian Galeazzo Sforza as early as 1480, which bestows upon the trusted physician-astrologer an annual stipend of a hundred golden ducats (the same amount stated by Cardano) for his services and his loyalty to his uncle Ludovico.[4] From 1491 onwards, Varesi became a member of Milan's influential Privy Council.[5] As we shall see, over the following two decades the relationship between Ludovico and Varesi intensified dramatically, becoming increasingly vital to Ludovico's daily actions in the years preceding and following the French descent.

Of course one could wonder if Cardano's judgment was not too harsh and his portrait of Ludovico's credulity exaggerated. Even a cursory examination of Ludovico's correspondence with Varesi, however, reveals the duke's almost compulsive—certainly excessive—reliance on interrogations and elections, two astrological techniques that are generally harder to document and whose validity was increasingly cast into doubt among sixteenth-century astrologers.[6] Equally, one could reasonably question Cardano's opinion about Varesi's astrological competence. Cardano, after all, is notorious for having expressed harsh judgments on many other prominent astrologers of the time, and his targeting of Varesi is not exceptional.[7] Varesi's knowledge of astrology does not seem very different from that of many of his contemporaries—he possessed the necessary skills to draw a celestial figure correctly and offer an interpretation. Cardano's opinion, therefore, may refer more to the kind of astrology that Varesi practiced—not the kind of classical astrology favored by Cardano, where interrogations and elections had little place—than to Varesi's astrological skills in general.[8]

In the face of Cardano's accusations, moreover, we can only speculate about Varesi's intellectual integrity or lack thereof. It is certain that Ludovico's exclusive patronage brought him wealth and privilege, but there is no evidence of foul play or bad intentions on his part. The evidence seems to show, instead, that he did all he could to serve his duke faithfully and that he reaped the benefits arising from such a privileged position as any other courtier would. Ambition and greed may have played a part in his ascent, but his success was closely linked to that of his lord, and therefore he had every motivation to help Ludovico succeed and maintain power. It is also clear, however, that Varesi was increasingly forced to produce prognostications on the most disparate (and sometime bizarre) topics and that he himself often doubted the reliability and usefulness of some of his predictions. In the present chapter I shall illustrate the ways in which Ludovico used astrology for his personal and political ends and how his reliance on astrology was both a sign and a symptom of a heightened sense of uncertainty about the future that characterized his reign and grew exponentially in the period that led to the French descent of 1494 and the final occupation of Milan in 1499.

Where the Personal Meets the Political: Sforza Marriages and Astrology

Chapter 4 recounted how the inability of the young Gian Galeazzo Sforza to consummate his marriage with his newly wedded spouse, Isabella of Aragon, in 1489 was not simply a personal matter, but an issue of great political concern. The fact that a political ruler could not perform his marital duties was certainly problematic, but hardly a novelty in the panorama of Renaissance ruling classes, where every marriage was arranged and the bride and groom were either married at a very young age or were separated by a considerable age difference.[9] What made the issue particularly sensitive, however, was the fact that Ludovico had turned down Ferrante's offer to marry Isabella and become duke himself. This choice was justified by the fact that Ludovico had every confidence that his nephew would never produce an heir, thus palpably weakening Gian Galeazzo's role as head of the Sforza dynasty and allowing for Ludovico's own takeover. Ludovico's belief was probably justified by Gian Galezzo's poor health and inexperience, but we should remember also that the person in charge of the arrangements for the consummation of the marriage was none other than Ambrogio Varesi da Rosate. Therefore, Ludovico's confidence may have also come from the conviction that, without the help of astrology and magic, his nephew's chances of fathering an heir were largely diminished, if not totally nonexistent.[10]

The case of Isabella and Gian Galeazzo is possibly the earliest documented instance of Ludovico's use of astrology in the context of a planned dynastic marriage, but it was not the last. He apparently resorted to Varesi's advice regarding the best time to celebrate and consummate a marriage on three other important occasions: in planning his own marriage with Beatrice d'Este, the youngest daughter of Ercole d'Este, Duke of Ferrara, and Eleonora of Aragon; in planning the related marriage of his niece, Anna Sforza, to Alfonso d'Este; and finally, and most importantly, in organizing that of his other niece, Bianca Maria Sforza—she, too, a daughter of the late Galeazzo Maria—to the Holy

Roman Emperor, Maximilian I, himself a staunch believer in astrology.[11] All three occasions were politically and personally momentous for Ludovico: with the first two marriages he acquired an important ally in Italy at the time when his relationship with other traditional powers, such as Florence and Naples, was deteriorating; with the third he gained a crucial family relationship with the Holy Roman Emperor and the much sought-after imperial investiture.

Beatrice d' Este was a good marriage prospect and presented the advantage of being quite dear to her grandfather, Ferrante of Aragon. As a young girl, Beatrice had spent eight years at the Neapolitan court, getting to know well both Ferrante and her future sister-in-law, Ippolita Sforza, not to mention her cousin, Isabella, the future legitimate Duchess of Milan.[12] Originally, Ludovico had asked for the hand of Beatrice's older sister Isabella, then only six, but she had already been promised to Francesco II Gonzaga. With Bona's advice, Ercole d'Este, therefore, offered Ludovico his other daughter, Beatrice, who was five at the time. Ludovico, however, was already twenty-nine years old and had to wait quite a few years to formally marry his young bride.[13] Almost ten years had to pass, but on May 10, 1489, the Ferrarese ambassador in Milan, Giacomo Trotti, was sent back to Ferrara to underwrite the nuptial agreement for the marriage. This stipulated that the wedding was to occur the following year, in March, in a private ceremony (*alla domesticha*).[14] March passed, however, and nothing happened. Only in April did Ludovico send his secretary, Francesco Casati, to Ferrara. At this time, however, he had detailed instructions about the marriage: Ludovico was ready to welcome Beatrice to Milan. He stressed one more time that the marriage had to be celebrated privately, without ostentation, so as not to offend the sensibilities of the newly married duke and duchess, Gian Galeazzo and Isabella, with excessive celebrations.[15] Concerning the time and place of the celebration, he informed Casati (who had to relay the message to Ercole) that he had deemed it desirable to consult his astrologers. Among them no doubt was Ambrogio Varesi da Rosate, who had witnessed Ludovico's marriage contract ten years before.[16] Ludovico's astrologers, Casati would report, had advised him to marry on July 18, "a fortunate and prosperous day for his union with her [i.e., Beatrice]."[17] For reasons that remain unclear, however, the marriage had to be postponed once again, this time to the winter of 1491.[18] At this point, the Ferrarese ambassador received orders to pressure Ludovico not to delay the marriage any further.[19] Ludovico then reassured Trotti of his intention to maintain his commitment and discussed the practicalities of the marriage. He pointed out, however, that Ercole and Beatrice would be hard pressed to reach Milan by boat in January as the weather conditions at that time of the year were treacherous and the river low. He suggested, therefore, that to inconvenience Ercole the least, the marriage take place in Reggio Emilia; then he and Beatrice could travel to Parma by horse, "without lying with her in this place, however, to respect the time and place already determined astrologically."[20] Despite the modest nature of the ceremony, Ludovico reassured Ercole d'Este that he would organize a joust (*giostra*) and other festivities to celebrate the event once he reached Milan.

Even with the concessions, Ludovico's suggestion did not win Ercole's favor. The Duke of Ferrara must have insisted on sending the bride and her entourage by water to Lombardy to have the marriage celebrated there. This had the obvious disadvantage of costing Ludovico more money, as he would have had to provide for the bride and her retinue while they were en route to Pavia and to organize everything so that the Este had suitable lodgings and provisions. Ludovico had explained his initial proposal to have the marriage celebrated in a private manner in Reggio Emilia by his desire not to upset the Duke and Duchess of Milan, Gian Galeazzo and Isabella (who had recently married), by outdoing them with his wedding celebration. Trotti maliciously gossiped, however, that Ludovico probably wanted to save money.[21] Ludovico grudgingly accepted Ercole's requests and promised to look after the practical arrangements in the best way possible. He posed as his only condition, however, that the marriage be celebrated on January 18; otherwise, more days or months might have to pass before they could marry and consummate the marriage.[22] As Trotti indicated in a letter dated to the day before the marriage, the day was chosen to follow the astrologers' orders as to the best moment to marry and consummate the marriage: "tomorrow at the 16th hour," Trotti announced, "the Duchess of Bari will be blessed and married with much pomp *per puncto de astrologia,* that is, on a Tuesday, and that night Ludovico will lie down with her."[23] Ludovico did not want to leave anything to chance: the marriage had been planned to its minute details to ensure it was blessed by a favorable celestial configuration of the skies.

This was not, however, the only Sforza-Este marriage to have been planned. To secure the support of the Este against possible attacks to her power after Galeazzo's death, Bona of Savoy had previously secured the marriage of their daughter, Anna, to Ercole's young son, Alfonso, who was then only an infant.[24] Ludovico's marriage was intended to further secure the Sforza's ties with the Este and thus gain an important political ally on the peninsula. While the two parties did not initially discuss the details of the planned marriage between Anna Sforza and Alfonso d'Este, as the arrangements proceeded it was decided to have Anna's marriage coincide with the visit of the Este retinue accompanying Beatrice. Thus, Ludovico's marriage with Beatrice became part of an elaborate double marriage ceremony with the Este family. In November 1490, Ludovico decided to have the two marriages celebrated around the same time: his own marriage in a private fashion in Pavia and Anna's with more pomp in Milan. In this way, his own wedding did not need to be a modest affair, but could be made part of the wider celebrations for Anna's marriage to Alfonso. As Trotti dutifully reported, Ludovico had put his astrologers to work once again to determine the best day for Anna's marriage, but unfortunately the astrologers had suggested January 19, the day after his own planned marriage in Pavia. Ludovico explained how it was impossible for he and Beatrice to reach Milan before January 21 and that his astrologers had consulted among themselves once more and advised that the 22nd might be equally favorable. They could not say this for sure, however, as, while they possessed Anna's geniture, they did not have Alfonso's. Trotti, therefore, requested that this be sent as soon as possible to Milan to be examined by Ludovico's astrologers.[25] As in the case of Galeazzo and Dorotea Gonzaga, in this instance, too, the astrologers were asked to scrutinize the nativities of the future bride and groom, this time, however, to determine the best time for the marriage. We can safely assume, therefore, that the same had happened with the nativities of Beatrice and Ludovico. (What remains to be established, of course, is if such a practice was relatively common among Renaissance elites or was something peculiar to the Sforza court.)

Ludovico's wedding took place in Pavia on January 18, and three days later the newlyweds and their party entered Milan to celebrate Anna's wedding.[26] This double union was to be accompanied by fitting celebrations. Rooms in the Castle of Porta Giovia were lavishly decorated for the occasion: the walls of the *sala della balla* (ballroom) were painted with the *historia* of Francesco's deeds, and a blue drape with stars painted on it was suspended from the ceiling.[27] The events surrounding the double marriage are lavishly recounted in Tristano Calco's marriage oration.[28] Ludovico may have wanted to outdo the extravagant wedding festivities that had accompanied the marriage of Beatrice's parents, Eleonora of Aragon and Ercole d'Este, but, as noted, what seems to have driven the preparations for the wedding was the principle that the union of the two couples was to happen under the most auspicious sky. Astrological elections determined the timing of each movement, witness Giacomo Trotti, who lamented with Ercole d'Este (who did not attend the wedding) that they had to travel under extreme weather conditions, as Ludovico would only travel "per puncto de astrologia." For this reason, he reported, the entire party had to wake up before dawn despite the bitter cold to leave the nearby village of Binasco and reach the church of Saint Eustorgio in Milan (near Porta Ticinese) as originally planned.[29]

While we may think of Trotti's comments as the exasperated exaggerations of a disgruntled courtier, Ludovico's obsessive planning is reported separately, and thus confirmed, in Calco's own account of the wedding preparations: three days after the Este party arrived in Piacenza, Ludovico returned to Milan to complete the wedding preparations and "determined that the guests would enter the city in the 5th hour of the day." To make this possible, Calco added, the party was stationed at Binasco, where they had spent the night.[30] According to Calco, Ludovico's wedding was accompanied by a comet (*stella crinita*), which appeared for seven days in the skies of Milan, Venice, Florence, France, and the German lands. This Calco (and probably Ludovico) interpreted as a good omen, a sort of public form of political and dynastic legitimation (albeit directed not just to Ludovico but to the Sforza dynasty more broadly).[31] Celestial events of this sort were dutifully recorded and interpreted by contemporaries, especially if their significance could be related in some meaningful way to events they were experiencing at the time.

As these examples reveal, marriage à la Sforza was no simple affair, and certainly not one left to chance. Astrology played a crucial role in establishing the marriage rituals, in particular their place, date, and time. Considering the premium placed on producing a healthy male heir and Gian Galeazzo's own difficulties, it seems quite understandable that the day and time of Ludovico and Beatrice's union was determined with the favor of the heavens in mind. Barely more than a year after their marriage, on January 25, 1493, at the 23rd hour, Beatrice gave birth to a boy, thus fulfilling Ludovico's expectations and his desire to father a potential heir to the duchy. To celebrate his own dynastic alliance with the Este, the child was aptly named after Beatrice's father, Ercole.[32]

Nothing about this birth seems to have been left unplanned: it was clear that Ludovico placed much hope on the birth of a boy who would confer on him a stronger sense of legitimacy in declaring himself duke once his nephew died. As recounted by Isabella d'Este's lady-in-waiting, Teodora Angeli, who had been sent by the marchioness to assist her sister, Beatrice, at Ercole's birth, Ambrogio Varesi's quarters in the castle were very close to where Beatrice was; her living quarters were next to the room used by Ludovico to summon the Privy Council, on one side of which, in the middle, stood Varesi's astrolabe, a vivid reminder of the astrologer's acquired prominence within Ludovico's group of counselors. Without Varesi, Teodora added emphatically, nothing was decided.[33] It is clear that Varesi was heavily involved in Beatrice's care and wellbeing, at least as much as he had been in choosing the date of her marriage and the day and time of its consummation. This is clearly indicated by the fact that even the *levatura di parto*—the moment when a woman left her bed after having given birth—was determined after consulting our astrologer. Given the difficulties of childbirth in early modern times, this moment was naturally considered particularly important for the new mother—the *levatura* signaled a woman's survival from childbirth (something that could not be taken for granted given the mortality rate of Renaissance women at childbirth), her return to full health, and her ability to resume her role as wife.[34] Not surprisingly, religion played an important role in the process: as Teodora recounted, on February 23, 1493, Beatrice d'Este and the Duchess of Milan, Isabella of Aragon (who had given birth to a daughter only a few weeks before), left their beds to attend a mass at Santa Maria delle Grazie and thank God for having safely delivered their children. What is remarkable, however, is that once again Varesi had been put in charge of establishing the best time for Beatrice to leave her bed: this, Teodora recounted, was established "per puncto d' astrologia," without which, she stressed once more, nothing could be decided.[35]

This example is clearly revealing of the trust that Ludovico placed on Varesi's astrological advice. His dynastic and personal ambitions were increasingly punctuated by astrological consultations that were requested to ensure the success of his political enterprise and of his marriage with Beatrice. Varesi, who had gained Ludovico's trust by saving his life, increasingly obtained a position of privilege within the ducal court, and the location of his rooms so close to those where the Privy Council was held is indicative of the close relationship of astrology and politics at this particular time.

On the Move: Astrology, Diplomacy, and Travel

Astrology's influence on Beatrice and Ludovico was not limited to their marriage ceremony or the delivery of a healthy boy, but extended to all other aspects pertaining to their persons where astrology was deemed relevant. Their health and travels certainly fell under this rubric. For this reason, when the couple travelled with Eleonora of Aragon and a large retinue to Ferrara a few months after Ercole's birth, Varesi's letters of astrological advice determined the day and time for their actions. The party left Milan on March 4, as established by Varesi after Ludovico consulted him.[36] Around the same time, Varesi enthusiastically predicted to Ludovico (presumably with the help of astrology) that his wife would bear him a second male child before a year had passed, generating some anxiety among the Este court that Beatrice could outdo her sister Isabella in producing the legacy required of all noble women. Ludovico put so much trust in Ambrogio Varesi's prediction that he had already invited Beatrice's mother, Eleonora of Aragon, to that birth![37] After having stopped in Vigevano and Cremona, the Sforza-Este retinue headed toward Ferrara. Ludovico joined his wife, Beatrice, in Parma, and the contingent proceeded toward the city, entering it on May 18 "at the astrologically propitious time." Ludovico's rapid and unexpected decisions as to his travels, apparently, left Isabella puzzled, seemingly indicating that such overreliance on astrological elections was not common among the Este, and it raised eyebrows among members of the court.[38] By the time Beatrice and Ludovico arrived, Isabella d'Este had already left for Venice, where the Este owned a palace on the Canal Grande, but the Milanese contingent was met by the marquis, Francesco Gonzaga.[39]

The Sforza retinue remained in Ferrara for a week. On May 25, the Sforza-Este party left the city, once more at the astrologically propitious time.[40] While Ludovico took off with Ercole d'Este for Belriguardo, near Ferrara, the rest of the contingent proceeded toward Venice. The Doge of Venice and other patricians of the city received Beatrice and the rest of the party in spectacular fashion.[41] The impressive size of the Sforza-Este retinue—1,200 people in total—already suggests that the visit was not simply a pleasure trip.[42] A long-standing opponent of both the Sforza and the Este, and particularly suspicious of Ludovico's intentions, Venice had not looked favorably on the Sforza-Este political union. Pressured by Ludovico, however, on April 22, 1493, Venice was forced to join the league with Milan and the Pope against Naples.[43] The trip thus had the clear diplomatic purpose of celebrating the recent political alliance, and Beatrice, with her charm and beauty, was instrumental to the success of this mission.

From Venice the young bride sent affectionate and informative letters to her husband detailing her daily activities and the many people she met in her busy social life. During the grand party that was organized for the Este-Sforza contingent by the Serenissima a few days after their arrival, however, Beatrice reported that she felt unwell, suffering from a headache and a niggling sore throat that forced her to take an hour's leave from the party to rest in one of the palace's rooms.[44] As her personal physician Luigi Marliani wrote the same day, she suffered from a common cold, largely due to the change of climate, her travels, and her lack of sleep.[45] Almost daily letters from her physician updated Ludovico on Beatrice's health, thus giving us a vivid impression of the concern and care taken for her person.

Beatrice's cold improved, and by the time she left Venice to return to Ferrara she was on the mend. Once again, Ludovico sought Varesi's advice on the most suitable time for the retinue's trip, asking him to advise on the appropriate time for Beatrice's departure from Ferrara. His letter to Luigi Marliani reveals much about Ludovico's constant dependence on Ambrogio Varesi's advice. In it, together with expressing his joy for the improved condition of his wife's health, Ludovico also voiced serious concerns that Marliani may have ignored Varesi's advice to depart on Monday, thus leaving earlier than the day elected by the astrologer for the retinue's journey. Ludovico was glad, however, to receive news that Marliani had followed Ambrogio Varesi's advice and that the party was ready to set off from Ferrara the following Monday as advised:

Dear Magister Luigi, I could not be more grateful for what you have written to me about the good health of my spouse, because there is nothing that makes us happier than the fact that she is well and will return with that good health that she possessed when she departed from us. And as we understood that the time of your departure was Saturday, namely yesterday, and having seen the note by Magister Ambrogio that said it was to be Monday at three, that is tomorrow, I was very surprised that you did not follow the orders given by Magister Ambrogio. But having been informed later that you will leave tomorrow, I was satisfied. Despite the fact that I said that the right day was Saturday, we had said so thinking that this was what Magister Ambrogio had said. For this reason, now you have to be careful and leave Ferrara so that it is the proper hour of the day [. . .][46]

Clearly, for Ludovico and his familial entourage, timing was everything: Marliani was reminded that not only the day of Beatrice's departure had to be right, but the time, too.

From other correspondence, we gather that, at the time, Varesi was in charge of Ercole's care and resided at the Castle of Porta Giovia, or, occasionally, at Pavia.[47] Yet, he constantly dispensed travel advice to both Beatrice and Ludovico from afar and followed similar rules when arranging for their children's travels, as documented by a letter dated March 16, 1494, in which Varesi informed Ludovico that the young Ercole, his personal retinue (including the physician-cum-educator Nicolò Cusano), and Varesi himself were on their way to Milan when they had to stop in Abiategrasso for a day to wait for Saturn's unfavorable aspect to pass.[48] As these examples clearly show, by the early 1490s there was virtually no occasion in which Ambrogio Varesi was not consulted. The relationship between Ambrogio Varesi and Ludovico had become symbiotic: Varesi was in charge of every important aspect related to Ludovico's family and actively used astrological elections to ensure his lord that all actions were undertaken at the most propitious time. This use of elections, I would argue, was not common, and this fact gives some credit to Cardano's bitter critique years later that Ludovico was foolishly following his astrologer's advice even when the conditions were clearly not favorable, like mounting on a horse in the middle of a storm.

As already mentioned, Beatrice's visit to Venice had a marked diplomatic character. She was there to promote her husband and his duchy to the Venetians and to help legitimate his role as *de facto* ruler of Milan as he was preparing himself to confront the Aragonese with the threat of a French descent into Italy. We know that at this stage Ludovico was already playing a double game with the Holy Roman Emperor, Maximilian I, with whom he was negotiating for an imperial investiture, and with the French king, Charles VIII, from whom he wanted to obtain protection against the Aragonese of Naples. When pressure from Naples to restore Gian Galeazzo and Isabella to their legitimate roles of Duke and Duchess of Milan grew in 1490, Ludovico sought a formal renewal of his alliance with France and renewed his efforts to gain the investiture of the fief of Genoa.[49] To accomplish this, however, Ludovico needed to give his ally something in exchange. When the French ambassadors travelled to Milan in January 1492 to join the league and grant their investiture, they asked, therefore, that Ludovico return some lands that the Milanese had expropriated to the Marquis of Monferrat, a close ally of

France.[50] Even in this case, Ludovico turned to his astrologers, and the advice he received was that January 16 was the most favorable day to join the league. In this case, however, Ludovico's choice could not be fulfilled, as the French ambassadors refused to join the league until the lands and castle that Ludovico had taken away from the Marquis of Montferrat had been returned. We know that the league was finally ratified on January 24, but it is not clear if in this case, too, astrology played a part in choosing the day and time.[51]

Be that as it may, it is clear that astrology played a prominent role in Ludovico's daily political practice and shaped his diplomatic activities. Other examples can aptly illustrate this point. At the end of February 1492, for instance, Ludovico sent his own embassy to Paris to thank the King of France and to congratulate him on his recent unexpected marriage to Anne of Brittany. The embassy was led by Gianfrancesco Sanseverino, Count of Caiazzo, and included Carlo Barbiano di Belgioioso, Girolamo Tuttavilla, Galeazzo Visconti, and Agostino Calco. Ludovico's entourage travelled with meticulous instructions regarding what they should say (and even wear) in the king's presence.[52] The aim was both to ratify and extend the alliance negotiated with the envoys a month earlier so that not only Duke Gian Galeazzo, but also his uncle Ludovico were guaranteed French protection if they were attacked by a foreign power.[53] Not surprisingly, in this instance, too, the entry of the Milanese retinue into Paris was determined astrologically: Ambrogio Varesi advised Ludovico that the most propitious day was March 28.[54] The ambassadors reported, however, that upon consideration, they "had found that on the 27th and 28th the Moon was combust, and it was not wise to seek a meeting with the king on those days," so they tried to arrange it for March 29. The king, however, had determined to leave Paris for two or three days to have some leisure and informed the envoys that he was keen to see them on March 28 after dinner. Things became further complicated when the party was briefed that the Duke of Bourbon had left the city and would not manage to return in time. The visit, in the end, was delayed to the morning of March 29, which, the Milanese envoys reported, was what they had preferred all along "because of the combustion of the Moon" of the previous days.[55] This example seems to suggest that, not only did Ludovico abide by the principles of astrology, but so did his most trusted ambassadors (it also shows that the French king and his entourage did not seem to abide by the same principles and did not mind meeting when the Moon was combust). However, at least in this specific case, Varesi's astrological advice was called into question, if only to say that he must have ignored or miscalculated the combustion of the Moon. Could it be that our astrologer did not take into consideration the longitude of Paris? Or was he pressed for time and made a mistake? This cannot be known. In any case, these documents suggest that astrological information was relatively accessible not only to Ludovico, but also to other members of the court, and that they, too, made use of astrological elections in their diplomatic activities. It cannot be excluded, moreover, that an astrologer or a physician was part of the Milanese contingent that travelled to Paris. In any case, it is clear that in the minds of many Milanese courtiers, daily political praxis and astrology were closely intertwined.

Despite the fact that the entry happened under the most propitious sky, the result was not quite as favorable as Ludovico and his envoys expected. Agostino Calco, the secretary who had accompanied the Milanese contingent, dutifully reported their grand entry into the city and their first assembly with the king, giving detailed information about their dress on each occasion, their conduct, and the welcome they received at the court.[56] After the first public oration, delivered in Latin by Galeazzo Visconti, the ambassadors were again received by the king. To stress the advantages of a French alliance with Milan, Ludovico had instructed the Count of Caiazzo to reveal secret plans from the King of England and the Holy Roman Emperor to invade the south of France. Ludovico's help had been requested for the enterprise, but he refused. Ludovico then instructed the Count of Caiazzo to reveal these letters to the king in private and to stress how he had refused an alliance with the Holy Roman Emperor to side with the King of France.[57] Ludovico also instructed the count to receive reassurance that, in case of Milan's intervention in support of his Majesty against England, Spain, and the Holy Roman Empire, he would be protected by his Majesty in return and included in the terms of a future peace. He further instructed the Count of Caiazzo to verify the information that the King of Naples had sent his ambassador to France "to convince his Majesty to move against our government" and, if necessary, to make clear to the king the great disadvantages he would receive if he were to embark on such a military *impresa*.[58]

The ambassadorial reports were mixed: the French envoys who had signed the peace a few months earlier were very unhappy about the presents they had received at their departure from Milan, lamenting that the Marquis of Monferrat had given to each of them twice what they had received collectively from Ludovico.[59] The message was clear: any form of alliance with the king had to be facilitated with generous donations. Like his rival Maximilian I, Charles VIII was always short of cash, and, playing the part of the lion, he tried to exert pressure on Ludovico to obtain military and monetary support.[60] Finally, on April 29, 1492, Charles admitted Ludovico Sforza, in his own right, as a member of the league concluded on January 16 of the same year with the legitimate Duke of Milan, Gian Galeazzo.[61] This was an important aspect of Ludovico's negotiations, for in this way, the King of France was implicitly recognizing Ludovico as a rightful political ally and the true lord of Milan, or so, at least, Ludovico was led to believe.[62]

To contrast effectively Ferrante's threat of intervention in favor of Gian Galeazzo, Ludovico had to strengthen considerably his ties with other national and international powers. As we shall see shortly, this included the Pope, who could exert great influence on the Kingdom of Naples (officially a papal fief) and was an essential ally in Italy. But as Milan was—officially, at least—a fief of the Holy Roman Empire, Ludovico's best chance of being accepted as its legitimate ruler resided in the imperial investiture that had been sought, albeit unsuccessfully, by both his father and brother before him. It was therefore essential for him to gain the full support of Maximilian I, the future Holy Roman Emperor.[63]

The Viper and the Eagle:
Ludovico, Maximilian I, and Astrology

In the same week Beatrice left for Venice in May of 1493, Ludovico sent his trusted ambassador, Erasmo Brasca, not to France, but to Germany, this time to meet the King of the Romans (and future Holy Roman Emperor) Maximilian I with two important propositions for him: first was the hand of his niece, Bianca Maria Sforza, which would come with the extremely generous dowry of 300,000 ducats; second was the request for his investiture to the Duchy of Milan. He was prepared to add another 100,000 ducats to Bianca's dowry to receive it. It is obvious that the two offers went hand in hand, and Ludovico was buying his investiture by offering Maximilian a wife and, more importantly, a substantial dowry that would solve many of Maximilian's financial problems. The offer, as Ludovico probably knew, was timely. Maximilian's first wife, Mary of Burgundy, had died in 1482, leaving him a widower but allowing him to inherit from her the Bugundian Netherlands. This, however, engendered a war with France for possession of the

Burgundian lands, and the war continued for almost a decade, drying up Maximilian's finances and occupying much of his time.[64] The emperor sought a solution by proposing his daughter, Margaret, in marriage to Charles VIII and becoming himself engaged to Anne of Brittany, whom he married by proxy in 1491. The official marriage ceremony, however, never happened, as Charles VIII's sister, Anne of France, then the Dauphin's regent, arranged a marriage between Charles and Anne of Brittany, who was forced to repudiate her marriage to Maximilian on the ground that it had not been consummated. After the events, Maximilian found himself without a wife, with an unmarried daughter, and with some rather poor finances. Ludovico's marriage proposal solved many of his problems and provided dynastic ties with a powerful neighboring Italian state.

Besides powerful foreign allies, Ludovico was aware that he needed support in Italy, so his internal politics, much of which was put in place with the help of his brother, Cardinal Ascanio, reflected such a necessity. As noted, after April 1493 Venice was no longer a menace.[65] The League of Saint Mark, as it was called, would have been unthinkable only a few years before when Venice, Milan, and Rome occupied opposite poles of the Italian political spectrum. Because it contained the threat of foreign intervention, however, the new situation was exceptional and induced the Italian states to seek new political alliances. The League of Saint Mark required that all states intervene in support of the others if attacked, and it was clear that its main aim was to discourage Naples from attacking Milan.[66] But such an alliance included a much more revealing clause: Ludovico insisted on the possibility that the King of France could join the league. Ludovico saw in Charles VIII the perfect ally against the threats of Naples and Venice. What he obviously did not realize, or otherwise recklessly decided to ignore, was the danger to which he exposed his own duchy and the entire peninsula by allowing the European powers to enter Italian diplomacy in this manner.

Ludovico's responsibility in the French descent has received much attention among political historians. An old historiography based on Venetian and Florentine accounts tried to make him out as virtually the only culprit behind the French expedition.[67] More recently, however, scholars have provided a more nuanced view, suggesting that, although Ludovico's wavering political diplomacy and his unjustified sense of superiority was certainly an important factor, other elements contributed greatly to Charles VIII's decision to intervene.[68] One should not forget that at various stages other political powers had

invited the king to settle the scores among the Italian powers: these included
Venice, Naples, and the papacy, all of whom had used the threat of a French
intervention as a way to exert more power in the peninsula.[69] Ludovico was
merely the last of a long line to do so, even if he admittedly remained blind to
the possible consequences of his actions. Nor was Ludovico the only one to
carry on with dangerous double games: Alexander VI's diplomacy clearly
reveals much duplicity of intent and an attempt to keep a door open both to
Naples and the French king.[70] What gave concrete impetus to the French
descent, however, was, more simply, Charles VIII's ambition coupled with his
attempt to distract the French lords from internal affairs. The catalyst, instead,
was the treaty of Senlis of May 23, 1493, which concluded the peace between
the Holy Roman Emperor and the King of France after a decade of war. Such
an event, it has been pointed out, shattered Ludovico's plans and exposed
Milan to the possibility of a French occupation.[71]

It was actually during Beatrice's visit to Venice that Ludovico received
news of the conclusion of the peace at Senlis from Count Belgioioso, his ambassador in Paris. By signing the peace, the King of France had made quite clear
his intention of invading Italy, and Ludovico was now forced to revise his
foreign and domestic policy accordingly. His negotiations with the King of
the Romans had to intensify further, and it was now absolutely crucial that he
receive the imperial investiture to protect his title from Louis of Orléans's
claims. Although Ludovico had obtained reassurance from the King of France
that his state would not be threatened, he knew all too well that these promises were fleeting. For this reason, when Beatrice met the Doge of Venice to
bring her husband's message to him, she had now to stress Ludovico's good
relationships with both the King of France and that of the Romans before
lying before the doge and Senate Belgioioso's letter informing Ludovico of
the forthcoming French descent. The letter said that Charles VIII was ready
to send envoys to Rome, Venice, and Milan to seek support for the French
impresa. At this point Beatrice also informed the doge that Ludovico's negotiations for his investiture were very advanced.[72] The scope of such a diplomatic maneuver was to show that Ludovico was in a strong position to remain
Duke of Milan, and that Venice, therefore, had to remain close to her ally
and follow a similar policy concerning the French descent.

Ludovico's best chance now was to hasten Bianca's wedding preparations.
Once Ludovico obtained Maximilian's consent to marry Bianca in August
1493, things proceeded relatively quickly for the new bride-to-be. By early

October, Ludovico was anxiously awaiting the arrival of the German envoys. Once more, Ludovico resorted to his astrologer's foresight to gain precious information. From a long letter Varesi wrote to his duke on October 8, 1493, we can gather that Varesi had received prior astrological interrogations from Ludovico about the date of the envoys' arrival; one interrogation, he specified, was dated September 1. Varesi had initially replied that the skies indicated that he would receive some news on October 3 or 4 and that the envoys would actually arrive by October 25. In a second letter, Varesi was quick to point out to his lord that, as he had predicted, the duke had received news on October 3 indicating that the ambassador would be there within ten days. While the first part of his prediction was correct, however, the second about their arrival before October 25 was a bit off the mark. At this point, by his own admission, Varesi had gone back to consult his books, studying again those letters and the interrogations they contained. After providing an elaborate explanation based on the position of the planets at the time of the last interrogation, he confirmed that the envoys were en route, that the letter was sealed, that it carried good news, and that the envoys would reach Milan in a day or two:

> That [the envoy] is on his way, is indicated by the Sun, lord of the tenth house, in the ascendant; by Venus, significatrix of the interrogation, in the ascendant; and by the Moon, in the eleventh house, together with the distancing of the Moon from Mercury. These things indicate without a doubt that the messenger has departed and is on his way, that the letter is sealed, and that, because Mercury is moving away from Venus, which is the fortune of the sky, in this letter there will be things pleasing to Your Illustrious and Excellent Lord.[73]

Even if the envoys did not reach Milan within a day or two, Varesi added, he hoped that the retinue would at least be within Milanese territory by then and thus would be there by Friday or Saturday of that week at the latest.[74] As to the earlier prediction regarding October 25, Varesi explained how he had come to that date by examining the revolution of Ludovico's nativity and that of the horoscope of his tutorship of Bianca (*tuttela*). Varesi believed, therefore, that this other date could indicate the arrival of other envoys or that around those days they would be able to clinch the deal.[75] Varesi was undoubtedly aware of the conflicting and confusing nature of his predictions, cautiously adding that "in the sky I cannot—and no-one can—see in its particulars how things below are."[76] Despite Ambrogio Varesi's reassurance, the two German envoys, Gaspar Melchior, Bishop of Brixen (Bressanone), and Jean Bontemps did not arrive in Milan with news and instructions for the marriage until November 7, well after the two proposed dates, and yet we have no documentation indicating that Varesi was questioned further or reproached for his inaccurate predictions.[77] What this correspondence reveals, instead, is that Varesi was asked by his patron to use astrological interrogations to pry into an increasingly uncertain and fleeting future. Varesi was aware, however, that this was not easy, stressing with caution that knowing "particular" events was hard and at times impossible. In this statement one could possibly read Varesi's self-doubt as to the soundness of these techniques, which, after all, did not find complete consensus among his peers and were generally less practiced than natal and natural astrology. But such a rhetorical statement, that may have served to a certain extent as a stopgap, contained a grain of truth: according to Aristotelian natural philosophy, particular effects had particular causes, and yet astrological influences inclined but not necessitated men to act in a certain way. For this reason, most astrologers believed that predicting men's particular behavior with absolute certainty was not possible. There was, of course, the obvious additional complication of the numerous factors that astrologers had to account for, especially in complex circumstances like this one, which made a judgment *par force* subjective and conjectural. This clearly made interrogations and elections less certain than other astrological practices.[78]

By the time of Bianca's proxy marriage in Milan in November 1493, Emperor Frederick III had died after a long illness, and Maximilian was believed by everyone to be the natural candidate for the title of Holy Roman Emperor.[79] Once the envoys arrived in the city, the court started preparing itself for the grand event. The marriage date was delayed for a further week, however, to allow the French envoys to reach Milan and participate in celebrating the newly acquired Milanese allies. On November 30, 1493, the young daughter of Galeazzo Maria Sforza stood next to her uncle Ludovico in the piazza facing the Castle of Milan, ready to proceed to the Duomo where the wedding ceremony was celebrated. That day she was married by proxy to her distant German husband by receiving the marriage band and by having the imperial crown placed on her head. The marriage, as Beatrice Sforza recounted to her sister, Isabella d'Este, who did not attend, was celebrated with the greatest pomp and without sparing any expense.[80] Three days later, Bianca started a long journey to meet the husband she had never seen and to start her new life in a foreign country. Reaching Innsbruck just before Christmas, she was received by Maximilian's uncle, Archduke Sigismund of Austria, and spent Christmas with him awaiting Maximilian's arrival. Bianca had to wait three more months, however, before meeting her husband. Finally, on March 9, 1494, the King of the Romans made his appearance in Innsbruck.[81]

What prompted Maximilian to wait so long before joining his bride? One likely hypothesis is astrology. Historian Franz Stuhlhofer has demonstrated that Maximilian's father, Frederick III, delayed the moment for consummating his marriage to Eleonor of Portugal to make it coincide with the most propitious time established by his astrologers. He suggested that the crucial factor—what he calls the "Habsburg rule"—may have been the conjunction of Venus with the Sun, which was common to a number of Habsburg marriages.[82] It has recently been suggested that Maximilian's delay in reaching Innsbruck may have been motivated by similar considerations and that the date and time of the marriage was established astrologically.[83] The fact that Ludovico had arranged all other marriages astrologically makes this explanation even more plausible. To do this, Maximilian's astrologers would have had to possess Bianca's horoscope, something very simple to obtain given that these were routinely cast for the Sforza children and had been used for previous marriages. Yet a close look at the dates of Maximilian's marriages—first to Mary of Burgundy and then to Bianca—reveals that while he loosely followed the so-called "Habsburg rule" in marrying Mary on August 19, 1477, he did not do the same in marrying Bianca. Stuhlhofer dismissed the question by arguing that this simply confirmed how little Maximilian cared about the marriage, but this may have been for lack of a better explanation.[84] Indeed, it is doubtful that a man as steeped in astrology as Maximilian, who had arranged his first marriage after receiving astrological counsel, would have completely dismissed such advice for his second marriage. When we add to this that the bride was a Sforza, such a possibility becomes even less likely. Therefore, it is more probable that, having seen how badly his first marriage with Mary had ended—with all of the challenging political consequences that had followed—he (or somebody else for him) may have wanted to avoid the very same conjunction that had sealed that union. Thus he may have thought that it was better to attempt a different "astrological strategy" (or perhaps, more simply, such conjunctions were so far distant in time that Maximilian would have had to delay meeting his spouse for an even more emarrassingly long period of time).[85]

What other factors, therefore, could have made the astrologers choose March 16 as a favorable date? Over the years, Maximilian I's geniture was cast and analyzed by numerous astrologers, most famously by Regiomontanus (who also copied that of Maximilian's mother, Eleonor of Portugal) and the other Nuremberg astronomer Johannes Schöner, who illustrated his own work with multiple interpretations of Maximilian's chart.[86] Cardano copied his chart of Maximilian from the latter and offered a brief interpretation that stressed how the trine aspect between the Sun and Mars indicated that Maximilian would be invincible at war. Cardano neglected to say, however, that the Sun and Mercury, which were conjunct in Aries, were also in a square with Saturn, and that this was hardly a favorable placement.[87] Similarly, Schöner stressed that Mars was the lord of the geniture, with Venus conjoining it, and in harmonious relationship with the Sun. Considering that Mars is the planet of war, this was an excellent placement for the future Holy Roman Emperor.[88] Schöner added, moreover, that Maximilian would have many friends among the kings and princes of Europe since the lord of the triplicity of the eleventh house (the Sun) was exalted in the seventh house, that of marriage.[89] He would also, however, have many hidden enemies, as could be evinced by the fact that Saturn is the lord of the significator of enemies in a squared aspect with the lord of the ascendant (Mercury), which is also the lord of the twelfth house, the house of enemies.[90] All in all, Maximilian's chart was that of a powerful ruler apt at war, but with internal problems, which was not very far from the truth. In his chart, Mars, Mercury, and Saturn were singled out as particularly significant planets.

Although I have yet to locate a copy of Bianca Maria Sforza's horoscope, we know that she was born sometime between April 4 and 5, 1472, and this kind of information is sufficient for us to say that her horoscope would have been favorably received by Maximilian's astrologers: Bianca's Sun and Mercury were nicely placed in Aries like Maximilian's, and this, according to Schöner, promised them a solid marriage.[91] Saturn, which was in the eleventh house (that of friends), was in a favorable aspect (sextile) with Mercury, which was trine Bianca's ascendant. On March 16, 1494, the day of their wedding, Mars was once again in Aries, its own mansion and Bianca's and Maximilian's birth sign. The Sun was in Aries (thus, broadly interpreted, in its exaltation), while Mercury was conjunct Saturn so as to mitigate its negative influence once again. As noted, astrologers believed that Maximilian was crossed by Saturn, and for this reason we may speculate that the marriage date was chosen by paying particular attention to Saturn's placement. On the wedding day, Saturn was not opposed to any of Maximilian's beneficial planets, but it was, instead, in conjunction with Mercury. Ludovico shared with Maximilian the favorable placement of Mars and the Sun, respectively in their domiciles (Aries and Leo), and the favorable placement of Mercury, conjoined the Sun. They also shared, however, the unfavorable placement of Saturn, which in Ludovico's case was in the eighth house, thus "suggesting many evil things," as Cardano put it[92] (See Figure 14, Chapter 1). Astrologically, therefore, the alliance between Ludovico and Maximilian could have been successful.

This opinion seemed to be shared by the participants at the wedding and other members of the court. Remarks about the importance of astrology in celebrating the nuptials are also present in two other sources, Giasone del Mayno's marriage oration before the King of the Romans on his wedding day[93] and Tristano Calco's celebration of the event in his wedding oration.[94] The celebrated Pavian jurist Giasone del Mayno, who accompanied Bianca in her journey across the Alps, was not sparing of praise nor of classical parallels for everyone involved. In a feat of rhetorical acrobatics, he first drew a series of parallels between two important figures of Roman history, the senator-turned-dictator Fabius Maximus (ca. 280 BC–203 BC) and the leading general of the Roman Republic Publius Cornelius Scipio Aemilianus (185–129 BC), from whom Maximilian took his name (Maximilianus being the union of Maximus and Emilianus). Both, quite aptly, had defended the Roman Empire against the barbarians. Maximilian, likewise, was not only to be lauded for being the King of the Romans, but also for being the one to protect all Christians against the barbaric Turks.[95] In addition, Mayno skillfully connected Maximilian's marriage to Bianca Maria Sforza to that of one of his ancestors to a Visconti, thus suggesting that this recent family link was not forged anew, but was more simply rekindled: Leopold III, Duke of Austria (1352–1414), had married Bernabò Visconti's daughter Viridis, who bore him five children, among whom was Maximilian's grandfather, Ernst the Iron. This fact in itself—Mayno seemed to imply—was highly propitious for Maximilian's marriage with Bianca.

The oration included lengthy praise of the two spouses and their virtues, and, in Bianca's case, also a rather tedious list of all the men who had asked her hand, wishing to marry her. Finally, some words of praise had to be offered to Mayno's patron, Ludovico: "nobody ignores," Mayno declaimed, "the extreme prudence of the divine Ludovico, his exceptional wisdom, how excellent is his divine acumen, how abundant the magnificence of his soul, how great his knowledge of military matters, how strong his authority, how much his grace."[96] If this was not enough, Mayno described Ludovico as the "arbiter of all things Italian," he who controls war and peace.[97] Indeed, in reading Mayno's oration one cannot but be struck by the irony of his words and the remarkable exaggeration of Ludovico's importance. No doubt, however, Ludovico would have read it as an appropriate and fair description of himself and his political position.[98] Last but not least, Mayno waved into his praise a subtle astrological reference to his patron's horoscope and to Maximilian's:

> The planets do not usually set their powers in motion unless they are coupled; but when they conjoin—as Jupiter with Saturn, or the Sun with Mercury— then they portend great consequences on Earth. Through imperial Bianca the viper sustains the thundering eagle; and Ludovico as Mercury is conjoined with Jupiter, that is Maximilian, Lord of the Earth.[99]

In this passage, the importance of Mercury and Jupiter is clearly emphasized, and not by chance, of course. Mercury was the lord of the ascendant in Maximilian's chart, while Jupiter was the lord of the ascendant in Ludovico's chart.[100] In an overflowing of allegorical and astrological imagery, therefore, the political union of Ludovico with Maximilian was both symbolized by the heraldic image of the thundering eagle and the viper—the symbols of the two houses—and the astronomical conjunction of Jupiter with Mercury.

Tristano Calco, however, was even more explicit than Mayno when it came to the role of astrology in the Habsburg-Sforza marriage. Not surprisingly, he stated unequivocally that Varesi had established the consummation of the marriage astrologically:

> Kept away from joining in making love, they are finally united at the time established by Ambrogio da Rosate, the most famous astrologer and at the same time illustrious physician, who had advised for the afternoon hours. Not only had Ambrogio predicted many things, but as an oracle [*vate*] rich in the art of the ancient Apollo, he controlled the departures of the ambassadors and prognosticated the arrivals of messengers when dealing with political affairs.[101]

Calco's passage is particularly revealing: given his reference to the astrological predictions of the coming and going of envoys—attested, as we have seen, by archival documentation—it would be hard to discard his comment regarding the consummation of the marriage as fanciful. Further astrological elements were weaved into the celebration, as can be evinced by the medal that Ludovico Sforza commissioned from the Mantuan goldsmith Gianmarco Cavalli to celebrate the wedding. As Gregory Harwell has suggested, the representation of Maximilian on a horse, led by Mars with an unusual caduceus—Mercury's traditional symbol—is a clear reference to Maximilian's horoscope and the extremely favorable conjunction of the two planets in the sign of Aries.[102]

We do not know how long the imperial couple had to wait before the consummation took place, nor exactly what factors Varesi and Maximilian's astrologers singled out for choosing the day of the marriage and its consummation. For lack of better documentation, our hypothesis about the reasons why the date of March 16 was chosen necessarily remains speculative. Were they to follow the "Habsburg rule" suggested by Stuhlhofer, they would have had to wait until May 17, when Venus and Mercury were conjunct, or until June 1, when the Sun conjoined with Venus again.[103] Bernardino Corio reports, however, that they consummated the marriage on the Monday after Easter, and this would make it March 31.[104] While uncertainties remain, we can safely say that, on this occasion, too, things were not left to chance and that the marriage and the consummation were established following astrological counsel.

Despite the attempts on both sides to have this union blessed by the stars, the marriage between Bianca and Maximilian was not a happy one, and Maximilian's alliance with Ludovico hardly advantageous for the former and disastrous for the latter. Maximilian found little he liked in his new bride, tiring quickly of her capricious behaviour and her tendency to spend large amounts of money. No offspring issued from that marriage and Bianca spent most of her short life in isolation in Innsbruck while Maximilian was in Vienna. Maximilian certainly gained a considerable amount of money from the deal, but he was also dragged, once again and much against his will, into a war against the King of France.

Ludovico had to wait many more months before actually receiving the much-desired investiture from Maximilian. When it came, however, Ludovico actively used it as a means to expand his role as legitimate Duke of Milan. As Sanuto recounts, the streets from the Duomo to the Castle of Porta Giovia had been lavishly decorated and a platform had been erected in front of the Duomo, much like during Bianca Maria Sforza's proxy marriage. In a real display of magnificence, the platform comprised a draped sky with "hand-made" stars, supported by columns, while the internal space was decorated with golden tree branches, thirty-five silver statues of saints, and hundreds of silver vases. Ludovico's chair was placed on the right in front of a drape with the white and red Sforza colors, while the imperial ambassadors were seated on the left.[105] This was certainly a lavish display of power and magnificence, especially for a duke whose finances were now so dry.

The day of the investiture was chosen, once again, with the help of astrology. It took place first in secret in the ducal chambers of the Castle of Porta Giovia on Friday, March 22, 1495, at the time elected by Ambrogio Varesi, "from whose counsel he [Ludovico] never departs, and indeed he does everything *per ponto di astrologia*," Sanuto quipped.[106] The public investiture in the streets of Milan, however, was scheduled for Sunday, March 24. Yet, as Sanuto reports, this date did not meet the favor of Varesi, and for this reason it was decided to postpone it to Monday, March 25. That day, however, "it rained so much that it seemed the day would come to an end," and for that reason the ceremony had to be postponed once more, this time until Tuesday, March 26. In Sanuto's account—the rain, the mud, the unfortunate choice of day—it is not difficult to detect some of the irony we encountered in Cardano's account of Ludovico's reliance on Varesi with which I opened this chapter. His sarcasm, however, is not that of an astrologer questioning the competence of a colleague, but that of the skeptical political analyst who had the advantage of knowing the political events that unfolded after 1495.

Predicting Death:
Innocent VIII's Death and His Succession

By seeking an alliance with France and the support of Charles VIII against the Aragonese, Ludovico was expecting to create enough tension between the Italian states to avoid a conflict with Ferrante and gain time for his investiture. Ludovico seems to have believed that simply the threat of a French invasion was more then enough to ensure that Ferrante would not mingle in Milanese affairs. Yet there was another European political power that he needed on his side against the Aragonese, namely the papacy. The election of the Genoese Gianbattista Cibo as Pope Innocent VIII in 1484 was the result of Ascanio Sforza's compromise to avoid the election of the Venetian cardinal Marco Barbo, the initial favorite of the faction led by Cardinal Giuliano della

Rovere. While Ascanio had little choice in the matter—the real mover having been della Rovere—the election of Cibo to the papal throne displeased Ludovico, who did not place much trust in the Genoese Pope.[107] Indeed, Ludovico had good reasons to feel threatened. Only a year after the election, Innocent VIII began scheming with Roberto Sanseverino to oust Ludovico from power and restore Bona to the regency.[108] The diplomatic relations between the papacy and Ludovico, moreover, did not improve over time. Rather, they remained fraught with suspicion and deep mistrust. At first staunchly anti-Aragonese, Pope Innocent VIII eventually moved closer to the King of Naples, to the point that a peace between the two powers was eventually concluded in January 1492.[109] This, no doubt, displeased Ludovico, who was trying to isolate Naples. Furthermore, in an attempt to counterbalance France's support of Milan, the Neapolitan secretary Giovanni Pontano had successfully enlisted not only Innocent VIII, but also Cardinal Giuliano della Rovere for the Aragonese cause, thus assuring the full support not only of the Pope, but also of the vast majority of cardinals.[110]

The fact that Innocent VIII took sides with the Aragonese worried Ludovico deeply, as the balance of power within the peninsula was not in his favor. Therefore, when Innocent VIII fell ill in the summer of 1492 (only a few months before the death of Gian Galeazzo Sforza), Ludovico was anxious to know the outcome of his illness and his chances of having a more favorable candidate elected to the Holy See. In Rome, Ludovico's brother, Cardinal Ascanio, could provide useful information and, at the same time, garner support for a candidate favorable to the Sforza. Yet neither of them felt he had good command of the situation. The issue, however, was crucial to Ludovico's plans for maintaining power against his nephew, Gian Galeazzo. Not surprisingly, therefore, he once again resorted to his trusted astrologer, Ambrogio Varesi da Rosate, this time to inquire about the outcome of the pope's illness.

As already seen in Chapter 3, the practice of casting interrogations about the illness of political leaders was not unusual at the Sforza court: Galeazzo had requested a similar interrogation from Annius of Viterbo in 1475 when Ferrante of Aragon was seriously ill. In that instance, as noted, the outcome predicted by the interrogation was strikingly different from what actually happened.[111] Although Annius had predicted Ferrante's death, Ferrante lived on. Indeed, at the time of Ludovico's interrogation about Innocent VIII, Ferrante was *still* alive. Varesi, however, was more successful, at least in this instance. Writing to him, Ludovico asked whether and when the pope would die. His response—which comes down to us in a lengthy and detailed letter—met Ludovico's hopes and expectations:

> In order to satisfy your question regarding the pope, which you posed to me in a letter handed in to me yesterday at the 21st hour, I have since studied, pondered and revolved (*revoltato*) the placement of the celestial bodies in the sky and their influence at the time of the question, which was cast the day before yesterday at 2 at night, when the letters arrived. Having examined your interrogation (as I do not know the pope's nativity), and having revolved the great significator of the Pastor and leader of the Christian faith in the revolution of the years of the world—which, according to some is the Sun with Mars, and, according to others, Mercury with Jupiter and the Sun—I find, in relation to your own interrogation, that Mars was the significator with Jupiter, as Mars is in his fall and Jupiter is under the Sun's rays; the Moon is under the Earth, in a wide aspect with the Sun (*coniuncta per generale aspecto*), which is lord of the house of infirmity, and Jupiter, lord of the ascendant (and unfortunate), in twenty-five days will be conjunct in the interrogation with Mars, lord of the house of death: therefore, the pope will die. The same can be inferred from the interrogation [sic: revolution] because Venus, significator of the year at the time of the revolution [because Libra is ascendant in the revolution], is combust and conjunct with the significator of death [i.e., Mercury], which denotes death in the present year. The same I find through the revolution of the years of the world, which started on March 10, in which the significator of the Pastor of the Christian Faith [i.e., Mercury], together with Venus, was damned by the combustion of the Sun in the house of infirmities, for which reason the death of the Pastor of the Christian Faith was indicated. As to when He should die, I find from the present interrogation that His death should happen in twenty-two days, which will be either the tenth or the eleventh of August, because of the conjunction of Mars with Jupiter in the interrogation, or in 15 days, because of the square aspect of the Moon with Venus, lady of the house of death in the interrogation [sic: revolution]. So, the interrogation shows me that He cannot reach the end of August, if Our Lord wishes and allows, for the private and the public good.[112]

As even a cursory reading of this letter indicates, either Ludovico was an expert practitioner or Varesi was showing off his astrological knowledge to impress his patron. As the first hypothesis is not supported by the evidence, it seems more plausible that Varesi's tactics were, once again, both to impress Ludovico and to give an aura of great complexity to the interrogation, without failing to stress, of course, that it would have been better to have the nativity of the pope and that God may wish differently, two caveats, implicitly indicating that his prognostication was tentative and fallible.

Which astrological principles were behind Varesi's convoluted interpretation? To answer his patron's query, Varesi had cast a chart for the moment the interrogation was formulated, which he compared with the chart for the revolution of the years (Figures 18 and 19). In the chart cast on July 18, 1492, around 9:30 p.m., the significators were Mars, which was in its fall (Cancer), and Jupiter, the ruler of the ascendant (which was Pisces). In the same chart, the Moon was under the Earth (the chart, therefore, was nocturnal), in a wide (sextile) aspect with the Sun, lord of the infirmities (as the sixth house, that of infirmities, was ruled by Leo), and Jupiter was the lord of the ascendant and debilitated by being under the Sun's rays (not a favorable placement). Furthermore, a progression of the chart indicated that Jupiter and Mars would become conjunct in twenty-two days. But because in this chart the lord of the house of death (the eighth) is Mars (which rules Scorpio), then Varesi predicted that the pope would die. Furthermore, this was confirmed in the revolution chart, to be cast for March 10 just after 8 p.m., the day of the Sun's entry into Aries (i.e., the spring equinox).[113] Since Venus, which was the significator of the year at the time of the revolution (because Libra is ascending in this chart), was combust and conjunct the significator of death (Mars), this confirmed that the pope would die that year. Looking once more at the chart of the revolution, moreover, Varesi noticed that the significator of the Pastor of the Christian Faith (i.e., Mercury) was conjunct Venus, which was "damned" because burned by the Sun's rays (*combusta*), and in the house of infirmities (the sixth). Once more, if needed, this confirmed that the pope would die. Asked when this would happen, Varesi confirmed that the conjunction of Mars and Jupiter on August 10 and 11 was ominous, but equally dangerous was the Moon's squared aspect with Venus on August 4, as Venus was lady of the house of death in the revolution (because Taurus is the sign on the cusp of the eighth house in the chart).

To sum up, all of these complex factors led Varesi to conclude that the pope would die sometime between August 4 and 11. As we know, Innocent VIII died on July 25, earlier than Varesi had predicted, but close enough to satisfy Ludovico. The same letter also contained reassurance of a favorable outcome for the election of the future pope: "Your Highness will be loved and will benefit from the succession," Varesi explained, "because the Moon moved

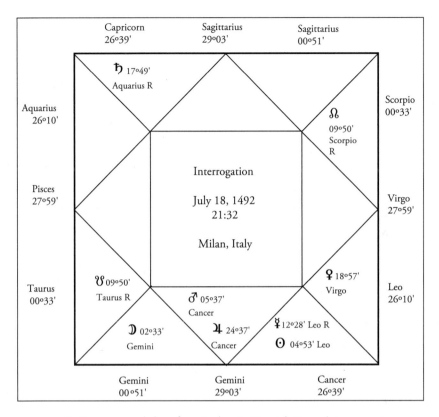

FIGURE 18. Reconstructed chart from Ambrogio Varesi da Rosate's interrogation about the possible death of Innocent VIII dated July 18, 1492, derived from Varesi's letter to Ludovico Sforza in ASMi, *Autografi,* Medici 219, Ambrogio Varese da Rosate to Ludovico, Milan, July 20, 1492.

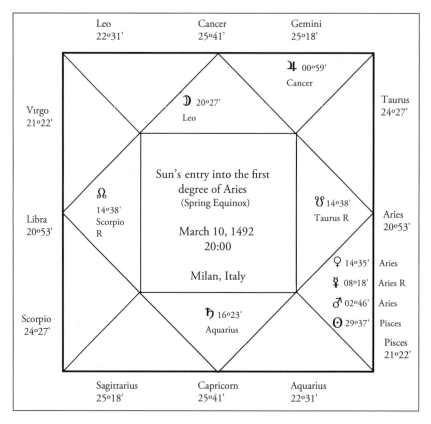

FIGURE 19. Reconstructed chart of the Sun's entry into the first degree of Aries (Spring Equinox), derived from Varesi's letter to Ludovico Sforza in ASMi, *Autografi,* Medici 219, Ambrogio Varese da Rosate to Ludovico, Milan, July 20, 1492.

away from Jupiter, and was received by the Sun in sextile to his house, which denotes that Your Highness will be loved by the successor as a friend or a true kin."[114] Even in this second instance, however, Ludovico and Ascanio did not favor the same candidate. While formally supporting Cardinal Oliviero Carafa, Ascanio secretly intended to promote Rodrigo Borgia's election. Ludovico, instead, favored Cardinal Ardicino della Porta, who presented the advantage of being a Lombard and thus more easily influenced. Eventually Ascanio chose to align himself officially with his brother, yet he never ceased to secretly pursue his original plan in support of Cardinal Borgia.[115]

Why did Ludovico resort to this type of interrogation? Much as in the case of Ferrante discussed in Chapter 3, the pope had been ill for some time, and the situation was unnervingly uncertain. Resorting to astrological interrogations, therefore, seemed a legitimate and useful way to plan a course of action. Machinations to steer the conclave to vote for one's favorite candidate started at least two months before Innocent VIII's death, and the two factions involved were prepared to use any means available to achieve the desired results. Neither Ludovico nor Ascanio doubted the soundness of Varesi's art: for this reason, when the news of the pope's death reached the court of Milan, Ludovico asked Varesi for another, more precise prognostication about the future election. They both clearly trusted that astrology could provide valuable guidance on how to best manage the delicate situation of papal succession. Thus Ludovico wrote to Ascanio telling him that Varesi's further investigations regarding Innocent's death had yielded fresh insights into the conclave's possible outcome. He was eager to communicate these results to his brother, warmly encouraging him to follow Varesi's advice and be liberal and generous toward other cardinals and cautious in choosing his allies:

> [Varesi] answered me that, at the time indicated for the Pope's death, Cancer was ascending, and this is the ascendant of His Reverend Highness, and that, in the revolution of the years of the world, Cancer was in the midheaven (*medium coeli*), which is the royal house. These two things clearly foretell a happy outcome for His Reverend Highness's enterprise. It is true, however, that in the hour of my interrogation, Saturn was aspecting His Reverend Highness's significator, which makes it possible that His Highness could be damaged, because of avarice or infidelity. Regarding avarice, however, I believe the risk is minimal, because Your Reverend Highness has good means, and you know you can use ours. I also believe that you should not allow stinginess—when the expense seems necessary

and could bear fruit—to get in the way of your plans. Infidelity is much more dangerous, however, as a man can be tricked also by his own men, and because people are no more tied to promises by faith than by goodwill, of which there is little.[116]

What Ludovico did not know was that Ascanio's favorite candidate was not Ardicino della Porta but the Spanish Borgia. In any case, it was essential for Ludovico that the new pope was not pro-Aragonese. Cardinal Borgia, in this respect, was an ideal candidate, being a long-term enemy of both Naples and Cardinal Giuliano della Rovere.

With the death of Innocent VIII, Ludovico had gained an important advantage over Ferrante, and he had no intention of losing it. Yet influencing the Curia was no easy matter. The situation before the conclave was deeply strained. Despite their formal alliance with Milan, both Venice and France sided with the Neapolitan favorite, della Rovere. So did Genoa and Florence.[117] This put Milan in a difficult position. Over the years, however, Ascanio had gained considerable experience in papal politics: his alliance with Borgia three years before could now bear its fruit. Despite Ludovico's advice to spare no necessary expenses, it was Ascanio himself who received ample compensation for his support from Cardinal Borgia. This is clearly symbolized by the four mules carrying loads of silver that—so it is told—Borgia, then vice-chancellor, had sent him the day before the conclave asking him to keep them safe.[118] This was a clear sign of the economic and political benefits awaiting Ascanio at Borgia's election to the pontificate. Indeed, Ascanio did not need to draw from his modest resources: vice-chancellor Borgia himself ensured that the wheels of the complex mechanisms were well greased before the conclave, in what remained a memorable election for the scale of its corruption and simony.[119]

The conclave of early August 1492 concluded with Borgia's election to the Holy See and with Ascanio's election to the office of vice-chancellor. Stefano Infessura recounts how all but five cardinals obtained rewards for their votes from the newly elected pope. Once again, the astrological forecast was accompanied by a series of celestial events that were recorded by contemporary chroniclers. Stefano Infessura describes how, on August 30, 1492, three weeks after the election, three suns—large, clear, and bright—were seen in the sky above Saint Peter's and were interpreted as bad omens.[120] Once again celestial phenomena were read against events that unfolded on Earth. These were celestial signs that cried for suitable interpretation: reading these signs was primarily the realm of the astrologer, but the chroniclers of the time were often remarkably attentive to their appearance and often posthumously ventured to offer their own interpretation of their significance. With the election of Rodrigo Borgia, Ascanio had his brief moment of glory and seemed the most likely candidate for the tiara at Alexander VI's death. Unfortunately for him, however, the events following the French descent of 1494, on the one hand, and the rise to the French throne of Louis of Orléans, on the other, eventually forced Ascanio to leave Rome to join his brother Ludovico in a desperate attempt to save Milan from foreign invasion. As we shall see, these attempts ultimately failed, and yet Ascanio, largely thanks to his being cardinal, escaped the bitter fate of his brother, Ludovico, and the other members of his family.

Waging War: Ludovico, Military Astrology, and the French Descent into Italy

Borgia's election to the pontificate put Naples once again at a disadvantage. Ludovico was thus confident that the outcome of the conclave and the threat of French intervention would now neutralize Ferrante's desire to interfere in Milanese affairs and would lead to the legitimation of his role as ruler of Milan. Meanwhile, through his trusted friend Cardinal Giovanni Battista Savelli, Ascanio tried to facilitate Ludovico's plans by sending false messages to the King of France on behalf of the pope, inviting him to conquer the Kingdom of Naples. Ferrante did not give up hope, however, and his next move was to attempt to draw a secret alliance with the pope.[121]

As suggested so far, the political situation at the end of 1493 was certainly tense; yet it seems that in Italy political rulers did not take the French menace very seriously. Even Ludovico, who had made gestures of invitation, had done so convinced that Charles VIII's military *impresa* would never happen. Unfortunately, however, a series of events that neither Ludovico nor his trusted astrologer had predicted precipitated the situation: one of these events, as noted, was the Peace of Senlis reached between Charles VIII and Maximilian in May 1493. With this peace, the possibility of a French invasion of Italy became, all of a sudden, remarkably concrete. Such an invasion, moreover, had the potential of having not only political consequences, but quite possibly even religious ones: Charles's *impresa* particularly worried the pope, who was a hostage to Ascanio's threats of conniving with Charles VIII for a gallican council that would oust him.[122] At the same time Ludovico was negotiating his investiture with Maximilian: by August 1493, he had obtained the investiture and had succeeded in marrying his niece Bianca to the Holy Roman Emperor, all at the staggering price of 400,000 ducats. Equally, and quite importantly, Ludovico did not predict the death of Ferrante. The Neapolitan King died on January 25, 1494, leaving the throne to his son, Alfonso, who had cultivated a much stronger sense of distrust and hatred towards Ludovico and who was naturally more protective of his daughter Isabella's rights. This fact is clearly signaled in a letter by the Milanese ambassador in Naples, who warned his duke, Ludovico, that the new king had sent some assassins to Milan and that he would need to keep his eyes open.[123]

By the autumn of 1493, therefore, the French invasion seemed virtually inevitable. Ludovico made gestures of reassurance toward Florence, informing Piero de' Medici of his mediation with the King of France to save his state, but intimating that he had to abandon his alliance with Naples and the papacy to be spared.[124] Ludovico was clearly confident of being fully in command of the situation and of having won the role of Italy's peacemaker. He seems to have truly believed that he could outwit both the King of France and the Holy Roman Emperor and that he finally held the destiny of the Italian states in his own hands.[125] By the spring of 1494, Charles VIII had sent the Milanese ambassador, Belgioioso, back to Italy to inform Ludovico that he was at the borders of his kingdom ready for the *impresa*. At this point Ludovico was forced to offer military support to Charles VIII.[126] He was aware, however, of the problematic nature of his rule while the legitimate Duke of Milan was still alive. For this reason, he never stopped actively seeking Maximilian's full support. Even after being promised the investiture, Ludovico did not feel completely confident about becoming the legitimate duke.

Gian Galeazzo's death after the visit of his distant relative Charles VIII must have provided a huge relief to Ludovico, who could now justify his ascendance as a matter of necessity and not of choice. Needless to say, few deaths were as timely as this one, and Ludovico did not miss the opportunity of being elected duke straightaway. Therefore, as Sanuto recounts, "On Wednesday, October 22, at the 17th hour—which Ludovico elected with the help of his astrologer, Magister Ambrogio—he mounted on a horse, dressed in gold, with a sword carried forward by Galeazzo Visconte, himself dressed in gold, who shouted: Duca! Duca! Moro! Moro!"[127] If this was not enough, in a sort of magic ritual that revealed a certain amount of superstition, he also astrologically elected the best time to write down his full title on paper: he wanted to be addressed as *Ludovicus Maria Sfortia Anglus Dux Mediolani Papiae Angleriaeque Comes ac Genuae et Cremonae Dominus*.[128] A few days later, having suspended some taxes and conceded some privileges to gain popular consensus, Ludovico left Milan with Beatrice to meet the King of France near Piacenza. Not surprisingly, the moment of their departure and their return were once more determined astrologically.[129] Once again, therefore, Varesi features prominently in the political and personal decisions of the Duke of Milan. Ludovico relied extensively on Varesi for astrological advice, but this advice had clear political applications, and as such it had a double nature. The physician-astrologer had also become political and personal counselor. Everybody, Sanuto included, was aware of the increasingly unique influence exerted by Varesi on Ludovico.

There is no doubt that Varesi's social and political ascent was closely connected to Ludovico's rise to the duchy. The more the duke relied on him, the more Varesi's competences expanded. Ludovico used astrology to gain legitimacy and to reinforce his power: he planned important dynastic marriages, he sounded his own chances of success in view of the conclave of 1492, and planned the consummation of his own marriage so as to guarantee for himself the best lineage possible. In a more mundane fashion, astrology dictated the time of his travels and those of members of his court and was used to treat Ludovico and his family members when ill. Ludovico's applications of astrology, and Varesi's duties, however, did not end here. They also invested other areas of political praxis, and, as we shall see shortly, war was one of them.

Once the war started, Ludovico often became directly engaged in military campaigns that took him away from Milan. It is difficult to determine if Varesi followed his duke in any of these instances. Given the documentation, however, it seems more likely that he continued to dispense his advice from a distance, often residing with Ludovico's children at the Castle of Porta Giovia and sending letters of advice to Ludovico when requested. This can be gathered by the ducal correspondence from those years. In November 1495, for example, we find Varesi in Milan at the bedside of the milanese nobleman Monsignor Daniele da Birago, brother of the ducal counsellor, Francesco.[130] Varesi had paid a visit to Birago at Ludovico's express request. He reported to Ludovico how Birago's own doctors had assured him that the worst was over, but that they still believed the case was particularly serious and that the patient had a temperature and much phlegm. Given the rainy weather and the forthcoming conjunction of the Moon—Varesi predicted—the situation, while not deadly, required close monitoring.[131] Similarly, in August 1498, when Ascanio fled Rome for Milan to support his brother in the campaign against King Louis XII (Charles VIII had died in April that year and Louis of Orlèans had become King of France), Varesi wrote to Ludovico from Milan to inform him of the ill health of his brother, who had recently undergone bloodletting and received some *manna* (a purgative).[132] Some of these letters clearly indicate that Varesi and the team of physicians he was directing were monitoring the progress of Ascanio's illness according to Galen's theory of the critical days, paying particular attention to the lunar phases.[133] They also suggest, once more, that Varesi remained mostly in Milan in the Castle of Porta Giovia and did not follow Ludovico in his campaigns. Other, more prosaic letters by Varesi simply reassure the duke about the good health of his children and briefly update him on the ailments of other courtiers—once again they were almost always sent from either Milan or nearby Vigevano.

Varesi's penchant for astrological medicine, already seen in his adherence to the theory of the critical days, extended also to the field of surgery: in a letter written on October 18, 1494, he informed Ludovico of the best time to be elected by his colleague Gabriele Pirovano and a team of surgeons for operating on Nicolò Orsini, Count of Pitigliano, himself a man not adverse to astrology.[134] A few days later two doctors reported to Ludovico how the Count of Pitigliano's wound was healing and that pieces of the chain and armor had been expelled from the wound along with some fabric. Believing, moreover, that more material would be expelled in the following days, they happily announced that the count was feeling reasonably well. He could stand up and move around, and they assumed he would continue to improve.[135] As on other occasions, Varesi functioned as the ultimate authority when it came to astrological elections, and for this reason even expert physician-astrologers like Pirovano referred back to him for advice.

Despite his full-time job as Ludovico's chief physician-astrologer, Varesi's role as political advisor grew over time, as he was increasingly asked to provide advice on military and diplomatic matters. By 1494, Varesi was regularly asked to determine the most appropriate day and time for the most important events: this could vary from the best time to make Ludovico's *castellani* swear allegiance to the duke,[136] to the appointment of his captains,[137] to the best time for his ambassadors to travel,[138] to the most favorable moment to sign an alliance or launch a military campaign. These last two instances are particularly significant, as they reveal a little studied aspect of diplomatic and military history. The very moment chosen by Ludovico to join the Italian League against Charles VIII, for instance, was not left to chance, but was carefully chosen with the help of astrology. Once again Ludovico resorted to astrological elections: "Today is not a good day to join the league," Varesi reported sometime in mid-April 1495, advising his duke to join when Taurus, a fixed and stable sign and the domicile of Venus, was in the ascendant.[139]

Similarly, astrology was used in planning battles. In January 1495, when Ludovico's alliance with the French king started to vacillate but his hostility had not yet been openly declared, the Gonzaga ambassador in Milan reported to his lord how he had been present at a dinner conversation between Ludovico and his two physician-astrologers, Gabriele Pirovano and Ambrogio Varesi da Rosate, about clashes between the Aragonese and French armies. Ludovico had asked Pirovano to forecast when the two armies would meet in battle again and who would win. In this circumstance, Pirovano confidently replied that the battle would take place around February 18 and that Alfonso of Aragon would be superior. This forecast, moreover, was seconded by Varesi.[140] The result, however, was not as predicted: Charles VIII entered Naples in late February, Alfonso abdicated, and his son Ferdinand was forced to flee to Ischia to save himself.[141]

Apparently, these incorrect predictions did not undermine Ludovico's faith in the predictive value of astrology. A few months later, in June 1495, the Duke of Milan resorted once more to astrological counsel, this time to use it actively as a military tool to plan his attack against a retreating French army. Ludovico once again placed confidence in his favorite astrologer. As Sanuto reports, Varesi, "from whose counsel he [Ludovico] never departs, and indeed he does everything *per ponto di astrologia*," had indicated to the duke that the King of France would be defeated on June 29.[142] This piece of astrological intelligence was obviously shared among Ludovico's Italian allies, Venice included. Once again revealing his own mild skepticism about Varesi's predictions, Sanuto added, however, that the French army was defeated by the Venetians only on July 6, at the battle of Fornovo, and that the astrologer thus erred a little in his prediction.[143] In any case, Ludovico trusted Varesi's advice completely: preparing himself for war, on June 28 at the 12th hour, he gave orders to his army to leave the ducal town of Vigevano for Cassolo, a smaller town near Piacenza, in Emilia Romagna.[144] With the benefit of knowing Ludovico's fate, a doubtful

Sanuto could not help but quip how "amazing it is that he [Ludovico] trusted him [Varesi] so much."[145] Regardless of Sanuto's negative opinion on Ludovico's gullibility, however, the Venetians, too, seem to have followed astrological counsel (possibly provided by Ludovico himself). The famous battle of Fornovo that took place on July 6, in fact, was similarly planned astrologically. This time with a heightened sense of national pride, Sanuto recounted how the retreating forces of the French king met the Venetian army on the river Taro near Parma and were forced to stop. This allowed the allied forces to regroup, albeit only partially, around the enemy, and on July 6, 1495, "the day chosen because of the celestial disposition to wage war against France," the mighty Venetians launched their attack against the French and succeeded in defeating them.[146] Unlike Sanuto, in recounting the events, Guicciardini did not make any reference to astrological counsel, and yet he linked the battle with the ominous meteorological event of a violent storm with heavy rain, thunder, lightening, and terrifying cracklings and flashes, "which many prognosticated as foretelling the most dire outcome," thus expressing the popular view that celestial and metereological portents were signs to be interpreted.[147] Astrology was thus never too far away from contemporaries' thoughts.

The battle of Fornovo remains memorable in the annals of military history for many reasons: besides often being considered the battle that marked the beginning of the Italian Wars, it is often celebrated by contemporary sources as a great success for the Venetian army and the Italian allies, despite the fact that Charles VIII escaped alive and was able to reach Asti safely.[148] More significantly, it was the first battle where many men on both sides sacrificed their lives, thus marking the end of the "paper wars" of the Italian powers that had characterized the earlier period. Modern military historians, however, seem to have remained generally oblivious to the fact that the battle itself was planned astrologically.

A few months later when the king was back in France, Ludovico tried to gain further insight into the king's future by asking Varesi once again to make a prognostication on his own and the king's fate. In his reply, the astrologer confidently assured the duke of success for his realm, at the same time making negative predictions about the King of France, his wife, and their chances of conceiving a male heir:

> Your Illustrious Highness will see the good success of your reign as I have predicted to you numerous times this year, and you will have heard of the loss of the Queen of France, of which I forecast that whenever the first

child died she would have few others. And this other prediction has become true and I affirm that either she will not have any, or even if she has them, they will die; in the end, she will have few children, and this is because of the Most Christian King's nativity [i.e., the French king's nativity].[149]

From this letter we can evince that Varesi possessed Charles VIII's geniture and that he used it to produce astrological predictions for his duke on all sorts of political issues, including the sensitive issue of the king's heir. As noted earlier, this was a subject on which Varesi had pronounced himself previously, possibly attempting to hinder Gian Galeazzo's success of conceiving an heir by means of astrology. The possibile death of Charles VIII's heir, however, should have alarmed Ludovico, who should have realized—and thus predicted—that if the king died without an heir, the title to the French crown would pass to his cousin Louis of Orléans, thus considerably weakening Ludovico's position as legitimate heir of the Visconti-Sforza domains. Varesi's advice, however, was not limited to the King and Queen of France. In the same letter he also predicted that the Florentines would not fare well, falling in the French king's hands, which surely must have pleased Ludovico.[150] Beyond its political content, it is evident from this letter that by 1494 Varesi's quasi-prophetic astrological role spanned the entire spectrum of European politics and diplomacy. The statement that nothing was decided without first consulting him was not far from the truth.

With the French invasion of Italy, the political scenario between 1494 and 1499 became extremely volatile, fraught with changing alliances and mounting uncertainty. Within this climate astrology continued to play a dominant role: as we shall see, Ludovico resorted to astrological advice in military and diplomatic matters right until the very end of his rule as Duke of Milan. After Charles VIII's death in 1498 and the ascent of Louis XII to the throne, the Italian powers quickly returned to their old political positions. Florence attempted almost imediately to reconquer Pisa, which, however, found in Venice a powerful ally. In this case, too, Ludovico resorted to Varesi, who was asked to pronounce himself both on the issue of Pisa and on the favorable outcome of Maximilian I's battle against the Swiss army, a war that, in Ludovico's eyes, was distracting the emperor from offering due support to Milan in the battle against the French. When the Milanese ambassador, Augusto Somenzi, shared the information contained in Varesi's forecast with the Holy Roman Emperor, however, he was met with an unexpected reaction: Maximilian laughed about it, saying that he had little faith in astrologers. The emperor did, however, use the prediction to reassure the other German princes and his court that his victory was almost certain. For this reason, Somenzi encouraged his lord to send other astrological advice of this sort to Maximilian, convinced that he would appreciate it.[151] Possibly following Varesi's astrological predictions, Ludovico unwisely lent his support to Florence, thus rekindling the hatred and distrust of his long-term nemesis, the Serenissima, which eventually sided with the French against Milan. Thus, instead of reuniting against a common foreign enemy as Machiavelli would later recommend in his *Prince,* the Italian powers soon became absorbed in old territorial disputes, almost oblivious to the fact that the new French king, Louis XII, was preparing himself for a new descent.[152]

Ludovico's position by 1499 was very different from that of 1494, when he had encouraged the French army to invade Italy and conquer the Kingdom of Naples. He was no longer his nephew's guardian and *de facto* ruler of Milan; he was now its legitimate lord. Yet the death of Charles VIII had weakened his position considerably, and now, on the eve of an attack from Louis XII, he found himself with his monetary resources depleted, his spirit fading, and his allies wavering.[153] In the early months of 1499, the earlier confidence in being "arbitro d'Italia"—he who controlled everything in Italy—turned to despair and dejection. Only his faith in astrology did not seem to waver. In the summer of 1499, while inspecting his fortifications in the west of Milan, he met his brother Ascanio near Melegnano: according to the Ferrarese ambassador in Milan, Antonio Constabili, the meeting was, once again, arranged astrologically.[154]

Ludovico maintained his faith in astrology until the very end. The stressful and dire conditions of war, however, seem to have undermined the duke's health, and it was reported that he had to take medication to ride to Santa Maria delle Grazie, where he often sought refuge and spiritual advice.[155] On August 22, 1499, Varesi once again produced a military forecast, affirming that the French army would be particularly dangerous in the period August 13 to 23, but that on the following day, the twenty-fourth, the duke would face the French army.[156] Varesi's prognostication for August was not very favorable, but he had every confidence that September would be victorious for the duke.[157] Varesi's relationship with his duke was symbiotic, his own success mirroring that of his duke. He had all the reasons to try and remain optimistic that his lord would ultimately succeed. Other forecasts, however, were much less optimistic. Ascanio's astrologer—whose identity remains unknown—was reported as saying that he could bet his life on the fact that the duke would lose his state,[158] while a nun whom Ludovico had asked to pray for him had allegedly told him that she had prayed to God, but she also prophesied that God's will was that he no longer be Duke of Milan.[159]

Ludovico continued to hope for Maximilian's help, but the emperor's internal affairs were much more pressing. Maximilian thus limited himself to sending his ambassador to Venice to discourage the Serenissima from persevering in its plan to attack Milan, hoping to persuade Venice, instead, to turn her formidable energies against the mighty Turk threatening her borders.[160] This invitation, however, was not met with favor, and Venice continued to fight against the Milanese army in support of France. By August Ludovico was "desperate," confessing to Antonio Constabili that he would rather die Duke of Milan than flee.[161] Ludovico's faith in Ambrogio Varesi's political counsel, in the meantime, started to waver. Constabili reported how Ludovico had consulted a group of astrologers about the current situation, and their advice was in stark contrast with what Varesi had said before. Worst still, now Varesi agreed with them, and this greatly displeased Ludovico:

> The astrologers that this Illustrious Lord received here this evening agree that even if His Lord was not the prince of this state, the influences of the heavens are of such a nature that this state will suffer. But they state that in the end His Highness will be able to defend himself and win. Magister Ambrogio has fallen out of favor, as when he disputed with these astrologers he agreed with what they said; His Excellency says that before then he had predicted exactly the opposite. For what is possible to know, it is a good thing that this lord was stopped from doing what he had planned and believed necessary. And if Your Highness had the chance of summoning Pietro Buono [Avogario] to ask him his opinion, I believe that His Highness would be nothing but grateful, even if he did not expressedly ask me this. Some of these astrologers say that this influence will last the whole year, other for the whole month of October, and others until September 27.[162]

As Ludovico's fate grew increasingly uncertain, he decided to resort not only to Varesi's counsel, but also to that of other astrologers, and the fact that Varesi was contradicting himself irritated the duke deeply. The other astrologers, however, were not in agreement as to the duration of the malignant influence exerting itself over Ludovico's duchy, and this must have added greatly to Ludovico's anxiety. On his part, Constabili, aware of Pietro Buono

Avogario's great reputation, suggested to Ercole d'Este that he ask his opinion on the matter so that his could be reported back to Ludovico. As this example clearly exemplifies, in this instance doubts were cast not on the practice of astrological counsel *per se,* but more simply on the skills (and possibly even the good intentions) of those who practiced it.

Constabili's letters at the end of August depict a situation in rapid evolution, with Maximilian promising help but not sending it and the French army gaining ground over the Milanese forces. The duke, Constabili reported, was preparing for a siege. Negotiations with the King of France and with Venice continued.[163] Fearing for his life, on September 3 Constabili fled Milan for Ferrara.[164] By early September, having lost most of his former allies and without enough men to defend the city, Ludovico lost all hope of maintaining his state. He arranged for Ascanio and his sons to leave Milan in the direction of Como to find refuge in the Tyrol. Ludovico followed them on September 2, and Gian Giacomo Trivulzio entered the city the same evening.[165] With Ludovico's fall, Varesi's splendid career came to a close. He was captured near Lecco while attempting to flee the ducal lands with one of his sons.[166] Having lost the duke's patronage and thus his privileged status, Varesi soon fell into disgrace: immediately after his capture, his fief and possessions were taken from him. Ludovico, who had escaped capture and was on his way to Bolzano, now put his trust in a new astrologer by the name of Ermodoro.[167] Meanwhile in Milan, Varesi was accused by Isabella of Aragon of having poisoned her beloved husband, Gian Galeazzo, Ludovico's behalf. Things could have not gone worse for the now disgraced astrologer. The new Ferrarese ambassador Ettore Bellingeri recounted how Giovanni Gonzaga had told him that the Duchess Isabella had revealed to him that Ambrogio Varesi had confessed to the murder. According to this account, Varesi had poisoned Gian Galeazzo with a syrup with the help of Ludovico's pharmacist (*spetiale*).[168] With the fall of Ludovico, Ambrogio Varesi's career came to an abrupt halt. His privileged relationship with the Duke of Milan had marked his ascent to power, money, and prestige. His fall was equally spectacular.

The year 1499 effectively signaled the fall of the Sforza dynasty. Ludovico's attempt to reconquer Milan in 1500 failed miserably and he was captured, ending his days in the Castle of Loches, where he died in 1508. With it ended also a splendid period for Milan: deprived of a sumptuous court, the city lived more modestly under the French and Spanish occupations that characterized the sixteenth and seventeenth centuries. For a brief time, Ludovico's sons,

Ercole Massimiliano and Francesco, were appointed Dukes of Milan again by European powers, but the title did not translate into the legitimate right of ruling the city like their father did. Their role was more symbolic than real.[169] Astrology in the duchy, however, continued to thrive and develop. Astrological consultations dated to the sixteenth century are still present in the Milanese archives, and, judging by the names of Milanese nobles included in Girolamo Cardano's *Liber de exemplis centum geniturarum,* it seems clear that horoscopes were still met with favor among the Milanese elites. It was, therefore, the end of the Sforza dynasty, but not the end of astrology in Renaissance Milan.

Epilogue

In 1494 Ludovico Sforza finally acquired the much desired imperial investiture from the Holy Roman Emperor, Maximilian I. The death of Charles VIII and the rise to the throne of France of Louis of Orléans, however, transformed the European scenario dramatically for the Sforza. As the grandson of Valentina Visconti, Louis XII of Orléans had legitimate rights over Milan, and he was determined to exercise them to the full. With the help of Gian Giacomo Trivulzio, Louis XII conquered Milan and eventually convinced Maximilian to withdraw Ludovico's investiture. The power of the Sforza dynasty crumbled. Ludovico fled, was captured, and ended his days in prison.

The House of Sforza came back to power briefly in the early sixteenth century, but at this stage they were only puppets in the hands of European powers. The Swiss and Venice briefly instated into power Ludovico's son Massimiliano when the French were defeated at the Battle of Ravenna in April 1512, but his rule over Milan was short lived. The defeat of Milanese and Swiss forces at the Battle of Marignano in October 1515 signaled the end of his rule, and the French regained control of the duchy. The next phase of the war between France and the Holy Roman Empire was fought by Francis I, the newly elected King of France, and Charles V, Maximilian's nephew. Milan was conquered by imperial forces on November 20, 1521: on that occasion, Charles V installed Francesco II Sforza, Ludovico's youngest son, as Duke of Milan. The battle between French and Imperial troops over Lombard soil protracted for another four years, to be concluded with the spectacular capture of Francis I at the Battle of Pavia (February 24, 1525). Francesco II ruled over Milan until his death in 1535, but always under the vigilant watch of the Holy Roman Emperor. More significantly, Francesco did not leave any heirs to the duchy. As Giorgio Valla had predicted, the Sforza dynasty was extinguished.[1]

Sixteenth-century Milan was a very different place than Milan in the previous century. Without its splendid court and Ludovico's patronage, the arts and sciences suffered. University teaching at Pavia was heavily disrupted by the Italian Wars. This signaled a decline of the Pavian *Studium,* a decline that only grew during the sixteenth century.[2] Significantly, the Milanese Girolamo Cardano did not obtain his degree in medicine from his local university, as he would have done had he lived two decades before, but from Padua. Cardano returned to Pavia to teach briefly, only to move to Bologna after his son was sentenced to death for having killed his unfaithful wife.[3] This was certainly not the only reason for the move, however. Bologna undoubtedly represented a much more attractive prospect for the teaching and study of medicine and astrology. Yet we know that Cardano remained well connected with the Milanese nobility. Many of the horoscopes he collected, and later reprinted in his works, were obtained through these channels, by copying from collections of genitures that were owned by other physicians, astrologers, and learned men in Milan. This is one of the reasons so many horoscopes of Milanese personalities populate his works.[4] It is clear that astrology did not die with the Sforza. It did not, however, find again a comparably fertile environment in which to thrive and remained relegated more and more in the semi-private sphere of the Milanese nobility. It never again obtained such a place of pride as a tool of political counsel.

In this book I have argued that in the fifteenth century astrology held prime of place among the predictive disciplines patronized at court. Arguably, in courts like that of Ferrara this place was contended with other predictive arts, one of them being female prophecy. This was not the case for the Sforza court: there astrology grew to reign supreme on the "predictive market." Although all the Sforza dukes were surrounded by physicians and astrologers eager to offer their services, some of them seem to have been keener than others to avail themselves of the astrolgers' skills. Among them, Ludovico

Sforza certainly emerges as the most dedicated patron of astrology and its more avid consumer. Unlike his father and his brother Galeazzo, he chose to privilege one astrologer among others, Ambrogio Varesi da Rosate. Varesi's career trajectory as a trusted counselor grew exponentially to reach its peak with Ludovico's investiture. His fall from favor, however, was even more rapid than his rise. As Ludovico's ambitions were quashed by the more powerful France, so were Varesi's. Varesi's case goes to show how delicate these relationships could be: scientific patronage, like any other form of patronage, was subject to various external forces and to the vagaries of time. Fortunes were made but could also be lost, and lost fast, especially in the Renaissance.

In this book I have striven to illustrate how astrology was put to practice within the Sforza court. My aim has been to investigate why astrology was so attractive to Italian lords, which astrological techniques may have been favored by Italian rulers, and the degree to which astrological advice responded to the needs and tastes of single individuals. What emerges clearly is that in Renaissance Italy astrology was never too far from the sites of power, particularly in time of crisis. As uncertainty grew, so grew the power of astrology. Responding to one of the most natural human desires, that of gaining knowledge of the future, astrology could provide men with some sort of answers. Renaissance political leaders who were unable to determine otherwise what the future had in hold for them often resorted to astrology. The Sforza were not alone in doing so. The case of Milan provides genuine new insights into astrology's appeal in Renaissance Italy. More research will need to be carried out on other courts to determine how exceptional were the Sforza when compared with other Italian and European lords. The evidence summoned so far suggests that astrology may have played a bigger role in political praxis than we may have generally assumed. While avoiding generalizations about the role of astrology at court, the evidence presented here clearly indicates that in the Renaissance astrology was neither considered a purely academic subject, nor believed to be irrelevant to the everyday life of Renaissance elites. Rather, it played an important role in areas as different as planning dynastic marriages, forecasting the outcome of political strife, treating diseases, and deciding when to sign a peace treaty or wage war. All these social and political dimensions find a place in the stories recounted in this book.

Abbreviations

Notes

Bibliography

Acknowledgments

Index

Abbreviations

AG	Archivio Gonzaga
APD	Archivio Pisani-Dossi
ASF	Archivio di Stato, Florence
ASI	*Archivio Storico Italiano*
ASL	*Archivio Storico Lombardo*
ASMa	Archivio di Stato, Mantua
ASMi	Archivio di Stato, Milan
ASMo	Archivio di Stato, Modena
BAM	Biblioteca Ambrosiana, Milan
BAR	Biblioteca Alessandrina, Rome
BAV	Biblioteca Apostolica Vaticana
BC	Biblioteca Corsiniana, Rome
BL	British Library, London
BLO	Bodleian Library, Oxford
BLYU	Beineke Library, Yale University
BLF	Biblioteca Laurenziana, Florence
BMP	Bibliothèque Mazarine, Paris
BMV	Biblioteca Marciana, Venice
BNCF	Biblioteca Nazionale Centrale di Firenze
BNF	Bibliothèque Nationale de France, Paris
BRB	Bibliothèque Royale, Bruxelles

BSB   Bayerische Staatsbibliothek, Munich
BT   Biblioteca Trivulziana, Milan
Cardano, *OO*   Girolamo Cardano, *Opera Omnia,* 10 vols. (Lyon: Jean
         Antoine Huguetan & Marc Antoine Ravaud, 1663). [repr.
         Stuttgart-Bad Canstatt, 1966]
Clm   Codices latini monacenses
COM   *Carteggio degli Oratori Mantovani alla corte Sforzesca (1450–*
         1500), coordinated and directed by Franca Leverotti (Rome:
         Ministero per i beni e le attività culturali, 1999-).
CUL   Cambridge University Library
DBI   *Dizionario Biografico degli Italiani* (Rome: Istituto della
         Enciclopedia Treccani, 1960).
DSB   *Dictionary of Scientific Biography,* ed. Charles Coulston
         Gillespie, 18 vols. (New York: Scribner's, 1970–1990)
GUL   Glasgow University Library
GW   *Gesamtkatalog der Wiegendrucke* (Leipzig: K. W. Hiersemann,
         1925–) [reprinted Stuttgart: A. Hiersemann/New York: H. P.
         Kraus, 1968-]
IGI   *Indice Generale degli Incunaboli delle biblioteche d'Italia,* 6
         vols. (Rome: Libreria dello Stato, 1943–1981).
ISTC   *Incunabola Short Title Catalogue* (now via BL Portal: http://
         www.bl.uk/catalogues/istc/index.html).
JWCI   *Journal of the Warburg and Courtauld Institutes*
Klebs   Arnold C. Klebs, *Incunabula scientifica et medica* (Bruges: The
         Saint Catherine Press, 1938) [first ed. "Incunabula scientifica
         et medica," *Osiris* 4 (1938): 1–359].
Kühn   Karl Gottlob Kühn, ed., *Claudii Galeni Opera Omnia,* 20
         vols. in 22 t. (Hildesheim: Olms, 1964–1965).
Lemay   Albumasar, *Liber introductorii maioris ad scientiam*
         judiciorum astrorum, ed. Richard Lemay, 9 vols. (Naples:
         Istituto universitario orientale, 1995–1996).
MAP   Medieco Avanti il Principato
MEFRM *Mélanges de l'École Française de Rome, Moyen Age*
ODNB   *Oxford Dictionary of National Biography,* 60 vols. (Oxford:
         Oxford University Press, 2004).
RIS²   *Rerum Italicarum Scriptores. Raccolta degli storici italiani*
         dal Cinquecento al Millecinquecento, ed. Giosuè Carducci,
         Vittorio Fiorini, and Pietro Fedele (Città di Castello: S. Lapi,
         1900-) [first ed. Milan: 1723–1751].
RQ   *Renaissance Quarterly*
SPE   Sforzesco, Potenze Estere
SPS   Sforzesco, Potenze Sovrane
Studi PA Studi, Parte Antica

Thorndike, *HMES*   Lynn Thorndike, *A History of Magic and Experimental
Science,* 8 vols. (New York: Columbia University Press,
1934–1958).

TK   Lynn Thorndike and Pearl Kibre, eds., *A Catalogue of Incipits
of Mediaeval Scientific Writings in Latin,* rev. and augm.
(London: Mediaeval Academy of America, 1963).

Notes

Introduction

1. The expression "minor intellectuals" has been used recently to describe the great number of minor figures who populated Italian courts. See Chiara Crisciani and Gabriella Zuccolin, eds., *Michele Savonarola: Medicina e cultura di corte* (Florence: SISMEL-Edizioni del Galluzzo, 2011). I am aware that the term carries a certain negative connotation. I want to stress, however, that these figures are considered "minor" only according to a particular set of criteria that should not be accepted uncritically.

2. For an analysis of the concept of celestial influence in Machiavelli's writing, see Anthony J. Parel, *The Machiavellian Cosmos* (New Haven and London: Yale University Press, 1992). To this one should add Raffaella Castagnola, ed., *I Guicciardini e le scienze occulte. L'oroscopo di Francesco Guicciardini, lettere di alchimia, astrologia e cabala a Luigi Guicciardini* (Florence: Olschki, 1990). On celestial influence more generally, see Edward Grant, "The Influence of the Celestial Region on the Terrestrial," in *Planets, Stars, and Orbs: The Medieval Cosmos, 1200–1687* (Cambridge: Cambridge University Press, 1994), 569–617; and John D. North, "Celestial Influence—The Major Premise of Astrology," in Paola Zambelli, ed., *"Astrologi Hallucinati": Stars and the End of the World in Luther's Time* (Berlin: Walter de Gruyter, 1986), 45–100.

3. Eugenio Casanova, "L'astrologia e la consegna del bastone al capitano generale della Repubblica fiorentina," *ASI* 7 (1891): 134–143.

4. The works that have explored the intellectual context of astrology, many in languages other than English, are too numerous to be listed here. Among the most significant studies that have appeared in English are: Keith Thomas, *Religion and the Decline of Magic* (London: Weidenfeld & Nicolson, 1971); Bernard Capp, *Astrology and the Popular Press: English Almanacs, 1500–1800* (London: Faber, 1979); Zambelli, *"Astrologi Hallucinati";* Patrick Curry, ed., *Astrology, Science and Society: Historical Essays* (Woodbridge: Boydell, 1987); Laura A. Smoller, *History, Prophecy, and the Stars: The Christian Astrology of Pierre D'Ailly, 1350–1420* (Princeton, NJ: Princeton University Press, 1994); Anthony Grafton, *Cardano's Cosmos: The Worlds and Works of a Renaissance Astrologer* (Cambridge, MA: Harvard University Press, 1999); Steven Vanden Broecke, *The Limits of Influence: Pico, Louvain, and the Crisis of Renaissance Astrology* (Leiden and Boston: Brill, 2003); Günther Oestmann, H. Darrel Rutkin, and Kocku von Stuckrad, eds., *Horoscopes and Public Spheres* (Berlin: Walter de Gruyter, 2005), and, very recently, Robert S. Westman, *The Copernican Question: Prognostication, Skepticism, and Celestial Order* (Berkeley and London: University of California Press, 2011). In Italian, see also Paola Zambelli, *Una reincarnazione di Pico ai tempi di Pomponazzi* (Milan: Il Polifilo, 1994); Germana Ernst, *Religione, ragione e natura* (Milan: FrancoAngeli, 1991); and Ornella Pompeo Faracovi, *Gli oroscopi di Cristo* (Venice: Marsilio, 1999). In French see Jean-Patrice Boudet, *Entre science et nigromance. Astrologie, divination et magie dans l'Occident médiéval (XIIᵉ–XVᵉ siècle)* (Paris: Publications de la Sorbonne, 2006); and Maxime Préaud, *Les astrologues à la fin du Moyen Age* (Paris: Lattès, 1984). In German, see Dieter Blume, *Regenten des Himmels: astrologische Bilder im Mittelalter und Renaissance* (Berlin: Akademie Verlag, 2000); Claudia Brosseder, *Im Bann der Sterne. Caspar Peucer, Philipp Melanchthon und andere Wittenberger Astrologen* (Berlin: Akademie Verlag, 2004); and Gerd Mentgen, *Astrologie und Öffentlichkeit im Mittelalter* (Stuttgart: Anton Hiersemann, 2005). Among the works that have focused more clearly on the political dimension, see Jean-Patrice Boudet, "Les astrologues et le pouvoir sous le règne de Louis XI," in Bernard Ribémont, ed., *Observer, lire, écrire le ciel au Moyen Age. Actes du Colloque d'Orléans (22–23 avril 1989)* (Paris: Klincksieck, 1991), 7–61; Jean-Patrice Boudet and Therese Charmasson, "Une consultation astrologique princière en 1427," in *Comprendre et maîtriser la nature au Moyen Age. Mélanges d'histoire des sciences offerts à Guy Beaujouan* (Geneva: Droz, 1994), 255–278; Jean-Patrice Boudet and Emmanuel Poulle, "Les jugements astrologiques sur la naissance de Charles VII," in Françoise Autrand, Claude Gauvard, and Jean-Marie Moeglin, eds., *Saint-Denis et la royauté. Études offertes à Bernard Guenée* (Paris: Publications de la Sorbonne, 1999), 170–178; Hilary M. Carey, *Courting Disaster: Astrology at the English Court and University in the Later Middle Ages* (London: Macmillan, 1992); Ann Geneva, *Astrology and the Seventeenth-Century Mind: William Lilly and the Language of the Stars* (Manchester: Manchester University Press, 1995), chaps. 7 and 8; Darin Hayton, "Astrology as Political Propaganda: Humanist Responses to the Turkish Threat in Early Sixteenth-

Century Vienna," *Austrian History Yearbook* 38 (2007): 61–91; idem, "Martin Bylica at the Court of Matthias Corvinus: Astrology and Politics in Renaissance Hungary," *Centaurus* 49 (2007): 185–198; idem, "Expertise ex Stellis: Comets, Horoscopes, and Politics in Renaissance Hungary," in Erik H. Ash, ed., *Expertise and the Early Modern State* (Osiris, vol. 25) (Chicago: University of Chicago Press, 2010), 27–46; Emmanuel Poulle, "Horoscopes princiers des XIVe et XVe siècles," *Bulletin de la société nationale des antiquaires de France* (1969): 63–69; and Paola Zambelli, "Astrologi consiglieri del principe a Wittenberg," *Annali dell'Istituto storico italo-germanico in Trento/Jahrbuch des italienisch-deutschen historischen Instituts in Trient* 18 (1992): 497–543.

5. Two noticeable exceptions are Sophie Page, "Richard Trewythian and the uses of astrology in late Medieval England," *JWCI* 64 (2001): 193–228, and Lauren Kassell, *Medicine and Magic in Elizabethan England. Simon Forman: Astrologer, Alchemist, and Physician* (Oxford: Oxford University Press, 2005).

6. Such a methodological approach is not novel, but it is still relatively unusual when it comes to the history of medicine and science, one noticeable exception being H. C. Erik Midelfort, *Mad Princes of Renaissance Germany* (Charlottesville and London: University Press of Virginia, 1999; originally published 1994).

7. Midelfort, *Mad Princes,* 7.

8. "De tal caxo havea maestro Pietro Bono Advogario, doctore e astrologo excellentissimo nostro citadino ferrarexe, previsto per simele parole, dicendo in lo iudicio de questo anno *quod aliquis magnus Dominus in ultima parte annni aut fer[r]o aut veneno interimeretur.* E cusì è sta' che subito fu amazato," in Bernardino Zambotti, *Diario ferrarese dall'anno 1476 sino al 1504,* ed. Giuseppe Pardi, *RIS²,* XXIV/7 (Bologna: Zanichelli, 1934), 29; and, "Se è verificato lo iudicio de maestro Piedrobono Avogario astrologo ferrarexe, qual diceva: *'Quod aliquis magnus dominus in ultima parte annni aut fero aut veneno interimetur.'* Et così se dice publice che dicte parole del suo iudicio sono verificate in lo duca de Milano, qual è sta morto. Et è verissimo, ut supra," in Girolamo Ferrarini, *Memoriale estense (1476–1489),* ed. Primo Griguolo (Rovigo: Minelliana, 2006), 58. I wish to thank Isabella Lazzarini for pointing me to these two chronicles.

9. Marilyn Nicoud, "Expérience de la maladie et échange épistolaire: les derniers moments de Bianca Maria Visconti (mai–octobre 1468)," *MEFRM* 112.1 (2000): 311–458; eadem, "Les pratiques diététiques à la cour de Francesco Sforza", in Bruno Laurioux and Laurence Moulinier-Brogi, eds., *Scrivere il Medioevo. Lo spazio, la santità, il cibo. Un libro dedicato ad Odile Redon* (Rome: Viella, 2001), 393–404; eadem, "La médecine à Milan a la fin du Moyen Âge: les composantes d'un milieu *professionel*," in Frank Collard and Evelyne Samama, eds., *Mires, physiciens, barbiers et charlatans. Les marges de la médecine de l'Antiquité au XVIe siècle* (Langres: D. Gueniot, 2004), 101–131; and eadem, "Les médecins à la cour de Francesco Sforza ou comment gouverner le Prince," in Odile Redon, Line Sallmann, and Sylvie Steinberg, eds., *Le Désir et le Goût. Une autre histoire (XIIIe–XVIIIe siècle)* (Paris: Presses universitaires de Vincennes, 2005), 201–217.

10. Other historians have investigated the relationship between client and practitioner in the pre-modern period. See Michael McVaugh, *Medicine before the Plague: Practitioners and their Patients in the Crown of Aragon, 1285–1345* (Cambridge: Cambridge University Press, 1993), and Gianna Pomata, *Contracting the Cure: Patients, Healers, and the Law in Early Modern Bologna* (Baltimore and London: The Johns Hopkins University Press, 1998). On Galeazzo's quality control, see Chapter 3; on Ludovico and Varesi, see Chapter 5.

11. One exception to this trend is Parel, *The Machiavellian Cosmos.*

12. Ferdinando Gabotto, "L'astrologia nel Quattrocento in rapporto colla civiltà: osservazioni e documenti storici," *Rivista di filosofia scientifica* 8 (June/July 1889): 377–413; and idem, *Nuove ricerche e documenti sull'astrologia alla corte degli Estensi e degli Sforza* (Torino: La Letteratura, 1891). Gabotto's references to ASMi, *Autografi, Astrologi* should be ignored as no longer valid as there is no longer such a *fondo Astrologi* in the Milanese State archives. Gabotto's transcriptions follow slightly different conventions from those adopted in this work. For this reason, unless specifically noted, all transcriptions are my own. The relocation of these letters at the turn of the century is problematic; it has not always been possible to relocate the material originally seen by Gabotto. The content of Gabotto's letters has been discussed briefly in H. Darrel Rutkin, *Astrology, Natural Philosophy and the History of Science, c.1250–1700: Studies Toward an Interpretation of Giovanni Pico della Mirandola's* Disputationes adversus astrologiam divinatricem (Ph.D. diss., Indiana University, Bloomington, 2002), chap. 4.

13. For England, see Carey, *Courting Disaster.* For France, see Poulle, "Horoscopes princiers"; Boudet and Charmasson, "Une consultation astrologique princière"; Boudet and Poulle, "Les jugements astrologiques sur la naissance de Charles VII"; Jean-Patrice Boudet, ed., *Le recueil des plus celebres astrologues de Simon de Phares,* 2 vols. (Paris: Honoré Champion, 1999); and, now, Luisa Capodieci, *Medicæa Medæa: Art, astres et pouvoir à la cour de Catherine de Médicis* (Geneva: Droz, 2011). For Burgundy, see Jan R. Veenstra, *Magic and Divination at the Courts of Burgundy and France: Text and Context of Laurens Pignon's* Contre les devineurs *(1441)* (Leiden: Brill, 1998). For the Holy Roman Empire, see Darin Hayton, *Astrologers and Astrology in Vienna during the Era of Emperor Maximilian I (1493–1519)* (Ph.D. diss., University of Notre Dame, 2004); idem, "Astrology as Political Propaganda," idem, "Martin Bylica at the Court of Matthias Corvinus"; and Mentgen, *Astrologie und Öffentlichkeit,* esp. 159–269.

14. There are, of course, articles dedicated to the subject, but they have failed to translate into academic monographs. One notable exception is the study of the Medici courts by Janet Cox-Rearick, *Dynasty and Destiny in Medici Art: Pontormo, Leo X, and the two Cosimos* (Princeton, NJ: Princeton University Press, 1984). For recent studies on aspects of court life that have been generally neglected, see Evelyn Welch, *Shopping in the Renaissance. Consumer Cultures in Italy, 1400–1600* (New Haven and London:

Yale University Press, 2005), and Evelyn Welch and Michelle O'Malley, *The Material Renaissance* (Manchester: Manchester University Press, 2007).

15. Westman, *The Copernican Question,* 62–63; 78–81. See also the remarks of Renaud Villard on Renaissance prophecies and political instability in *Du bien commun au mal nécessaire. Tyrannies, assassinats politiques et souveraineté en Italie* (Rome: École française de Rome, 2008), 424–442.

16. The literature on medieval and Renaissance prophecy is too vast to be summed up here. On these aspects, see, at least, Donald Weinstein, *Savonarola and Florence: Prophecy and Patriotism in the Renaissance* (Princeton, NJ: Princeton University Press, 1970); Cesare Vasoli, *Profezia e ragione. Studi sulla cultura del Cinquecento e del Seicento* (Naples: Morano Editore, 1974); idem, *I miti e gli astri* (Naples: Guida, 1977); Ottavia Niccoli, *Prophets and People in Renaissance Italy,* trans. Lydia G. Cochrane (Princeton, NJ: Princeton University Press, 1990); Gabriella Zarri, *Le sante vive: profezie di corte e devozione femminile tra '400 e '500* (Turin: Rosenberg & Sellier, 1990); and Marion Leathers Kuntz, *The Anointment of Dionisio: Prophecy and Politics in Renaissance Italy* (University Park, PA: Pennsylvania State University Press, 2001).

17. Together with the classic study of Franz Cumont, *Astrology and Religion among the Greeks and Romans* (New York and London: Putnam, 1912), see also Tamsyn Barton, *Ancient Astrology* (London and New York: Routledge, 1994); eadem, *Power and Knowledge: Astrology, Physiognomics, and Medicine under the Roman Empire* (Ann Arbor: University of Michigan Press, 1994); and David S. Potter, *Prophets and Emperors: Human and Divine Authority from Augustus to Theodosius* (Cambridge, MA: Harvard University Press, 1994).

18. This distinction is not maintained consistently in the primary sources, however. The medieval physician and astrologer Pietro d'Abano, for instance, did not embrace this division but included medical astrology and meteorology under judicial astrology. See Nancy Siraisi, *Arts and Sciences at Padua: The Studium of Padua before 1350* (Toronto: Pontifical Institute for Medieval Studies, 1973), 86. See also Rutkin, *Astrology, Natural Philosophy and the History of Science,* Introduction.

19. For a discussion of Renaissance medical terms and these concepts, see Heikki Mikkeli, *Hygiene in the Early Modern Medical Tradition* (Helsinki: Finnish Academy of Science and Letters, 1999), 19–23, 54–68; and Ian Maclean, *Logic, Signs and Nature in the Renaissance: The Case of Learned Medicine* (Cambridge: Cambridge University Press, 2002), 251–254.

20. Giuseppe Dell'Anna, *Dies critici: La teoria della ciclicità delle patologie nel XIV secolo,* 2 vols. (Galatina: Mario Congedo Editore, 1999), vol. 1, 9.

21. On the genre of the *regimen sanitatis,* see Mikkeli, *Hygiene in the Early Modern Medical Tradition,* and Marilyn Nicoud, *Les régimes de santé au Moyen Âge: Naissance et diffusion d'une écriture médicale, XIIIᵉ–XVᵉ siècle,* 2 vols. (Rome: École française de Rome, 2007).

22. Dell'Anna, *Dies critici,* vol. 1, 9; Maclean, *Logic, Signs and Nature,* 304. These themes are also discussed in Albumasar, *Introductorium in astronomiam Albumasaris abalachi octo continens libros partiales* (Augsburg: Erhard Ratdolt, 1489), Book I, chap. 1, sig. a2v–a5v. On this text, see Richard J. Lemay, *Abû Ma'shar and Latin Aristotelianism in the Twelfth Century. The Recovery of Aristotle's Natural Philosophy through Arabic Astrology* (Beirut: American University of Beirut, 1962).

23. Galen, *De diebus decretoriis,* in Karl Gottlob Kühn, ed., *Claudii Galeni opera omnia* (Hildesheim: Georg Olms, 1964–1965 [first ed. Leipzig: C. Cnobloch, 1821–1833]), vol. 9, 769–941. Galen rejected the possibility that the future course of a disease could be determined simply from the configuration of the heavens, arguing that it was not the configuration itself, but the consequent atmospheric changes that brought about diseases; therefore no single astral moment could encompass all the developments that occurred during the course of an illness. Despite Galen's qualms about the level of precision of astro-medicine in some of his works, the circulation of his *De diebus criticis* in the Middle Ages and the Renaissance made him possibly the most authoritative writer on astro-medicine of classical antiquity. On this classical tradition and Galen's take on astro-medicine, see Vivian Nutton, "Greek Medical Astrology and the Boundaries of Medicine," in Anna Akasoy, Charles Burnett, and Ronit Yoeli-Tlalim, eds., *Astro-Medicine: Astrology and Medicine, East and West* (Florence: SISMEL-Edizioni del Galluzzo, 2008), 17–31. For a revisionist take on Galen's astrological medicine, see now Glen M. Cooper, "Galen and Astrology: A Mésalliance?," *Early Science and Medicine* 16 (2011): 120–146. On its reception in Renaissance Italy, see Concetta Pennuto, "The Debate on Critical Days in Renaissance Italy," in Akasoy, Burnett, and Yoeli-Talim, *Astro-Medicine,* 75–98.

24. The theory of the critical days is very complex and varied, as the Hippocratic and Galenic traditions offered diverging views on how to calculate the *dies indicativi* and establish the *dies critici.* In general it was believed that the fourth, eleventh, and seventeenth day of an illness were *dies indicativi* and the seventh, fourteenth, and twentieth *dies critici.* Furthermore, if a crisis occurred on the sixth, eighth, or tenth day, it generally resolved itself negatively, while if it happened on the seventh, fourteenth, or twentieth, it resolved positively. The human body-celestial bodies relationship is fully developed in the third book of the *De diebus criticis.* For a thorough discussion of this theory, see Dell'Anna, *Dies critici,* 43–82, esp. 67–72; and the older study of Karl Sudhoff, "Zur Geschichte der Lehre von den kritischen Tagen im Krankenhaitsverlaufe," *Wiener medizinische Wochenschrift* 52 (1902): 210–275. For the debate generated by these inconsistencies, see Pennuto, "The Debate on Critical Days."

25. Dell'Anna, *Dies critici,* 71–73. The principles of astrological medicine are also elucidated in Book III of Marsilio Ficino, *Three Books on Life: A Critical Edition and Translation with Introduction and Notes,* ed. Carol V. Kaske and John R. Clark (Tempe, AZ: Arizona Center of Medieval and Renaissance Studies and The Renais-

sance Society of America, 2002 [first ed. Binghamton, NY: Medieval and Renaissance Text and Studies in Conjunction with the Renaissance Society of America, 1989]). A provocative study of Girolamo Cardano's ambiguous stance toward astrological medicine is offered in Anthony Grafton and Nancy Siraisi, "Between the Election and My Hopes: Girolamo Cardano and Medical Astrology," in William R. Newman and Anthony Grafton, eds., *Secrets of Nature: Astrology and Alchemy in Early Modern Europe* (Cambridge, MA: The MIT Press, 2001), 69–131. The same article neatly outlines the sixteenth-century debate that ensued around Galen's theory of the critical days. On the topic, see also Pennuto, "The Debate on Critical Days."

26. On this text, see Lynn Thorndike, "The Three Latin Translations of the Pseudo-Hippocratic Tract on Astrological Medicine," *Janus* 49 (1960): 104–129; and Pearl Kibre, "*Astronomia* or *Astrologia Ypocratis*," in Erna Hilfstein, Paweł Czartoryski, and Frank D. Grande, eds., *Science and History, Studies in Honor of Edward Rosen* (Wrocław: Polish Academy of Sciences Press, 1978), 133–156.

27. Fevers could be conceived both as symptoms and as diseases in their own right. On the problematic nature of the concepts of symptom and disease in the Renaissance, see Nancy G. Siraisi, "Disease and Symptom as Problematic Concepts in Renaissance Medicine," in Eckhard Kessler and Ian Maclean, eds., *Res et verba in der Renaissance* (Wiesbaden: Harrassowitz Verlag, 2002), 217–240, esp. 226–228; Maclean, *Logic, Signs and Nature*, 261–266.

28. As quoted in Kibre "*Astronomia*," 137. On the theory of fevers and astrology, see Dell'Anna, *Dies critici*, 9–11, 265–276, and, for an example, 200–201.

29. The basic principles are outlined in Alcabitius's *Liber introductorius*. See *Al-Qabisi (Alcabitius): The Introduction to Astrology. Editions of the Arabic and Latin Texts and an English Translation,* eds. Charles Burnett, Keiji Yamamoto, and Michio Yano (London and Turin: The Warburg Institute and Nino Aragno Editore, 2004), differentia I [25–48], 242–248. On *melothesia*, its classical origins, and the zodiac man, see also Charles W. Clark, *The Zodiac Man in Medieval Medical Astrology* (Ph.D. diss., University of Colorado, 1979); and idem, "The Zodiac Man in Medieval Medical Astrology," *Journal of the Rocky Mountain Medieval and Renaissance Association* 3 (1982): 13–38.

30. "Se adonque tu desideri sapere el tempo bono & accomodato alla diminution del sangue [. . .] [n]ota prima & attendi alle regole generale. Et primamente nel tempo della luna nuova & della luna piena non vale né è utile la diminution del sangue, anchora che la luna fossi in quel tempo in buono segno. Anchora tagliare alcuno membro con ferro quando la luna è nel segno che governa el dicto membro è da guardarse. Anchora quando la luna è in alcun segno o aereo o igneo più vale la operatione delle vene che se fosse la luna in alcun segno terreo overo aquatico. Per tanto li giovani si debono minuir el sangue nel crescere della luna, ma li vecchi quando è la luna in declinazione. Anchora nel tempo della primavera & della state si deve minuire el sangue della parte destra, nel tempo dello autunno & della invernata

della parte sinistra". Johannes de Ketham, *Fasiculo de medicina* (Venice: Giovanni and Gregorio De Gregori, 1493), sig. b4*v*.

31. Ibid, sig. b4*v*–b6*r*. Following Avicenna's *Canon,* which made no reference to surgical activity, bloodletting was sometimes defined more broadly as any type of evacuation that discharged the humors in excess much like purgative medicine. See Pedro Gil-Sotres, "Derivation and Revulsion: The Theory and Practice of Medieval Phlebotomy," in Luis Garcìa-Ballester, Roger French, Jon Arrizabalaga, and Andrew Cunningham, eds., *Practical Medicine from Salerno to the Black Death* (Cambridge: Cambridge University Press, 1994), 110–155, esp. 115–119; and especially Mary Katherine Keblinger Hague Yearl, *The Time of Bloodletting* (Ph.D. diss., Yale University, 2005).

32. See Chapter 1.

33. Grafton and Siraisi rightly argue that "generalizations about medical astrology in the Renaissance should be discouraged." Grafton and Siraisi, "Between Election and My Hopes," 110.

34. See Chapter 1. In addition, the long life of the ps.-Hippocratic *Astronomia Ypocratis,* which had wide manuscript circulation and was printed as early as 1485 in Venice, indicates that the principles of astro-medicine may have circulated widely. See Kibre, "Astronomia," esp. 146–156.

35. See Chapter 4.

36. The only sustained efforts so far have been carried out by French historian Marilyn Nicoud. See n. 9 above.

37. While the moon and other planets are not explicitly mentioned in the correspondence relating to Galeazzo Maria Sforza's illness as a child, the treatment of fevers described in the documents was based on the Galenic theory of the critical days and thus adhered to the principles of astrological medicine. For these documents and their discussion, see Nicoud, "Les médecins à la cour de Francesco Sforza," 204–209.

38. Fully documented instances of the illnesses of Sforza family members are few: that of Bianca Maria Visconti, wife of Francesco, has been discussed extensively in Nicoud's "Expérience de la maladie," which also includes a substantial number of documents. See also Giorgio Cosmacini, "La malattia del duca Francesco," in *COM,* vol. 3 (2000): 23–26, as well as letters on Francesco Sforza's illness included in the same volume.

39. See Chapter 5.

40. For instance, Girolamo Benivieni famously cast the horoscope of his friend Giovanni Pico della Mirandola. See Paul O. Kristeller, "Giovanni Pico della Mirandola and his Sources," in *L'opera e il pensiero di Giovanni Pico della Mirandola nella storia dell'Umanesimo,* 2 vols. (Florence: Istituto Nazionale di Studi sul Rinascimento, 1965), vol. 1, 35–142 (the horoscope is reproduced in a plate placed between pp. 112–113). The same horoscope has also been reproduced in Paolo Viti, ed., *Pico, Poliziano e l'Umanesimo di fine Quattrocento* (Florence: Olschki, 1994), fig. 42 (at the end

of the volume); and is discussed more extensively in Patrizia Castelli, "L'oroscopo di Pico," in Patrizia Castelli, ed., *Giovanni e Gianfrancesco Pico: L'opera e la fortuna di due studenti ferraresi* (Florence: Olschki, 1998), 17–43, esp. 19. Another example is provided by the annual horoscope (or "revolution") sent by the Pavian professor of natural philosophy Apollinare Oliffredi to Francesco Gonzaga "si per eterne obligatione como ancora per nostre nuperrime promisse." Oliffredi had already sent the marquis a booklet (*libello*). ASMa, AG 1632, fol. 501r, Lodi, November 4, 1498.

41. The most striking example of the Italian Renaissance is perhaps the horoscope of Agostino Chigi frescoed on the vault of Villa Farnesina, Rome. For recent, but contrasting, interpretations of the vault's astrological frescoes, see Mary Quinlan-McGrath, "The Astrological Vault of the Villa Farnesina: Agostino Chigi's Rising Sign," *JWCI* 47 (1984): 91–105; Kristen Lippincott, "Two Astrological Ceilings Reconsidered: The Sala di Galatea in the Villa Farnesina and the Sala del Mappamondo at Caprarola," *JWCI* 53 (1990): 185–207; and Quinlan-McGrath's own reply, "The Villa Farnesina: Time-telling Conventions and Renaissance Astrological Practice," *JWCI* 58 (1995): 52–71. Cosimo I de' Medici similarly used his ascendant, the sign of Capricorn, as a propaganda device to promote himself as a new Augustus. On Cosimo's use of the Capricorn as personal device, and Medici's astrological imagery more generally, see Cox-Rearick, *Dynasty and Destiny in Medici Art*, esp. 199–220, 269–291; Roger J. Crum, "'Cosmos, the World of Cosimo': The Iconography of the Uffizi Façade," *The Art Bulletin* 71 (1989): 237–253, esp. 246–251; and Henk Th. van Veen, *Cosimo I de' Medici and his Self-Representation in Florentine Art and Culture* (Cambridge: Cambridge University Press, 2006), esp. 20–26, 95–98, 138–139. On the use of the sign of Capricorn in Augustan iconography, see Barton, *Power and Knowledge*, 40–47.

42. Other evidence emerges from the genre of annual prognostications (*pronostici*), in which astrologers often based their comments for a given year upon the annual revolution of the nativity of various political leaders. See Monica Azzolini, "The Political Uses of Astrology: Predicting the Illness and Death of Princes, Kings and Popes in the Italian Renaissance," *Studies in the History and Philosophy of Biological and Biomedical Sciences* 41 (2010): 135–145, esp. 140–142.

43. See Monica Azzolini, "Refining the Astrologer's Art: Astrological Diagrams in Bodleian MS Canon. Misc. 24 and Cardano's *Libelli quinque* (1547)," *Journal for the History of Astronomy* 42 (2011): 1–25.

44. Both Cardano's and Gaurico's collections contain Sforza horoscopes. Girolamo Cardano, *Liber de exemplis centum geniturarum*, in Girolamo Cardano, *OO*, vol. 5, respectively geniture XLIII: Francesco Sforza (p. 482); XLIV: Galeazzo Maria (p. 482); IX: Ludovico Maria (p. 463). Ludovico Maria Sforza's horoscope is also included in Luca Gaurico's collection, *Tractatus astrologicus in quo agitur de praeteritis multorum hominum accidentibus per proprias eorum genituras ad unguem examinatis* (Venice: Curzio Troiano Navò, 1552), Tractatus tertius imperatorum regum principum schemata et apothelesmata, fol. 50r.

45. Examples of this sort of application are included in this chapter and in Chapter 5.

46. See Chapter 5 for examples.

47. On the need for consultation with other doctors, Giovanni Matteo Ferrari da Grado wrote as follows to his duke in relation to the health of his children: "A mi pare da fare domandare tri o quatro deli più excelenti pratici de questo collegio per visitare essi vostri figlioli, insieme cum noy dui [i.e. Giovanni Matteo Ferrari and Marco da Roma], non perché diffida non si proceda quanto vole la rasone et el debito circa li loro casi, et con ogni diligentia possibile havere, nec etiam perché le loro dispositione siano in mal termine—ymo ne spero de tuti duy sanità—ma solum perché non possa parere ad alchuno che me voglia arrogare tanto caricho cum solo Magistro Marcho [da Roma], quanto è la cura de duy figlioli de tale Signore, perché etiam de li figlioli de gentilhomeni o castelani a simili casi han[n]o a le volte uno col[l]egio de quatro o cinque medici." See ASMi, *Autografi,* Medici 215, Giovanni Matteo Ferrari da Grado to Galeazzo Maria Sforza, August 10, 1472. On the increasing professionalization of these medical practitioners, see Nicoud, "La médecine à Milan à la fin du Moyen Âge": 101–131, esp. 101–110.

48. Simon de Phares, *Le recueil,* vol. 1, 579–580. On Annius's astrology, see Chapter 3.

49. On the Gonzaga and astrology, see now Rodolfo Signorini, *Fortuna dell'astrologia a Mantova: Arte, letteratura, carte d'archivio* (Mantua: Sometti, 2007).

1. The Science of the Stars

1. Martin Kemp, *Leonardo da Vinci: The Marvellous Works of Nature and Man,* rev. ed. (Oxford: Oxford University Press, 2006), 206–207.

2. Ladislao Reti, "The Two Unpublished Manuscripts of Leonardo da Vinci in the Biblioteca Nacional of Madrid-II," *The Burlington Magazine* 110 (1968): 81–89.

3. It is impossible to identify which edition with any certainty. The text was first published in Ferrara in 1472 and was later republished numerous times in Venice, Brescia, Bologna, and Milan. The ISTC lists no fewer than sixteen editions for the period 1472–1500. The Milanese edition was printed by either Filippo Lavagna or Leonardus Pachel and Uldericus Scinzenzeler on April 1, 1478.

4. Francesco Sirigatti was a Florentine astronomer and astrologer, with a strong interest in astrology and vernacular translations, considering his translations of both Guido Bonatti's *Liber astronomiae* (now BLF, MS Plut. 30.30) and Lucio Bellanti's *Tractatus astronomiae.* He later dedicated his only published work, *De ortu et occasu signorum libri duo,* to Leo X, the Medici pope famous for his interest in astrology. See Carlo Pedretti, *Studi Vinciani: documenti, analisi e inediti leonardeschi* (Geneva: Droz, 1957), 120–121; and Gabriella Federici Vescovini, "Note di commento a alcuni passi del «Libro di Pittura». «L'astrologia che nulla fa senza la prospettiva . . . »," in

Fabio Frosini, ed., *Leonardo e Pico. Analogie, contratti, confronti. Atti del Convegno di Mirandola (10 Maggio 2003)* (Florence: Olschki, 2005), 99–129 (esp. 123–129).

5. In a passage of the *Libro della Pittura,* Leonardo (or his editor?) called judicial astrology "fallace." See Romano Nanni, "Le «disputationes» pichiane sull'astrologia e Leonardo," in Frosini, *Leonardo e Pico,* 53–98, esp. 69. In the same volume, see also Federici Vescovini, "Note di commento," esp. 122–129.

6. Leonardo similarly expressed reservations toward these disciplines, as well as alchemy and necromancy. See Patrizia Castelli, "Leonardo, i due Pico e la critica alla divinazione," in Frosini, *Leonardo e Pico,* 131–172.

7. Castelli, "Leonardo, i due Pico e la critica alla divinazione," 132, 157–158.

8. On this event, see Monica Azzolini, "Anatomy of a Dispute: Leonardo, Pacioli, and Scientific Entertainment in Renaissance Milan," *Early Science and Medicine* 9.2 (2004): 115–135.

9. Luca Pacioli, *De divina proportione,* introduction by Augusto Marinoni (Milan: Silvana, 1982), fol. iv. This is the facsimile edition of the copy now in the BAM. The manuscript was concluded by 1498.

10. Azzolini, "Anatomy of a Dispute," 119–123.

11. Ibid.

12. Olaf Pedersen, "The *Corpus Astronomicum* and the Traditions of Medieval Latin Astronomy," in Owen Gingerich and Jerzy Dobrzycki, eds., *Astronomy of Copernicus and its Background,* Colloquia Copernicana III (Wroclaw-Warszawa-Krakow-Gdansk: Ossolineum, 1975), 57–96.

13. A commendable step in this direction has been taken by David Juste, who is in the process of cataloging astrological manuscripts in a number of European collections. The first volume of this admirable enterprise is dedicated to the texts in the BSB. A companion volume for the BNF will appear shortly. See David Juste, ed., *Les manuscrits astrologiques latins conservés à la Bayerische Staatsbibliothek de Munich* (Paris: CNRS Éditions, 2011).

14. Considering the important contribution of the student population to the economy of university towns, most Italian states formally required their students to study in their own *Studia.* The dukes of Milan, at various points in time, issued decrees reminding their subjects of this obligation. See Agostino Sottili, "Università e cultura a Pavia in età visconteo-sforzesca," in *Storia di Pavia,* 5 vols. (Milan: Banca del Monte di Lombardia, 1990), vol. 3, t. 2, 418–420; and *Documenti per la storia dell'Università di Pavia nella seconda metà del '400,* 3 vols. (Milan and Bologna: Cisalpino, 1994–2010), vol. 1 (1994): (1450–1455), ed. Agostino Sottili, 11–18, esp. 14–17.

15. *Documenti per la storia dell'Università di Pavia,* vol. 1, 13–16. Unfortunately only one *rotulo* is extant for the period 1450–1455 and none for the period 1456–1460. The records are more complete for the period 1461–1499. See *Documenti per la storia dell'Università di Pavia,* vol. 2 (2003): (1455–1460), ed. Agostino Sottili and Paolo Rosso, x.

16. Ibid., vol. 2, viii.

17. Some of the promoters may have simply been prominent members of the profession. See Paul Grendler, "New Scholarship on Renaissance Universities," *RQ* 53 (2000): 1174–1182, esp. 1177.

18. Siraisi, *Arts and Sciences at Padua*, 11. The same approach has been followed in Katharine Park, *Doctors and Medicine in Early Renaissance Florence* (Princeton, NJ: Princeton University Press, 1985).

19. Historian of the book, Ennio Sandal, estimated that Venetian books constituted over 40 percent of the entire Italian output, Milanese less than 10 percent, and Pavian only a little more than 2 percent. This data is taken from Brian Richardson, *Printing, Writers, and Readers in Renaissance Italy* (Cambridge: Cambridge University Press, 1999), 5–6.

20. This is BL, MS Arundel 88. Another manuscript that may be linked to Pavia is BLYU, MS Mellon 13, *Iudicium astrologicum* for 1475. The anonymous author of this *iudicium* calls Giovanni Marliani "preceptor noster" (fol. 68v). He must have studied, therefore, at Pavia.

21. Michael Shank, "L'astronomia nel Quattrocento tra corti e università," *Il Rinascimento Italiano e l'Europa*, 6 vols. (Treviso: Fondazione Cassamarca & Angelo Colla Editore, 2005–2010), vol. 5: *Le Scienze*, ed. Antonio Clericuzio and Germana Ernst (2008), 5.

22. Siraisi, *Arts and Sciences at Padua*, esp. 67–94, 78, 83–84, 91–93; eadem, *Taddeo Alderotti and His Pupils: Two Generations of Italian Medical Learning* (Princeton, NJ: Princeton University Press, 1981), 139–145; Richard Lemay, "The Teaching of Astronomy in Medieval Universities, Principally at Paris in the Fourteenth Century," *Manuscripta* 20 (1976): 197–217. The exception was England, where, in the late Middle Ages and the Renaissance, astrology/astronomy was not officially taught at Cambridge and Oxford. See Carey, *Courting Disaster*.

23. Shank, "L'astronomia nel Quattrocento tra corti e università," 4–5.

24. H. Darrel Rutkin, "Astrology," in *The Cambridge History of Science*, 5 vols. (Cambridge: Cambridge University Press, 2002–2009), vol. 3: *Early Modern Science*, ed. Katharine Park and Lorraine Daston (2006), 541–561; idem, *Astrology, Natural Philosophy and the History of Science*, chap. 2 (on astrology and natural philosophy); and chap. 3 (on astrology, medicine, and mathematics).

25. Our main source of information regarding the subjects taught are the extant rolls with the annual list of the university professors who were contractually obliged to give their lectures in a given year. See Paul Grendler, *The Universities of the Italian Renaissance* (Baltimore and London: Johns Hopkins University Press, 2001), 415–426.

26. Rodolfo Majocchi's *Codice diplomatico dell'Università di Pavia*, 2 vols. (Bologna: Forni editore, 1971) [originally published Pavia: Fratelli Fusi, 1905–1915] covers the period 1401–1450 and contains transcriptions of a number of rolls (not all

have survived). At the end of the fourteenth century there were three professors of astrology at Pavia, a considerable number by comparison with other universities. Indeed, the rolls of 1399–1400 indicate "M. Blazio de Parma legenti philosophiam moralem, naturalem et astrologiam," "Magistro Iohanni de Catelonia legenti astrologiam," and "M. Francisco de Crispis legenti astrologiam." See Majocchi, *Codice diplomatico,* vol. 1, 421. These became two later on: under the rubric "Ad lecturam astrologie," the rolls of 1425 report "Magister Petrus de Montealcino" and "Magister Antonius de Bernadigio." That year, Antonio Bernareggi lectured also in *physica* (natural philosophy), while Pietro da Montalcino was extraordinary professor of practical medicine. See Majocchi, *Codice diplomatico,* vol. 2, t. 1, 221–222. By 1439 Bernareggi had risen considerably within the academic ranks and received 300 florins for his appointment "ad lecturam ordinariam medicine," but he was also still teaching astrology. Pietro da Montalcino no longer appears in the rolls, being replaced by "Stephanus da Faventia," who also taught astrology that year. See Majocchi, *Codice diplomatico,* vol. 2, t. 1, 395.

27. The rolls for 1448/49, 1469/70, 1471/72, 1473/74, 1474/75, 1475/76, 1477/78 (incomplete), 1480/81 are now in ASMi, Studi PA, 390. Giovanni Marliani lectured on astrology from 1469 to at least 1475/76. Concurrently, he held the chair of theoretical medicine (*ad lectura medicine de mane/ad lectura medicine theorice ordinarie de mane*) with a salary of around 750 florins, to which, in 1475, the duke added 250 more. No salary is indicated for the teaching of astrology per se, but presumably the salary indicated for medicine also took into consideration his other teaching. For a similar phenomenon in Padua, see Grendler, *The Universities,* 416.

28. A number of important works have been devoted to the teaching and practice of medicine in Italian universities, and these contribute to a relatively uniform picture. The Paduan medical curriculum prior to 1350 seems very similar to that of Bologna in 1405. See Siraisi, *Arts and Sciences at Padua,* 94. The University of Florence similarly modelled its curriculum on that of Bologna. See Park, *Doctors and Medicine,* 245. Ferrara also modelled its curriculum on Bologna's and Padua's. See Graziella Federici Vescovini, "L'astrologia all'università di Ferrara nel Quattrocento," in Patrizia Castelli, ed., *La rinascita del sapere: libri e maestri dello studio Ferrarese* (Venice: Marsilio, 1991), 293–306; and eadem, "I programmi degli insegnamenti del collegio di medicina, filosofia e astrologia dello statuto dell'università di Bologna del 1405," in Jaqueline Hamesse, ed., *Roma, magistra mundi. Itineraria culturae medievalis. Mélanges offerts au Père L. E. Boyle à l'occasion de son 75ᵉ anniversaire,* 3 vols. (Louvain-la-Neuve: Fédération Internationale des Instituts d'Études Médiévales, 1998), vol. 1, 193–223. For another example, namely that of Perugia, see Grendler, *The Universities,* 323. There is no reason to believe that things were substantially different for the teaching of natural philosophy or astrology and astronomy.

29. The statutes (in Latin) were first published in Carlo Malagola, *Statuti delle università e dei collegi dello studio bolognese* (Bologna: Nicola Zanichelli, 1888). The

reader interested in the relevant passage can access it, among other places, in my article "Reading Health in the Stars: Prognosis and Astrology in Renaissance Italy," in Günther Oestmann, H. Darrel Rutkin, and Kocku von Stuckrad, eds., *Horoscopes and Public Spheres* (Berlin and New York: Walter de Gruyter, 2005), 183–205, esp. 191, n. 19. An English translation is provided in Lynn Thorndike, *University Records and Life in the Middle Ages* (New York: Norton, 1975; first published 1944), 281–282, and in Grendler, *The Universities*, 410. Further discussions of the curriculum are also provided in Vescovini, "I programmi degli insegnamenti"; Grendler, *The Universities*, 408–429; Boudet, *Entre science et nigromance*, 288–290; and Rutkin, *Astrology, Natural Philosophy and the History of Science*, chap. 3.

30. On the Alphonsine tables, see Emmanuel Poulle, *Les Tables Alphonsines avec les Canons de Jean de Saxe* (Paris: Éditions du centre national de la recherche scientifique, 1984), and José Chabás and Bernard R. Goldstein, *The Alfonsine Tables of Toledo* (Dordrecht and London: Kluwer Academic, 2003).

31. Over the century, various authors have been credited with writing the *Theorica planetarum*: these include Gerardo da Cremona and Gerardo da Sabbioneta. According to Olaf Pedersen, none of the attributions is particularly convincing. On these texts, see Olaf Pedersen, "The *Theorica planetarum*-literature of the Middle Ages," *Classica et Medievalia* 23 (1962): 225–232; idem, "The Decline and Fall of the *Theorica planetarum*: An Essay in Renaissance Astronomy," in Erna Hilfstein, Pawel Czartoryski, Frank D. Grande, eds., *Science and History: Studies in Honor of Edward Rosen*, Studia Copernicana XVI (Wroclaw-Warszawa-Krakow-Gdansk: Ossolineum, 1978), 157–185; idem, "The Origins of the *Theorica planetarum*," *Journal for the History of Astronomy* 12 (1981): 113–123. Another *Theorica planetarum*, superior to that attributed to Gerardo da Cremona and Gerardo da Sabbionetta is that of Campanus of Novara, which circulated only in manuscript. For these two texts, see Carmody, *Arabic Astronomical and Astrological Sciences*, 167–168; Francis J. Carmody, ed., *Theorica planetarum Gerardi* (Berkeley: [s.n.], 1942), and *Campanus of Novara and Medieval Planetary Theory: Theorica planetarum*, ed. with an introduction, English translation, and commentary by Francis S. Benjamin, Jr. and G. J. Toomer (Madison: University of Wisconsin Press, 1971).

32. The *Sphaera* of Sacrobosco was often accompanied by commentaries and remained a staple text within the curriculum of Italian and European universities. On the *De Sphaera*, see Lynn Thorndike, *The Sphere of Sacrobosco and Its Commentators* (Chicago: The University of Chicago Press, 1949).

33. Carmody, *Arabic Astronomical and Astrological Sciences*, 23–24. Other anonymous treatises by this title exist, as well as works by medieval authors such as Andalò de Negro. See, for example, Enrico Narducci, ed., *Catalogo di manoscritti ora posseduti da D. Baldassarre Boncompagni*, 2nd edition (Rome: Tipografia delle scienze matematiche e fisiche, 1892). Item n. 51 on the list contains all the statutory texts for

the teaching of astronomy/astrology and others, too, like Andalò de Negro's *Practica astrolabii* and his *Tractatus scalae quadrantis sive astrolabii.*

34. This list was probably the result of consolidation of the university curriculum: a number of manuscripts dating to an earlier period contain many of these works in one single tome, suggesting that these were taught as a course of studies well before 1405. One example is BAM, MS M28 sup. (XIV cent.), which contains: Sacrobosco's *Sphaera* with commentary, his *Algoritmus,* Euclid's *Elementa,* two versions of the *Theorica planetarum,* an anonymous *De mensurationibus rerum secundum quadrantem,* Messahallah's *Practica astrolabii,* and Acabitius's *Liber introductorius.* For a brief description see now: http://ambrosiana.comperio.it/opac/viewdetail .php?bid=77769 [accessed August 19, 2010].

35. On the *De urina non visa,* see Roger K. French, "Astrology in Medical Practice," in Ballester, French, Arrizabalaga, and Cunningham, eds., *Practical Medicine,* 30–59 [reprinted in Roger French, *Ancients and Moderns in the Medical Sciences: From Hippocrates to Harvey* (Aldershot: Ashgate, 2000), 44–48], and, now, Laurence Moulinier-Brogi, *Guillaume l'Anglais le frondeur de l'uroscopie médiévale (XIIIe siècle). Édition commentée et traduction du* De urina non visa (Geneva: Droz, 2011).

36. For the 1405 medical curriculum of Bologna, see Park, *Doctors and Medicine,* 245–248. On the teaching of Galen's *On critical days,* see the discussion below.

37. Pedersen, "The *Corpus Astronomicum,*" 57–96.

38. Grendler, *The Universities,* 412–426; Shank, "L'astronomia nel Quattrocento," 8–9.

39. Thorndike, *The Sphere of Sacrobosco.* For an example of Renaissance innovation in the teaching of Avicenna's *Canon,* see Nancy G. Siraisi, *Avicenna in Renaissance Italy: The Canon and Medical Teaching in Italian Universities after 1500* (Princeton, NJ: Princeton University Press, 1987).

40. For a census of the numerous surviving manuscripts of the *Centiloquium Hermetis,* see Paolo Lucentini and Vittoria Perrone Compagni, *I testi e i codici di Ermete nel Medioevo* (Florence: Edizioni Polistampa, 2001), 27–32. For a census of the numerous manuscripts of ps.-Ptolemy's *Centiloquium* in the major European libraries, see Dell'Anna, *Dies critici,* 137, n. 2.

41. Brief mention of Boerio's employment can be found in John Noble Johnson, *The Life of Thomas Linacre: Doctor in Medicine, Physician to King Henry VIII* (London: Edward Lumley, 1835), 168–169. For information on Boerio's life, I rely on Giuseppe Portigliotti, "G. B. Boerio alla Corte d'Inghilterra," *Illustrazione medica Italiana* 5 (January 1923): 8–10; Silvana Seidel Menchi, "Boerio, Giovanni Battista," *DBI* 11 (1969), 126–127; and Peter G. Bietenholz and Thomas B. Deutscher, eds., *Contemporaries of Erasmus: A Biographical Register of the Renaissance and Reformation,* 3 vols. (Toronto and Buffalo: University of Toronto Press, 1985–1987), vol. 1, 158–159. None of these sources, however, lists MS Arundel 88 among Boerio's works.

42. BL, MS Arundel 88, *Canones primis mobilis Johannis de Lineriis*, fols. 39v–46v. Incipit: "Cuiuslibet arcus propositi sinum rectum invenire . . ." (TK 276); *De urina non visa, et de concordia astrologiae et medicinae*, fols. 67r–70v (incomplete). Incipit: "Ne vel ignorancie vel pocius . . ." (TK 906). As noted, these texts were generally studied in the second and fourth year of the four-year cycle: "In secundo anno primo legatur tractatus de sphera, quo lecto, legatur secundus geumetrie [sic] Euclidis, quo lecto legantur *canones super tabulis de linerijs*. [. . .] In quarto anno primo legatur quadripartitus totus, quo lecto, legatur *liber de urina non visa*" (Malagola, *Statuti delle università*, 276). Emphasis mine.

43. BL, MS Arundel 88, Johannes of Sacrobosco, *Computus*, fols. 38r–39r. Incipit: "Cum ex motu lune quemadmodum et solis plures . . ." (TK 298). Explicit (fol. 39r): "Explicit tractatus de illuminatione lune quo ad se et quo ad nos, et de divisione mensis lunaris in quatuor species, editus a Magister Johanne de Sacrobusco in computo suo. Scriptus per me Johannem Baptistam Boerium de Tabia artium et medicine scholarem nec non astrologiam in presenti audientem, filium spectabilis ac insignis illustris doctoris domini Bernardi qui obiit anno elapso die 20 Septembris scilicet 1483 [. . .] 1484 die 5 Augusti."

44. Indeed, Sacrobosco's *Computus* differed from earlier and simpler versions of the *computus* in the way it integrated more fully diagrams within the text. On this, see Jennifer Moreton's "John of Sacrobosco and the Calendar," *Viator* 25 (1994): 229–244, esp. 237–238. Two fine examples of this manuscript tradition are CUL, MS Ii.III.3 (diagrams at fols. 46v–47r) and MS Ii.I.15 (only one diagram at fol. 33v). For a Visconti-Sforza copy, see BNF, MS Lat. 7363, which contains the *Algoritmus* (1r–7r), the *Sphaera* (7v–19v), and the *Computus* (fols. 20r–39v); diagram at fol. 30r. I wish to thank Adam Mosley for pointing me to Moreton's article and the two Cambridge manuscripts.

45. BL, MS Arundel 88, *De chiromantia tractatus*, fols. 58v–66r. Incipit: "Chiromantia est ars demonstrans mores et inclinationes naturales hominis per signa sensibilia manus . . ." (TK 225). On the ps.-Aristotelian *Chiromantia*, see Roger A. Pack, "Pseudo-Aristoteles: Chiromantia," *Archives d'histoire doctrinale et littéraire du Moyen Âge* 39 (1972): 289–320. The image of the hands (fol. 58v) precedes the treatise that starts at fol. 59r. This text is preceded by a treatise on physiognomy, which is, apparently, a common combination. See Charles S. F. Burnett, "The Earliest Chiromancy in the West," *JWCI* 50 (1987): 189–195, esp. 191.

46. At fol. 66r it says: "Libellus de ciromantia [sic] excopiatus per me Johannem Baptistam Boerium de Tabia, artium et medicine studentem filosofie (?) ac astrologie de anno 1484 die 10 Septembris, qui [est] profogus a Papia propter pestem quam deus dignet mittigare. Resideo hic in Arce Valencie." Valenza was a two days' walk from Pavia.

47. BL, MS Arundel 88, fol. 26v: "Explicit Meseala de revolutione annorum mundi, excopiatus per me Johannem Baptistam Boerium de Tabia artium, medicine et astrologie studentem, 1484 die 24 Septembris."

48. BL, MS Arundel 88, *De revolutione anni mundi 1484 prognosticon per Gabrielem Priovanum,* fols. 28r–29v. At fol. 29v: "Explicit iudicium de 1484 editum a clarissimo artium et medicine doctore D. M. Gabriel de Pirovano ducali medico. Scriptum autem ab originali per me Johannem Baptistam Boerium artium et medicine studentem nec non etiam astrologie incumbentem."

49. Portigliotti, "G. B. Boerio"; Seidel Menchi, "Boerio, Giovanni Battista"; Carey, *Courting Disaster,* 161.

50. Elizabeth Lane Furdell, *The Royal Doctors, 1485–1714: Medical Personnel at the Tudor and Stuart Courts* (Rochester, N.Y.: University of Rochester Press, 2001), esp. 17–43. Despite covering the chronological period in which Boerio was employed at court, Furdell fails to include him among the Tudor doctors she discusses in the book.

51. Seidel Menchi, "Boerio, Giovanni Battista," 126.

52. Portigliotti, "G. B. Boerio."

53. BL, MS Arundel 88, fol. 95r–v.

54. Ibid., *Aphorismus secundum Ptolomeum explicatur,* fol. 55r, opens (roughly) with verbum 51 of Ptolemy's *Centiloquium,* "Locus lune in nativitate est ipse gradus ascendens de circulo hora casus spermatis, et locus lune hora casus spermatis est gradus ascendens hora nativitatis." The *Tractatus de physiognomia,* opens at fol. 57r. Some of these inconsistencies were clearly created by Boerio himself, who then inserted some personal reminders, but others like this one on the celestial configurations that bring about the plague may have been introduced at a later stage when the manuscript was bound.

55. Ibid., Johannes of Bruges, *De veritate astronomie,* fols. 1r–15r (incomplete). Incipit: "Quoniam, ut ait philosophus, in libro de natura et proprietatibus elementorum" (TK 1307). This version lacks the introduction and the concluding chapter (chap. 12) present in other versions. The British Library Catalogue of Manuscripts wrongly attributes this text to John Ashenden and the misattribution is reported in TK 1307, who, however, lists also an early printed copy with the same incipit and it credits it correctly to Johannes of Bruges (TK 1307; Klebs 556.1). I am aware of three other copies of this work: BMP, MS 3893 (fols. 65r–98r), BSB, MS Clm 51 (fols. 89ra–98va), and BRB, MS II.1685 (fols. 68r–93v). I have only looked at MS 3893, which counts twelve chapters preceded by an introduction. Thorndike claims to have consulted the following early printed edition: BNF, Rés. p.V.186 (Klebs 551.1), which I have not consulted. On John of Bruges, see Simon de Phares, *Le recuieil,* vol. 1, 563–564; Boudet, *Entre science et nigromance,* 321–322; and Laura A. Smoller, "Of Earthquakes, Hail, Frogs, and Geography: Plague and the Investigation of the Apocalypse in the Later Middle Ages," in Caroline Walker Bynum and Paul H. Freedman, eds., *Last Things: Death and the Apocalypse in the Middle Ages* (Philadelphia, Penn.: University of Pennsylvania Press, 2000), 156–187, esp. 185–186. I wish to thank David Juste for the information on the manuscripts in Munich and Bruxelles.

56. The explicit, which corresponds virtually verbatim to MS Arundel 88, reads (fol. 19v): "Millennarium omnino finiatur. Anno Christi 1096. Henricho imperatore in Alamania. Philippo Rege Francie regnante in Francia. Papa Urbano secundo viro egregio et bono fuit passagium in triplicitate tenea [in MS Arundel 88: terrea]. Deo Gratias amen."

57. *Judicium cum tractatibus planetariis* (Milan: Filippo Mantegazza, December 20, 1496), in gothic with illustrations (Klebs 566.1). I have consulted the following copy: BL, IA. 26847. The book is listed in Teresa Rogledi Manni, *La tipografia a Milano nel XV secolo* (Florence: Olschki, 1980), 154 (n. 529).

58. All of these authors are cited at least once in the treatise. A collection of Abraham ibn Ezra's works appeared in print in Venice in 1507: *Abrahe Avenaris Judei astrologi peritissimi in re iudiciali opera ab excellentissimo philosopho Petro de Abano post accuratam castigationem in Latinum traducta. Introductorium quod dicitur principium sapientie. Liber rationum. Liber nativitatum & revolutionum earum. Liber interrogationum. Liber electionum. Liber luminarium & est de cognitione diei cretici . . . Liber conjunctionum planetarum & revolutionum annorum mundi qui dicitur de mundo vel seculo. Tractatus insuper quidam particulares ejusdem Abrahe. Liber de consuetudinibus in judiciis astrorum & est centiloquium Bethen breve admodum. Ejusdem De horis planetarum* (Venice: Peter Lichtenstein, 1507). On Abraham ibn Ezra and his works, see Shlomo Sela, *Abraham Ibn Ezra and the Rise of Medieval Hebrew Science* (Leiden and Boston: Brill, 2003); and Renate Smithuis, "Abraham Ibn Ezra's Astrological Works in Hebrew and Latin: New Discoveries and Exhaustive Listing," *Aleph: Historical Studies in Science and Judaism* 6 (2006): 239–338.

59. These include Albumasar, *De magnis coniunctionibus, annorum revolutionibus, ac eorum profectionibus* (Augsburg: Erhard Ratdolt, 31 Mar. 1489; Klebs 39.1); Messahallah, *Epistola [Messahallah] in rebus eclipsis lune et in coniunctionibus planetarum ac revolutionibus annorum breviter elucidata,* in *Opera astrologica* (Venice: Bonetto Locatelli for Ottaviano Scoto, December 20, 1493; Klebs 814.2), fols. 148r–149r; and Abraham ibn Ezra, *Liber coniunctionum* (also called *De mundo vel seculo*), in *Abrahe Avenaris Judei astrologi peritissimi in re iudiciali opera,* fols. LXXVIr–LXXXVr.

60. John D. North, "Astrology and the Fortunes of Churches," *Centaurus* 24 (1980): 181–211; Eugenio Garin, *Astrology in the Renaissance: The Zodiac of Life* (London: Routledge & Kegan Paul, 1983).

61. As Luther himself did not know if he was born in 1483 or 1484, this allowed the astrologers to identify him with the prophet-monk (or pseudo-prophet, depending on their religious leanings) announced by the conjunction of 1484. On this issue, see the classic essay by Aby Warburg, "Pagan-Antique Prophecy in Words and Images in the Age of Luther," in *The Renewal of Pagan Antiquity: Contributions to the Cultural History of the European Renaissance,* intro. Kurt W. Forster, trans. David Britt (Los Angeles: Getty Research Institute, 1999), 597–697 [originally published as Aby

Warburg, Aby Warburg, *Heidnisch-Antike Weissagung in Wort und Bild zu Luthers Zeiten* (Heidelberg: C. Winter, 1920); as well as Dietrich Kurze, "Popular Astrology and Prophecy in the Fifteenth and Sixteenth Centuries: Johannes Lichtenberger," in Zambelli, *'Astrologi Hallucinati,'* 177–193; Ottavia Niccoli, *Prophecy and People in Renaissance Italy,* trans. Lydia C. Cochrane (Princeton, NJ: Princeton University Press, 1990), 136–139; and Anthony Grafton, *Cardano's Cosmos: The Worlds and Works of a Renaissance Astrologer* (Cambridge, MA: Harvard University Press, 1999), 33, 75–77, 101–102. Giovanni Pico della Mirandola discusses the upheaval caused by this conjunction in his *Disputationes adversus astrologiam divinatricem,* ed. Eugenio Garin, 2 vols. (Turin: Aragno, 2004), vol. 1, book V, cap. 1 [originally published Florence: Vallecchi, 1946–1952]. This conjunction was also associated with the appearance of syphilis. See Anthony Grafton and Nancy Siraisi, "Between the Election and My Hopes: Girolamo Cardano and Medical Astrology," in Newman and Grafton, *Secrets of Nature,* 83–84; and Darin Hayton, "Joseph Grünpeck's Astrological Explanation of the French Disease," in Kevin Siena, ed., *Sins of the Flesh: Responding to Sexual Disease in Early Modern Europe* (Toronto: Centre for Reformation and Renaissance Studies, 2005), 81–106.

62. Bruges's work comprises twelve chapters; both MS Arundel 88 and the Milanese incunabulum only eleven. In addition, MS Arundel 88 contains an anonymous chapter on the conjunction of Saturn and Jupiter in the fiery triplicity (fol. 17r). This is not part of Bruges's work but seems modeled on it.

63. BL, MS Arundel 88, fols. 1r, 7v, 8r. The conjunction of the two superior planets Saturn and Jupiter occurs every twenty years. On the principles of conjunctionist astrology, see North, "Astrology and the Fortunes of Churches," and Smoller, *History, Prophecy, and the Stars,* 20–22.

64. The note is on fol. 14v. D'Ailly's was another of Bruges's sources. See Smoller, "Of Earthquakes, Hail, Frogs," 185–186. For the role of conjunctionist astrology in Pierre d'Ailly's natural theology and the connection between apocalyptic notions of the end of the world and conjunctionist astrology in the later Middle Ages, see Smoller, *History, Prophecy, and the Stars.*

65. BL, MS Arundel 88, fols. 15r–16v. The incipit does not correspond to any of those given in Marjorie Reeves, *The Influence of Prophecy in the Later Middle Ages: A Study of Joachimism* (Notre Dame, IN: University of Notre Dame, 1993; first edition Clarendon Press, 1969), and it is possible that it was a uniquely Lombard creation or it was adapted for a Lombard audience.

66. On the intermingling of astrology and prophecy in the sybilline and Jochamite traditions, see now Laura A. Smoller, "Teste Albumasare cum Sibylla: Astrology and the Sibyls in Medieval Europe," *Studies in History and Philosophy of Biological and Biomedical Sciences* 41.2 (2010): 76–89; eadem, *History, Prophecy and the Stars;* and Jean-Patrice Boudet, "Simon de Phares et les rapports entre astrologie et prophétie à la fin du Moyen Âge," *MEFRM* 102 (1990): 617–648. On Lichtenberger, see Domenico

Fava, "La fortuna del pronostico di Giovanni Lichtenberger in Italia nel Quattrocento e nel Cinquecento," *Gutenberg Jahrbuch* 5 (1930): 126–148; Dietrich Kurze, *Johannes Lichtenberger: Eine Studie zur Geschichte der Prophetie und Astrologie* (Lübeck and Hamburg, 1960); idem, "Popular Astrology and Prophecy in the Fifteenth and Sixteenth Centuries: Johannes Lichtenberger," in Zambelli, *'Astrologi Hallucinati',* 177–193; and Giancarlo Petrella, *La Pronosticatio di Johannes Lichtenberger. Un testo profetico nell'Italia del Rinascimento* (Udine: Forum, 2010).

67. For fifteenth- and sixteenth-century circulation of ps.-Joachite prophecies in northern Italy, see Marjorie Reeves, *The Influence of Prophecy in the Later Middle Ages,* esp. 92–93 and 259–273; Roberto Rusconi, *Profezia e profeti alla fine del Medioevo* (Rome: Viella, 1999); Gian Luca Potestà, ed., *Il profetismo gioachimita tra Quattrocento e Cinquecento: atti del III Congresso internazionale di studi giochimiti, San Giovanni in Fiore, 17–21 Settembre 1989* (Genoa: Marietti, 1991). On the relationship between Charles VIII's impresa and prophetic writings, see Henri-François Delaborde, *L'expédition de Charles VIII en Italie: histoire diplomatique et militaire* (Paris: Firmin-Didot, 1888), 312–318 and Anne Denis, *Charles VIII et les Italiens: Histoire et mythe* (Geneva: Droz, 1979). For some general considerations and a list of some of the prophecies connected to Charles VIII, see Henri Hauser, *Les Sources de l'histoire de France: XVIe siècle* (1494–1610), 4 vols. (Paris: A. Picard, 1906–1915), vol. 1: "Sources Françaises", 107–109; Samuel Krauss, "Le Roi de France, Charles VIII, et les espérances messianiques," *Revue des études juives* 51 (1906): 87–95; Cesare Vasoli, "Il mito della monarchia francese nelle profezie fra 1490 e 1510," in Dario Cecchetti, Lionello Sozzi, and Louis Terreaux, eds., *L'aube de la Renaissance: pour le dixième anniversaire de la disparition de Franco Simone* (Geneva: Editions Slatkine, 1991), 149–165; David Abulafia, *The French Descent into Renaissance Italy, 1494–95: Antecedents and Effects* (Aldershot: Ashgate, 1995), 15–17; and Francesco Guicciardini, *The History of Italy,* trans., ed. with notes and an introduction by Sidney Alexander (Princeton, NJ: Princeton University Press, 1984 [originally published New York: Macmillan, 1969]) Book 1, chap. iv.

68. BL, MS Arundel 88, fols. 18r–26v; see also Messahallah's *De revolutionibus annorum mundi,* in *Opera astrologica,* fols. 149r–152r.

69. BL, MS Arundel 88, *Almansoris sententiae ad Saracenum regem,* fols. 89r–93r, and *Quinquaginta praecepta Zaelis,* fols. 94r–97r.

70. BL, MS Arundel 88, fol. 96v.

71. Tiziana Pesenti Marangon, "La miscellanea astrologica del prototipografo padovano Bartolomeo Valdizocco e la diffusione dei testi astrologici e medici tra i lettori padovani del '400," *Quaderni per la storia dell'università di Padova* 11 (1978): 87–106. I have not been able to consult the Valdizocco manuscript.

72. The full content is described in Pesenti Marangon, "La miscellanea astrologica," 88–93. On Nicolò Paganica and this work, see Nicolaus de Paganica, *Compendium medicinalis astrologiae,* ed. Giuseppe Dell'Anna (Galatina: Congedo, 1990).

On Albumasar's *De revolutionibus annorum mundi,* also called *De experimentis (Book of Experiments),* see Thorndike, *HMES,* vol. 1, 650–652. Thorndike alerts us to the fact that both contemporaries and present-day scholars could easily confuse this text with another by Albumasar entitled *De revolutione annorum in revolutione nativitatum.* A work by Albumasar entitled *De revolutionibus annorum mundi* is cited in BLYU, MS Mellon 13 (fol. 55r). The author of this manuscript had studied astrology at Pavia. Considering this is an annual *iudicium* for the year 1474, it is likely that the text referred to in MS Mellon 13 is the *De experimentis* and not the *De revolutione annorum in revolution nativitatum.*

73. Ibid. On Prosdocimo de' Beldomandi, see Cesare Vasoli, "Bedemandis, Prosdocimo," *DBI* 7 (1965), 551–554; on Nicolò de' Conti, see Augusto De Ferrari, "Conti, Nicolò," *DBI* 28 (1983), 461–462 and chap. 3.

74. *Liber quadripartiti Ptholemei. Centiloquium eiusdem. Centiloquium hermetis. Eiusdem de stellis beibenijs. Centiloquium bethem. [et] de horis planeta[rum]. Eiusdem de significatione triplicitatum ortus. Centu[m] quinquaginta p[ro]po[sitio]nes Almansoris. Zahel de interrogationibus. Eiusdem de electionibus. Eiusde[m] de te[m]po[rum] significationib[us] in iudicijs. Messahallach de receptionibus planetaru[m]. Eiusdem de interrogationibus. Epistola eiusde[m] cu[m] duodecim capitulis. Eiusdem de reuolutionibus anno[rum] mundi* (Venice: Bonetto Locatelli for Ottaviano Scoto, 20 Dec. 1493). Ottaviano Scoto gave up printing (that is, being a *tipografo-editore* who owned a press) for publishing and bookselling (thus becoming a *libraio-editore*). On the differences between these categories, see Richardson, *Printing, Writers, and Readers,* 34–35. Title pages are one feature that distinguishes manuscripts from printed books. The lack of title page is not so unusual in early printed books, but by the 1490s most volumes had title pages as these were seen as an attractive feature for potential purchasers (Richardson, *Printing, Writers, and Readers,* 131–132). For a description of this incunabulum's contents, see also IGI 8187, and Carmody, *Arabic Astronomical and Astrological Sciences,* 13. I have consulted the following copies: BNCF, B.1.21, BL, IB. 22900, and the digital edition housed at Gallica's permanent link: http://gallica.bnf .fr/ark:/12148/bpt6k596584.

75. Elisabeth Pellegrin, *La Bibliothèque des Visconti et des Sforza, ducs de Milan* (Paris: Publications du C.N.R.S., 1955), Inventory A, n. 129.

76. On this *bottega,* see Angela Nuovo, *Il commercio librario nell'Italia del Rinascimento* (Milan: FrancoAngeli, 1998), 77. On the Scoto family (also called Scotti or Scotto), see also Carlo Volpati, "Gli Scotti di Monza tipografi-editori in Venezia," *ASL* 59 (1932), 365–382; Jane Bernstein argues that by the 1490s the academic market had become the bread and butter of the Scoto Press. See Jane A. Bernstein, *Music Printing in Renaissance Venice* (Oxford: Oxford University Press, 1998), 29–34. On Venetian printers' and booksellers' attempts to enter the university market at Pavia (which included some books travelling in the other direction), see Martin Lowry, *Nicolas Jenson and the Rise of Venetian Publishing in Renaissance Europe* (Cambridge,

MA: Blackwell, 1991), 161, 172, 184–185; on the early speculative printing of scientific texts, see Ian Maclean, "The Reception of Medieval Practical Medicine in the Sixteenth Century: The Case of Arnau de Vilanova," in idem, *Learning and the Market Place: Essays in the History of the Early Modern Book* (Leiden and Boston: Brill, 2009), 89–106.

77. Lowry, *Nicolas Jenson*, 196.

78. For the source of comparison and conversion data, see Gregory P. Lubkin, *A Renaissance Court: Milan under Galeazzo Maria Sforza* (Berkeley: University of California Press, 1994), xix, 128–133, 288; and Grendler, *The Universities*, 84. For the cost of books in Milan, of which less is known, see Monica Pedralli, *Novo, grande, coverto, e ferrato: gli inventari di biblioteca e la cultura a Milano nel Quattrocento* (Milan: Vita e Pensiero, 2002), 190–192.

79. The BL copy IB. 22900 is bound together with editions of Johannes Sacrobosco's *Sphaera* (Venice: Simone Bevilacqua da Pavia, 1499) and Georg von Peurbach's *Theorica nova planetarum* (Venice: Simone Bevilacqua da Pavia, 1495) (separate shelf mark IB .23996), which may reinforce the hypothesis of a *corpus* of texts aimed at university professors and wealthy students.

80. Westman, *The Copernican Question*, 96–97.

81. Judging from the ISTC entries, Girolamo Salio seems to have specialized mainly in medical and astrological books and texts edited by university professors or targeted for a university audience. On Scoto's *bottega* in Pavia, see Angela Nuovo and Christian Coppens, *I Giolito e la stampa nell'Italia del XVI secolo* (Geneva: Droz, 2005), 27, n. 32. The medical book is Barzizza's *Introductorium ad opus practicum medicinae. Liber IX Almansoris* (Pavia: Antonio Carcano for Ottaviano Scoto, 1494) (GW3672; ISTC: ib00260000). Barzizza was professor of medicine at Padua. See, Paolo Sambin, "Barzizza, Cristoforo," *DBI* 7 (1970): 32–34.

82. The inventory has been published in Daniela Fattori, "Il dottore padovano Alessandro Pellati, la sua biblioteca e *l'editio princeps* del *De medicorum astrologia*," *La Bibliofilía* 110 (2008): 117–137 (inventory at 121–131).

83. On Bianchini's tables see José Chabás and Bernard R. Goldstein, *The Astronomical Tables of Giovanni Bianchini* (Leiden and Boston: Brill, 2009). Little is known of John of Lübeck's tables, but he claims to have been at Padua for a time, as indicated in the only work by him to be published by Valdizocco: Joahnnes de Lübeck, *Prognosticon super Antichristi adventu Judaeorumque Messiae* ([Padua]: Bartolomeo Valdizocco, [not before April 1474]). I have consulted the following copy: BL, IA. 29826. On Nicolò de' Conti, see also chap. 2.

84. Regiomontanus's *Kalendar* contains printed paper instruments: the *Instrumentum horarum inaequalium,* the *Instrumentum veri motus lunae* with two superimposed rotating discs held to the page by a piece of string ("volvelles"), the *Quadrans horologii horizontalis* and the *Quadratum horarium generale* with a brass pointer. A particularly fine example is in GUL, Sp Coll BD7-f.13. Significantly, this copy is

bound with Sacrobosco's *Sphaera*, Alcabitius's *Introductorius*, and Ibn Ezra's *De luminaribus et diebus creticis*.

85. Fattori, "Il dottore padovano Alessandro Pellati," 132–136. On this text, see Thorndike, "The Three Latin Translations," 104–129; and Kibre, "*Astronomia* or *Astrologia Ypocratis*," 133–156.

86. This is probably John Ashenden (Johannes Eschuid), *Summa astrologiae iudicialis* (Venice: Johannes Lucilius Santritter, July 7, 1489) of which a great number of copies are extant.

87. Hyginus, *Poetica astronomica*, ed. Jacobus Sentinus and Johannes Lucilius Santritter (Venice: Erhard Ratdolt, 1485), and Werner Rolewinck's *Fasciculum temporum* (Venice: Erhard Ratdolt, 1480 and 1481).

88. In Italian universities, the *Physiognomia* was taught by extraordinary professors of natural philosophy in the first year. There were a number of different works that went by this title, Scot's being one of them. Grendler, *The Universities*, 270. It is likely that the *Chiromantia* was taught by the same professors in the first or second year of the arts and medicine curriculum.

89. Bindo de' Vecchi, "I libri di un medico umanista fiorentino del sec. XV," *La Bibliofilía* 34 (1932): 293–301 (inventory at 297–301).

90. Ibid., 301.

91. Ibid., 298.

92. On these inventories, see Pedralli, *Novo, grande, coverto, e ferrato*, 88–91. None of these Milanese inventories is representative of an entire library collection. The inventory of Filippo Pellizzoni, for a time Filippo Maria Visconti's physician, reflects only a part of his library and lists fifty-eight books he had deposited at the Misericordia of Milan when he left the city for Bologna and ultimately Rome. Presumably, Filippo took the most important books related to his practice with him. The inventory is in Pedralli, *Novo, grande, coverto e ferrato*, 351–358. The inventory of the ducal physician and apostolic protonotary Ambrogio Griffi (c. 1420–1493), which has been published by Paolo Galimberti, reports the books donated by the physician on the occasion of the funding of a college for poor students at Pavia. The college was to house six to eight students studying either law or medicine. None of the books relates to astrology, but this does not mean Griffi did not possess any astronomical or astrological books. Of the four manuscripts belonging to him that have been identified, only one was listed in the inventory. Furthermore we know Griffi owned a copy of Fazio Cardano's edition of Peckham's *Perspectiva communis* and a copy of Ficino's *De vita sana* (and thus presumably the whole of the *De triplici vita*, famous for its astrological medicine, of which the *De vita sana* is the first book). None of this appears in the inventory, thus suggesting that this incomplete list is not reflective of Griffi's entire library. See Paolo M. Galimberti, "Il testamento e la biblioteca di Ambrogio Griffi, medico milanese, protonotario apostolico e consigliere sforzesco," *Aevum* 72 (1998): 447–483. On Giovanni Matteo Ferrari da Grado's library, see Henri-Maxime

Ferrari, *Une chaire de médecine au XVe siècle. Un professeur à l'université de Pavie de 1432–1472* (Paris: Félix Alcan, 1899), 83–93, and Tullia Gasparrini Leporace, "Due biblioteche mediche del Quattrocento," *La Bibliofilía* 52 (1950): 205–220 (this article includes partial inventories of the libraries of Ferrari da Grado and the little known doctor Baldassarre de Vemenia). Henri-Maxime Ferrari remarks with some surprise how medical astrology is completely absent from Giovanni Matteo's works (and indeed, I would add, from his library). On at least one occasion, however, he mentions the conjunction of the Moon (with the Sun) as relevant to the prognosis of the illness of one of Galeazzo Maria Sforza's sons. The letter is in ASMi, *Autografi, Medici* 215, Giovanni Matteo Ferrari da Grado to Galeazzo Maria Sforza, Castle of Pavia, August 16, 1471 (cited in *Une chaire*, 172–174). On Ferrari's funding of a college and the donation of other books, see Maria Luisa Grossi Turchetti, "La dotazione libraria di un collegio universitario del Quattrocento," *Physis* 22 (1980): 463–475.

93. The first college to be founded was the Collegio Castiglioni, in 1429. See Pedralli, *Novo, grande, coverto, e ferrato*, 321–323. On Pavia's colleges, see also Paolo Rosso, "Presenze studentesche e collegi pavesi nella seconda metà del Quattrocento," *Schede umanistiche* 2 (1994): 24–42.

94. Pedralli, *Novo, grande, coverto, e ferrato*, 324; Francesco Malaguzzi Valeri, *La corte di Ludovico il Moro*, 4 vols. (Milan: Hoepli, 1913–1923), vol. 4, 121–122.

95. The two lists are reproduced in Pedralli, *Novo, grande, coverto, e ferrato*, 329–333 and 336–337. I am inclined to believe that the *Liber aggregatoris* indicated at p. 332 is Alfraganus's *Liber de aggregationibus scientiae stellarum* rather than the *Aggregator* of Jacopo Dondi or Guglielmo Corvi da Brescia (less unlikely, in this context) or Serapion's *Liber aggregationis medicinarum simplicium* suggested by Pedralli (another text commonly adopted in university teaching). Without any further details, however, my choice is based only on the proximity of this text to other astrological works by Ptolemy and Haly Abenragel. On the now little known Giorgio Anselmi da Parma, and his works, see Thorndike, *HMES*, vol. 4, 242–246. Anselmi's works must have circulated widely within the Milanese court. His treatises on music were cited with admiration by the celebrated Milanese musician Franchino Gaffurio in his works. See Liliana Pannella, "Anselmi, Giorgio, senior," *DBI* 3 (1961), 377–378. A copy of his *Astronomia* by three different hands is in BAV, Vat. Lat. 4080, fols. 41r–87r. On Anselmi see also Nicolas Weill-Parot, *Les "images astrologiques" au Moyen Âges et à la Renaissance: Spéculations intellectuelles et pratiques magiques, XIIème–XVème siècles* (Paris: Honoré Champion, 2002), 622–636 and chap. 4.

96. BL, MS Arundel 88, *De prognosticatione morborum per crisim et alia signa*, fols. 71r–86v, as well as fol. 56r, which must have been misplaced at some point before the book was bound (TK 204). On the attribution of this text to Bartholomew of Bruges, see Cornelius O'Boyle, "An Updated Survey of the Life and Works of Bartholomew of Bruges († 1356)," *Manuscripta* 40.2 (1996): 67–95. MS Arundel 88 is

not among the manuscripts listed by Thorndike and Kibre and in the studies of Cornelius O'Boyle.

97. I owe much of the information that follows about the *Aggregationes de crisi et de creticis diebus* to the thorough study of Cornelius O'Boyle, *Medieval Prognosis and Astrology: A Working Edition of the* Aggregationes de crisi et creticis diebus, *with Introduction and English Summary* (Cambridge: Wellcome Unit for the History of Medicine, 1991).

98. Pearl Kibre, *Hippocrates Latinus: Repertorium of Hippocratic Writings in the Latin Middle Ages* (New York: Fordham University Press, 1985), rev. ed. with additions and corrections, 94–107; eadem, *"Astronomia* or *Astrologia Ypocratis,"* 133–156.

99. BL, MS Arundel 88, *Potiones quaedam medicinales,* fols. 107r–108r. Not much is known of this Francesco Benci (or Benzi). On his activities at court, see Giulio Bertoni, *La biblioteca estense e la coltura ferrarese ai tempi del Duca Ercole I (1471–1505)* (Turin: Loescher, 1903), 191; Angela Dillon Bussi, "I Benzi a Ferrara," in Paola Castelli, ed., *«In supreme dignitatis». Per la storia dell'Università di Ferrara (1391–1991)* (Florence: Olschki, 1995), 439; Primo Girguolo, "Professori di medicina senesi tra Ferrara e Padova: notizie dei Benzi e di Alessandro Sermoneta," *Quaderni per la storia dell'Università di Padova* 41 (2008): 135–150; and Alessandro Simili, "Alcuni documenti inediti intorno a Francesco Benzi," *Rivista di storia delle scienze mediche e naturali* 33 (1942): 81–102.

100. BL, MS Arundel 88, fols. 108v–122r. By Renaissance standards, this horoscope interpretation is reasonably long. Given that the text is so neatly written, it is possible that the intepretation was not written by Boerio, but rather that he copied it from an original now lost.

101. Ibid., fols. 123r–125v. "Finitum pronosticum super genitura insignis et ellegantis [sic] adulescentis domino Johannis Baptiste Colis, anno domini 1491 die 14 Maii, Cremone, per Baptista Piasium Cremonensem. Extractum ab originali per me, Johannem Baptistam Boerium Genuenssem [sic], artium et medicine doctorem." In the chart and in the explicit, Giovanni Battista Colis is called an *adolescens;* at the time of the horosocope, he was twenty years old.

102. Ibid., fols. 125v–133v. "Infrascriptum est pronosticum de universitate oxonii 1513 super Regia . . . missum suo elimosinario. Et quamvis in eo plura inepta contineantur, illud tamen transcripsi dicto anno essere primum Martii." This lengthy interpretation includes Henry VII's geniture with the chart of Henry VIII's conception (fol. 129r); the geniture of Henry VIII and the conception chart of his son (fol. 129v); the geniture of Henry VIII's son (fol. 130r); the "revolution" of Henry VIII's birth chart for the year 1512 (June 28) (fol. 131r); and another revolution of Henry VIII's birth chart for 1513 accompanied by a revolution of the year (*introitus solis in Arietem*) for the same year (fol. 132r). Again this is a rather long and composite interpretation for the time.

103. "Item horologium unum magnum cum theoricis septem planetarum." The 1488 inventory has been transcribed in Anna Giulia Cavagna, "«Il libro desquadernato: la carta rosechata da rati». Due nuovi inventari della libreria Visconteo-Sforzesca," *Bollettino della Società Pavese di Storia Patria* 41 (1989): 29–97. This was probably Giovanni Dondi's famous planetarium, which the Paduan physician-astronomer donated to Duke Gian Galeazzo Visconti in 1381. On this device see Silvio A. Bedini and Francis R. Maddison, "Mechanical Universe: The Astrarium of Giovanni de' Dondi," *Transactions of the American Philosophical Society* 56 (October 1966): 1–69 and Giovanni Dondi Dall'Orologio, *Tractatus astrarii,* ed. and trans. Emmanuel Poulle (Genève: Droz, 2003).

104. Three inventories were published in Pellegrin, *La Bibliothèque.* These are dated 1426 (*Consignatio* A), 1459 (*Consignatio* B), and 1469 (*Consignatio* C, which includes only the texts owned by Galeazzo Maria Sforza that were added to the ducal library). Two other inventories have been discovered more recently in the notarial archives at Pavia. These date to 1488 (Inventory D) and 1490 (Inventory E). On them, see Maria Grazia Albertini Ottolenghi, "La biblioteca dei Visconti e degli Sforza: gli inventari del 1488 e del 1490," *Studi Petrarcheschi* 8 (1991): 1–238 and Simonetta Cerrini, "Libri dei Visconti-Sforza. Schede per una nuova edizione degli inventari," *Studi Petrarcheschi* 8 (1991): 239–281. On Inventories D and E, see also Cavagna, "«Il libro desquadernato»." Inventories D and E both list 947 items. Inventory A, which lists 988 books, offers a very detailed description of the books and includes, quite unusually for the time, both incipits and explicits of most of them. All these details have contributed to the identification of a number of books in French and European libraries. Unless stated otherwise, in this chapter I refer only to Inventory A.

105. Pellegrin, *La Bibliothèque,* 136–137, 166, 287. The manuscripts containing the *Sphaera* are A. n. 290 (BNF, MS Lat. 7267); n. 409 (BNF, MS Lat. 7363); n. 971 (BNF, MS Lat. 7400). The copy with the *Liber de iudiciis in astrologia* (fols. 6vb–7rb), which has been identified as Raymond de Marseille, *Liber iudiciorum,* is now BNF, MS Lat. 7267. Other astrological texts included in this manuscript include Thebit Bencora's *De motu octavae sphaerae* and Albertus Magnus *De mineralibus.* I wish to thank David Juste for sharing with me his identification of Raymond de Marseille's text.

106. Ibid., 166, A. n. 409 (BNF MS Lat. 7363).

107. Ibid., 130. A. n. 255 and n. 256, respectively *Campanus super geometria Euclidis copertus corio pavonacio levi ad modum parisinum* and *Euclidis geometria cum planisphero Tholomei copertus corio rubeo levi* (now BNF, MS Lat. 7214).

108. Ibid., 138. A. n. 292 (BNF, MS Lat. 7258). *Almagestum Tolomei copertum corio rubeo levi.*

109. Ibid. 166. A. n. 410 (BNF, MS Lat. 7409: *Liber tabularum tolentinarum parvus copertus assidibus cum fondo rubeo*).

110. Ibid., 128 and 136. These are, respectively, A. n. 246 and A. n. 287. For the latter, the inventory of 1459 (*Consignatio* B) adds the name of the commentator John

of Saxony, leaving no doubt that we are dealing with Alcabitius's *Introductorius* and not any other work by the same author.

111. Ibid., 97. A. n. 112. This is the second book of Scot's untitled work on astrology. The first book, the *Liber introductorius,* is a rambling introduction to astrology, while the third, the *Liber physionomiae,* deals with the art of physiognomics (but includes also a good deal of astrology). Scot expressed much admiration for the Arabic authors Albumasar, Alcabitius, and Alfraganus. Scot was famously put in Dante's *Inferno* as a charlatan and diviner by Dante (*Inferno,* 20. 115–117). On Scot, see Lynn Thorndike, *Michael Scot* (London: Nelson, 1965), Thorndike, *HMES,* vol. 2, 307–337, and Piero Morpurgo, "Scot, Michael (*d.* in or after 1235)," *ODNB* (http://www.oxforddnb.com /view/article/24902, accessed July 28, 2011).

112. Pellegrin, *La Bibliothèque,* 128–129. A. n. 246 and A. n. 251 (MS Lat. 7307).

113. Ibid., 170. A. n. 431. Possibly now BNF, MS Lat. 7038, a manuscript version of Taddeo Alderotti's commentary on the *Isagoge.* On its content see Dell'Anna, *Dies critici,* 17–18.

114. Cavagna, "«Il libro desquadernato»," 45 (n. 156), 74 (n. 177).

115. Pellegrin, *La Bibliothèque,* 136. A. n. 288.

116. Ibid., 294. B. n. 103.

117. Ibid., 294. B. n. 105.

118. Cavagna, "«Il libro desquadernato»," 48. On this rare text attributed to Alfodhol, see Lynn Thorndike, "Alfodhol and Almadel: Hitherto Unnoted Medi-aeval Books of Magic in Florentine Manuscripts," *Speculum* 2 (1927): 326–331; idem, "Alfodhol de Merengi Again," *Speculum* 4 (1929): 90; and idem "Alfodhol and Almadel Once More," *Speculum* 20 (1945): 88–91. The text was also present in the library of Francesco Gonzaga, Marquis of Mantua (Thorndike, "Alfodhol and Almadel Once More," 90).

119. Malaguzzi Valeri, *La corte di Ludovico,* vol. 4, 149–151.

120. Pesenti notes how the inventories of the libraries of the Paduan physicians Antonio Cermisone and Cristoforo Barzizza and the surgeon Nicolò Raimondi contain virtually no astronomy/astrology books. Pesenti Marangon, "La miscellanea astrologica," 96.

121. Among those authors who were less favorable toward medical astrology one can count the famous fifteenth-century Ferrarese physician Nicolò Leoniceno (1428–1524). Other medical humanists, including his own student Giovanni Mainardi, shared his skepticism. This may have been part of a broader anti-Arabist stance. See Peter Dilg, "The Antiarabism in the Medicine of Humanism," in *La diffusione delle scienze islamiche nel Medio Evo Europeo* (Rome: Accademia Nazionale dei Lincei, 1987), 269–289, esp. 272–274; Daniela Mugnai Carrara, *La biblioteca di Nicolò Leoniceno. Tra Aristotele e Galeno: cultura e libri di un medico umanista* (Florence: Olschki, 1991), 73–74; and eadem, "Fra causalità astrologica e causalità naturale. Gli interventi di Nicolò Leoniceno e della sua scuola sul morbo gallico," *Physis* 21 (1979):

37–54. On Mainardi, see now Pennuto, "The Debate on Critical Days." See also note n. 92 in this chapter for Giovanni Matteo Ferrari da Grado.

122. Siraisi, *Arts and Medicine at Padua*, 77–94; Graziella Federici Vescovini, *Astrologia e scienza: la crisi dell'aristotelismo sul cadere del Trecento e Biagio Pelacani da Parma* (Florence: Vallecchi, 1979); eadem, *Il "Lucidator dubitabilium astronomiae" di Pietro d'Abano: opere scientifiche inedite* (Padua: Programma e 1+1 Editori, 1988); eadem, "La medicina astrologica dello «Speculum phisionomie» di Michele Savonarola," in Castelli, *«In supreme dignitatis»*, 415–429; Tiziana Pesenti, *Marsilio Santasofia tra corti e università. La carriera di un «Monarcha Medicinae» del Trecento* (Treviso: Antilia, 2003); Siraisi, *Taddeo Alderotti*, 139–145, 179–186; Tommaso Duranti, *Mai sotto Saturno: Girolamo Manfredi medico e astrologo* (Bologna: CLUEB, 2008); and more generally, Grendler, *The Universities*, 415–426.

123. For a broad, scholarly treatment of the way celestial influence was conceived, see North, "Celestial Influence—The Major Premise of Astrology," 45–100; and Grant, "The Influence of the Celestial Region on the Terrestrial," *1200–1687* (Cambridge University Press, 1994), 569–617. See also Barton, *Ancient Astrology* (Routledge, 1994), 102–111.

124. See, for example, Ashenden, *Summa astrologiae iudicialis,* fol. 67v: "Dixit itaque Ptolomeus in primo Quadripartiti, capitulus quintus quod planetae bonae sive fortunae sunt Iupiter, Venus, et Luna. Planetae malivolae sive infortunae sunt Saturnus et Mars. Mediocres vero sunt Sol & Mercurius." For these astrological commonplaces in what follows I draw largely from Barton, *Ancient Astrology,* and J. C. Eade, *The Forgotten Sky: A Guide to Astrology in English Literature* (Oxford: Clarendon Press, 1984).

125. For an outline of these properties, see *Al-Qabisi (Alcabitius): The Introduction to Astrology*, differentia II [1–43], 267–292. See also differentia III [30–31], 313–314.

126. *Al-Qabisi (Alcabitius): The Introduction to Astrology*, differentia II [45–48], 293.

127. *Al-Qabisi (Alcabitius): The Introduction to Astrology*, diifferentia I [57–68], 255–260.

128. On houses and house division, see John D. North, *Horoscopes and History* (London: Warburg Institute, 1986), 1–69 and Eade, *The Forgotten Sky*, 41–51 and 73–76. Eade explains how to erect a celestial figure at pp. 51–59.

129. For these planets and their relationships with houses and *partes* (*pars fortunae, pars mortis*, etc.), see *Al-Qabisi (Alcabitius): The Introduction to Astrology*, differentia V, [4–17], 351–358.

130. The issue of free will and the will of God are part and parcel of most attacks against astrology from the Middle Ages onward. See Grant, "The Influence of the Region on the Celestial," and Stefano Caroti, ed., "Quaestio contra divinatores horoscopios," *Archives d'histoire doctrinale et littéraire du moyen age* 43 (1976): 201–310; idem, "La critica contro l'astrologia di Nicole Oresme e la sua influenza nel Medioevo

e nel Rinascimento," *Atti della Accademia Nazionale dei Lincei (Classe di Scienze Morali, Storiche e Filologiche)* 23 (1979): 545–685; idem, "Nicole Oresme's Polemic against Astrology in his *Quodlibeta*," in Curry, *Astrology, Science and Society: Historical Essays,* 75–93; and George W. Coopland, *Nicole Oresme and the Astrologers: A study of his* Livre de divinacions (Liverpool: Liverpool University Press, 1952).

131. Barton, *Ancient Astrology,* and Eade, *The Forgotten Sky,* explain roughly how to construct a chart, and I refer the reader to these texts for guidance.

132. A basic explanation is provided in *Al-Qabisi (Alcabitius): The Introduction to Astrology,* differentia IV [4–6], 319–325. See also chap. 3.

133. The system of "dignities" was quite complicated and required the astrologer to consider a great number of factors that cannot all be discussed here. For a reasonably exhaustive description, see Eade, *The Forgotten Sky,* 59–88 (for dignities) and 88–95 (for hyleg and alchocoden).

134. On these branches of astrology, see Thomas, *Religion and the Decline of Magic,* 286–287; Siraisi, *Arts and Sciences at Padua,* 85–86; Charles S. F. Burnett, "Astrology," in Frank. A. C. Mantello and Arthur G. Rigg, eds., *Medieval Latin: An Introduction and Bibliographical Guide* (Washington, D.C.: Catholic University of America, 1996), 369–382, esp. 375–376; Jean-Patrice Boudet, "Astrology," in Thomas F. Glick, Stephen J. Livesey, and Faith Wallis, eds., *Medieval Science, Technology and Medicine: An Encyclopedia* (New York: Routledge, 2005), 61–64; and Rutkin, *Astrology, Natural Philosophy and the History of Science,* esp. chap. 4. For an expanded version of this taxonomy in Nicole Oresme, see Hilary M. Carey, "Judicial Astrology in Theory and Practice in Later Medieval Europe," *Studies in History and Philosophy of Biological and Biomedical Sciences* 41 (2010): 90–98, esp. 91.

135. For the analysis of one particular example, see Monica Azzolini, "The Political Uses of Astrology: Predicting the Illness and Death of Princes, Kings and Popes in the Italian Renaissance," *Studies in the History and Philosophy of Biological and Biomedical Sciences* 41 (2010): 135–145, esp. 141–142.

136. For a lucid explanation, see North, "Astrology and the Fortune of Churches," 181–211.

137. Over 60 authors dedicated over 160 pamphlets to this conjunction. See Kurze, "Popular Astrology and Prophecy," 177–193. On this phenomenon, see also Thorndike, *HMES,* vol. 5, 178–233; Warburg, "Pagan-Antique Prophecy," 597–697; Ottavia Niccoli, "Il diluvio del 1524 fra panico collettivo e irrisione carnevalesca," in *Scienze, credenze occulte, livelli di cultura. Convegno Internazionale di Studi (Firenze, 26–30 giugno 1980)* (Florence: Olschki, 1982), 369–392; eadem, *Prophecy and People,* chap. 6; Paola Zambelli, "Fine del mondo o inizio della propaganda? Astrologia, filosofia della storia e propaganda politico-religiosa nel dibattito sulla congiunzione del 1524," *Scienze, credenze occulte, livelli di cultura,* 291–368; eadem, *'Astrologi Hallucinati';* and Robin Bruce Barnes, *Prophecy and Gnosis: Apocalypticism in the*

Wake of the Lutheran Reformation (Stanford: Stanford University Press, 1988), esp. 141–181.

138. Albumasar, *De magnis coniunctionibus.*

139. Smoller, *History, Prophecy and the Stars,* and Boudet "Astrology," 63.

140. Boudet, "Astrology."

141. The existence of unfair competition from these low-life practitioners did not escape the haughty Cardano, who remarked on this in relation to both medicine and astrology. Grafton and Siraisi, "Between the Election and My Hopes," 105–106.

142. See Cardano's opinion in ibid., 106.

2. The Making of a Dynasty

1. ASMi, SPE Mantova, 390, Antonio da Camera to Francesco Sforza, Mantua, February 4, 1452. It seems that in early January, da Camera had requested a release from Ludovico Gonzaga so that he could be a free man and offer his services to other people, not only Ludovico. This request was granted. According to da Camera himself, Sigismondo Pandolfo Malatesta requested that he compile a work (presumably an astrological work, an *operetta*) for him, together with some calculations (*certo calchulo*), and for this reason da Camera travelled to Ferrara to consult some books and then to Mantua to do his calculations. Little is known of this astrologer. A *iudicium* by da Camera addressed to Ludovico Gonzaga regarding the apparition of a comet in 1456 is now preserved in BNF, MS Lat. 16021, fols. 10v–11v, and in ASMi, *Sforzesco,* Miscellanea 1569. A prognostication for the year 1464 written in Pisa for Piero de' Medici exists also in two copies: BNF, MS Lat. 7336, fol. 365r–v (incomplete), and BAR, MS 102, fols. 50r–52r. Simon de Phares seems to believe he was originally from Florence. See Simon de Phares, *Le recuil des plus célèbres astrologues,* 574. On Antonio da Camera, see also Thorndike, *HMES,* vol. 4, 438, n. 2.

2. ASMi, SPE Mantova, 390, Antonio da Camera to Francesco Sforza, Mantua, February 27, 1452. Originally quoted in Gabotto, *Nuove ricerche,* 9–10.

3. ASMi, SPE Mantova, 390, Antonio da Camera to Francesco Sforza, Mantua, March 15, 1452. Originally quoted in Gabotto, *Nuove ricerche,* 11–12. Antonio da Camera wrote to Cosimo de' Medici (the Elder) from Milan on March 17, 1458, referring to a *iudicium* for that year that he had sent to him earlier. He invites Cosimo to send a copy of it to "his people" in Rome. See, ASF, MAP 12, doc. 274. According to one source, Antonio da Camera was Jewish and was called by Pope Paul II to his court. See Anna Maria Corbo, *Paolo II Barbo: dalla mercatura al papato, 1464–1471* (Rome: Edilazio, 2004), 61.

4. The three signs form the fiery triplicity. See Eade, *The Forgotten Sky,* 65–66.

5. Gabotto, *Nuove ricerche,* 12.

6. Ibid.

7. ASMi, *Sforzesco,* Miscellanea 1569. Antonio da Camera to Francesco Sforza, Rome, March 1, 1453.

8. Ibid. Antonio da Camera to Francesco Sforza, Rome, June 14, 1457. This letter is transcribed in Gabotto, *Nuove ricerche,* 12–13.

9. Ibid. Antonio da Camera to Francesco Sforza, Castellaccio (Grottaferrata, near Rome?), October 29, 1457. See Gabotto, *Nuove ricerche,* 13.

10. ASMi, *Sforzesco,* Miscellanea 1569, *Copia iudicii Antoni de Camera.* In this *iudicium,* da Camera briefly cites authors as varied as Eusebius, Aristotle, Ptolemy, and Haly Avenrodoan (Ali Ibn Ridwan) on the nature and properties of comets. The rest of the *iudicium* includes a description of the comet's position in the sky, its physical properties, and its effects on the sublunary world, all framed within astrological discourse. Another copy in BNF, MS Lat. 16021, fols. 10v–11v. See Jane L. Jervis, *Cometary Theory in Fifteenth-Century Europe* (Dordrecht-Boston-Lancaster: Reidel Publishing, 1985), 51–54, for a fragment of Avogario's treatise on the 1456 comet.

11. ASMi, *Sforzesco,* Miscellanea 1569. The document is very damaged. Probably composed originally of a series of folios bound together, it must have come apart and now some of the folios are missing. The name of Giovanni Lorenzo de Fundis, however, is clearly legible on the recto of the last folio. On Giovanni de Fundis, lecturer in astronomy/astrology in Bologna from 1428–1473 and author of numerous astronomical and astrological works, see Thorndike, *HMES,* vol. 3, 423; vol. 4, 232–242. Thorndike calls him *Lauratius,* but at least in the Milanese document, we find the form *Laurentius.* On his works, see also Emmanuel Poulle, *La Bibliothèque scientifique d'un imprimeur humaniste au XVe siecle: Catalogue des manuscrits d'Arnaud de Bruxelles à la Bibliothèque Nationale de Paris* (Geneva: Droz, 1963), and Fabrizio Bònoli and Daniela Piliarvu, *I lettori di astronomia presso lo Studio di Bologna dal XII al XX secolo* (Bologna: CLUEB, 2001), 104–106.

12. Malatesta Malatesta detto dei Sonetti (1366–1429), son of Pandolfo II Malatesta, lord of Pesaro, was famous for his interest in the arts and his study of literature, medicine, philosophy and astrology. See Anna Falcioni, "Malatesta, Malatesta," *DBI* 68 (2007), 77–81. Copies of de' Conti's *De motu octavae spherae* are now in BLF, MS Ashburnham 134 (208–140), and BAV, MS Vat. lat. 3379, fols. 1r–4v.

13. On Nicolò de' Conti, see Thorndike, *HMES,* vol. 4, 250–255, and De Ferrari, "Conti, Nicolò," 461–462; on Regiomontanus's oration, see now James Steven Byrne, "A Humanist History of Mathematics? Regiomontanus's Padua Oration in Context," *Journal of the History of Ideas* 67 (2006): 41–61.

14. On these themes in the Renaissance, see now Craig Martin, *Renaissance Meteorology: Pomponazzi to Descartes* (Baltimore, MD: The Johns Hopkins University Press, 2011), which, however, does not cover our period.

15. Thorndike, *HMES,* vol. 4, 252–255.

16. Gabotto, *Nuove ricerche*, 15.

17. ASMi, *Sforzesco,* Miscellanea 1569. In referring to Francesco's astrologers, De' Conti sounded inquisitive ("de quali la Vostra Illustrissima Signoria ne diè essere copiosissima"). The letter, dated Monselice (*Montiscilicis*), Padua, January 25, 1455, is transcribed in Gabotto, *Nuove ricerche,* 15–16.

18. ASMi, *Sforzesco,* Miscellanea 1569.

19. Ibid.

20. On the issue of private versus public prognostications and the problems related to the circulation of this type of information, see my remarks in "The Political Uses of Astrology," 135–145.

21. Pier Candido Decembrio, *Vita di Filippo Maria Visconti* (Milan: Adelphi, 1983), 51. The bibliography on the Visconti dynasty and its rise to power, especially in Italian, is vast and cannot be summed up here. For general orientation, see Francesco Cognasso, *I Visconti* ([Milan]: Dall'Oglio Editore, 1966); idem, "L'unificazione della Lombardia sotto Milano," in *Storia di Milano,* vol. 5, *La signoria dei Visconti (1310–1392)* (Milan: Fondazione Treccani, 1955), 1–567; and idem, "Il Ducato Visconteo da Gian Galeazzo a Filippo Maria," and "Istituzioni comunali e signorili di Milano sotto i Visconti," in *Storia di Milano,* vol. 6, *Il Ducato visconteo e la Repubblica ambrosiana (1392–1450)* (Milan: Fondazione Treccani, 1955), 1–383 and 451–544. On Filippo Maria's father, see Daniel M. Bueno de Mesquita, *Giangaleazzo Visconti, Duke of Milan (1351–1402): A Study in the Political Career of an Italian Despot* (Cambridge: Cambridge University Press, 1941). On the institutional aspects of the Visconti rule, see now Francesco Somaini, "Processi costitutivi, dinamiche politiche e strutture istituzionali dello Stato visconteo-sforzesco," in *Storia d'Italia,* vol. 6, *Comuni e signorie nell'Italia settentrionale: La Lombardia* (Turin: UTET, 1998), 681–786, and 809–825 (bibliography), and Jane Black, *Absolutism in Renaissance Milan. Plenitudo of Power under the Visconti and the Sforza 1329–1535* (Oxford: Oxford University Press, 2009).

22. Decembrio, *Vita,* 120–123.

23. Ibid., 123. Pietro Lapini da Montalcino was for a time professor of pratical medicine and astrology at the *Studium* of Pavia; Stefano Fantucci, *physicus et astronomus,* was also professor at Pavia, and so was Antonio Bernareggi; Luigi Terzaghi was ducal physician from 1428 onward; nothing is known of this Lanfranco, but Elia Ebreo is presumably the former doctor of Popes Martin V and Eugene IV. See *Vita,* 212–213. On the university appointments of Lapini and Fantucci at the university (in practical medicine and astrology or solely in astrology), see Alfonso Corradi, *Memorie e documenti per la storia dell'Università di Pavia e degli uomini più illustri che v'insegnarono* (Pavia: Stabilimento Tipografico-Librario Successori Bizzoni, 1877–78), 107 and 110. Corradi does not indicate a Luigi Terzaghi, but only an Antonio Terzaghi, who taught astrology and medicine in Pavia in the years 1441–1450 (at 112). On Pietro Lapini's diplomatic missions and his teaching in Pavia, see Gigliola

Soldi Rondinini, "Ambasciatori e ambascerie al tempo di Filippo Maria Visconti," *Nuova Rivista Storica* 49 (1965): 313–344, esp. 333 (despite the fact that he is recorded as Pietro da Montalcino di Siena, Rondinini, however, does not identify him with the Pietro from Siena mentioned by Decembrio). On Terzaghi, see also Majocchi, *Codice diplomatico,* vol. 2, t. 2, 432, 470, 481, 485, 496, 519. On Elia Ebreo, Corradi, *Memorie e documenti,* 113.

24. Decembrio, *Vita,* 123.

25. On this device, see Chapter 1.

26. Decembrio, *Vita,* 124–127.

27. See Vittorio Zaccaria, "Sulle opere di Pier Candido Decembrio," *Rinascimento* 7 (1956): 13–74, esp. 18–20.

28. In this work Decembrio makes reference to "Marcatius avuus meus, vir phisice peritus." Pier Candido Decembrio, *De genitura hominis* (Rome: Stephan Plannck, around 1495), sig. a2r. I have consulted BL, I.A. 18718 and a copy of the same edition in the BNCF. On Decembrio's grandfather, see Paolo Viti, "Decembrio, Pier Candido," *DBI* 33 (1987), 488–498.

29. Viti, "Decembrio, Pier Candido," 492. See also Mario Borsa, "Pier Candido Decembrio e l'umanesimo in Lombardia," *ASL* 20 (1893): 5–75, 358–441; Zaccaria, "Sulle opere di Pier Candido Decembrio," 18–20. On Arcimboldi, see Nicola Raponi, "Arcimboldi, Nicolò," *DBI* 3 (1961), 779–781. On the humanistic *cenacolo* revolving around him, see James Hankins, *Plato in the Italian Renaissance,* 2 vols. (Leiden and New York: Brill, 1991), vol. 1, 148–153.

30. Three editions appeared in Rome between 1474–1475 (printed by the German printers, Georgius Sachsel and Bartholomaeus Golsch, Johannes Gensberg, and Bartholomaeus Guldinbeck, respectively); a further Roman edition appeared in 1485 from the press of Eucharius Silber; one more sometime before 1487 from the press of Stephan Plannck, who published it again around 1495, as did the Roman-based printer Johann Besicken. Another edition appeared in Vicenza around 1490, a later one in Augsburg in 1498, and a last one in Geneva in 1505, in a collectanea entitled *Arcana medicinae,* which included also *De conservatione sanitatis* by the physician Benedetto Reguardati da Nursia, Pietro d'Abano's *De venenis,* and other short medieval medical treatises. See Klebs, 327.1–9. For manuscripts containing only the *De genitura hominis,* see BNF, MS Nouv. acq. lat. 315, fols. 1r–11r (copy of an earlier printed version, c. XVI cent.), and BC, MS 232 (36 D 28), misc. XV cent., fols. 9r–17r.

31. Borsa, "Pier Candido Decembrio," 13–18 and 370–379; and Marcello Simonetta, *Rinascimento Segreto: Il mondo del segretario da Petrarca a Machiavelli* (Milan: FrancoAngeli, 2004), 142, n. 62, who discusses and dismisses the prevalent historiographical view of Decembrio as a republican. Similarly, following Malaguzzi Valeri, it has often been argued that Decembrio was *persona non grata* to the Sforza, but also this view should be dismissed as it is only representative of a very brief period of

Decembrio's life. If he seems to have been only mildly successful in attracting Francesco's direct patronage, this was possibly because of the acrimonious war of words with his fellow humanist, Francesco Filelfo, or simply because Francesco found Decembrio more useful where he was, away from Milan.

32. Simonetta, *Rinascimento Segreto*, 140–149, and esp. 146–149.

33. In 1454, for instance, he sends her a little box with some images of the "agnus dei" blessed by the Pope. Ibid., 144; and Borsa, *Pier Candido Decembrio*, 384.

34. This is confirmed, among other things, by his successful sponsorship of the humanist and Hellenist scholar Costantino Lascari, who Decembrio promoted against Demetrio Castreno from Constantinople, Filelfo's chosen candidate. Lascari was elected to the chair of Greek on July 24, 1463. See Malaguzzi Valeri, *La corte di Ludovico il Moro*, vol. 4, 108.

35. ASMi, *Autografi*, Uomini celebri, scienziati e letterati, 215, fasc. 17, P. C. Decembrio to Bianca Maria Visconti, Milan, July 26, 1460: "Mi vene a mente che essendo io a Roma intese dala sancta memoria de Nicolao Papa Quinto essere due picole historie nela Bibia cioè quella de Joseph e quella de Tobia le quale havevano tanta virtute e iociunditate in se che quando se legesseno ad una persona afflicta de infermitate o d'alchuna melanconia e le intendesse bene la liberariano indubitatamente; tanta era la virtù de quella lectione per le quale parole per potire giovare non solo a me ma ad altre persone digne mi nusse a tradurle tuti due de litra in vulgare tanto proprie e consimile quanto si possa dire." Decembrio is referring probably to the Book of Job and the Book of Tobit (or Tobias), the latter considered apocryphal by the Protestants.

36. Jole Agrimi and Chiara Crisciani, *Medicina del corpo, medicina dell'anima: note sul sapere medico fino all'inizio del secolo XIII* (Milan: Episteme Editrice, 1978), esp. 17–24; Joseph Ziegler, *Medicine and Religion c. 1300: The Case of Arnau de Vilanova* (Oxford: The Clarendon Press, 1998); Peter Biller and Joseph Ziegler, eds., *Religion and Medicine in the Middle Ages* (York: York Medieval Press, 2001).

37. See Pellegrin, *La Bibliothèque*, 338–339, 347, 447–448.

38. On the importance of generation in fourteenth- and fifteenth-century medical discourse, see Romana Martorelli Vico, *Medicina e filosofia. Per una storia dell'embriologia medievale nel XIII e XIV secolo* (Milan: Guerini e associati, 2002); Katharine Park, *Secrets of Women: Gender, Generation, and the Origins of Human Dissection* (New York: Zone Books, 2006); and Monica Azzolini, "Exploring Generation: A Context to Leonardo's Anatomies of the Female and Male Bodies," in Domenico Laurenza and Alessandro Nova, eds., *Leonardo da Vinci's Anatomical World: Language, Context and "Disegno"* (Venice: Marsilio, 2011), 79–97.

39. Decembrio, *De genitura*, chaps. xxxv–xxxviii: "Qualiter homo fortunatus ex corpore et anima sumit ab astris accidentia ad utrumque/untrinque"; "De influentia omnium planetarum in generatione fetus"; "De lune influentia et quare natus in septimo mense vivat, moriatur in octavo, et de statu lune quadripartito et quod eius

dispositio in oppositione sit deterior"; "De virtute planetarum generaliter et influentia eorum ad utrumque/untrinque." Interestingly enough, the only part to have marginal notes in the Florentine copy examined is the one about the monthly rulership of the planets on the fetus.

40. According to medieval and Renaissance psychology, the soul possessed three different faculties (or kinds of soul): the vegetative, the sensitive, and the intellective. See Katharine Park, "The Organic Soul," in Charles B. Schmitt, Quentin Skinner, and Eckhard Kessler, eds., *The Cambridge History of Renaissance Philosophy* (Cambridge: Cambridge University Press, 1988), 464–484, esp. 466.

41. The planetary scheme illustrated here was first described in Constantinus Africanus's *De humana natura* and was later popularized in a number of medieval texts, including ps.-Albertus Magnus *De secretis mulierum*. See Charles S. F. Burnett, "The Planets and the Development of the Embryo," in Gordon R. Dunstan, ed., *The Human Embryo: Aristotle and the Arabic and European Traditions* (Exeter: University of Exeter Press, 1990), 95–135. Textual evidence suggests that Decembrio may have lifted his material from a manuscript version of the very popular ps.-Albertus Magnus, *De secretis mulierum cum commento [Henrici de Saxonia]* ([Rome?] Venice: Simone Bevilaqua, July 8, 1499). I have consulted BL, IA. 23989. Cf. Decembrio, *De genitura*, sig. a6r–v and ps.-Albertus Magnus, *De secretis*, sig. b5v–c3v. For an English translation of these passages in Albertus Magnus, see Helen R. Lemay, *Women's Secrets: A Translation of Pseudo-Albertus Magnus'* De secretis mulierum *with Commentaries* (Albany, NY: State University of New York Press, 1992), 84–86.

42. Decembrio, *De genitura*, sig. a6v. Here Decembrio reports that this is the opinion of the doctors, as opposed to that of Aristotle, who believed that the heart was formed in the first month of the fetus's life. The passage, once again, is a loose paraphrase of ps.-Albertus Magnus. Cf. *De secretis*, sig. c1v. This debate reflects the discrepancies existing between the two traditions, the natural philosophical of Aristotle and the medical of Hippocrates and Galen. See Martorelli Vico, *Medicina e filosofia*.

43. Most medical sources from antiquity agreed that the eighth month was mostly fatal for premature babies. The small chance of survival for premature babies in their eighth month was already noted in the Hippocratic corpus and constituted common knowledge at the time. See Ann Ellis Hanson, "The Eight Months' Child and The Etiquette of Birth: *Obsit Omen!*," *Bulletin of the History of Medicine* 61 (1987): 589–602.

44. Thorndike, *HMES*, vol. 4, 238. An extended form of astrological embryology is also present in Cecco D'Ascoli's commentary to the *Sphaera*, suggesting that this type of information was routinely included in basic university teaching of astrology at Bologna and probably at other Italian universities. See Thorndike, *The* Sphere *of Sacrobosco*, 371–372, 384–385, 402–405.

45. Again, this section of Decembrio's work is an abridgement of the same passage in the *De secretis mulierum*. Cf. Decembrio, *De genitura*, sig. a6v–a7v and

ps.-Albertus Magnus, *De secretis,* sig. c7r–d2v. For a translation of the corresponding passages in ps.-Albertus Magnus, see, Lemay, *Women's Secrets,* 91–95. Presumably here ps.-Albertus and Decembrio mean the ruling planets (i.e., the planet that rules the sign of the zodiac on the ascendant at birth).

46. These principles resonate strongly with those of physiognomics. On the relationship between astrology and physiognomics, see Jole Agrimi, *Ingeniosa scientia nature. Studi sulla fisiognomica medievale* (Florence: SISMEL Edizioni del Galluzzo, 2002).

47. See North, "Celestial Influence—The Major Premise of Astrology"; and Gad Freudenthal, "The Medieval Astrologization of Aristotle's Biology: Averroes on the Role of the Celestial Bodies in the Generation of Animate Beings," *Arabic Sciences and Philosophy* 12 (2002): 111–137 [reprinted in idem, *Science in the Medieval Hebrew and Arabic Traditions* (Aldershot: Ashgate, 2005), 111–137]. See also Chapter 1.

48. Pico della Mirandola, *Disputationes adversus astrologiam divinatricem,* vol. 1, Book III, chap. 1. For the discussion that occurred among European intellectuals after Pico's publication, see Zambelli, "Fine del mondo o inizio della propaganda?"; Broecke, *The Limits of Influence,* esp. chaps. 1 and 3.

49. BNF, MS Italien 1586, fol. 15r–v, Pier Candido Decembrio to Cicco Simonetta, Naples, January 14, 1457: "Magnifice vir e maior honorandissime, quando mi partite da Milano per venire in queste parte, non credeva trovare tante impedimenti quanti ho ritrovati; non voglio replicare la difficultate dela via, che may non si vide la più dolorosa de fangi, de pluvie continue et inundatione de fluvii, e quali sono usciti fuor d'ogni mensura. Solo mi basta a scrivere la terribile novitate del movimento dela terra, dala quale tuta questa regione è desolata, che niuno vivente may non vide. Io per gratia de Maestro Antonio da Bernadigio, el quale a questo puncto ho experimentato bono astrologo per me, mi partite un pocho più tarde da Milano, che de certo si più presto mi partiva, m'era necessario vedere ultimum terribilium, ma era anchora a Roma, quando el terremoto fu a Napoli, e pocho o niente l'ò sentito. Quelli che si ritrovareno a Napoli videno de quanto periculo fu la chosa, ch'io tremo ad udirla. Qui dopoi a jovedi a xxx del mese passato fu alquanto de terrimoto, ma molto dissimile dal primo, perché assai dela materia era resoluta, pur non fu che non mi facesse correre molto bene con gli altre brigate. Ma per venire ala conclusione pensate che terremoto fu quello che dura per spacio de duy paternostri e che se durava el tertio non si ritroveria vestigio de Napoli." Transcription mine. The letter is now published in *Dispacci sforzeschi da Napoli,* vol. 1 (Salerno: Carlone, 1997): (1444–2 Luglio 1458), ed. Francesco Senatore, 477–479. I wish to thank Nerida Newbigin for drawing my attention to this document. On the contents of a good part of the Sforza documents now at the BNF, see Giuseppe Mazzatinti, "Inventario delle carte dell'Archivio Sforzesco contenute nei codd. Ital. 1583–1593 della Biblioteca Nazionale di Parigi," *ASL* 10 (1883): 222–326.

50. BNF, MS Italien 1586, fol. 15r–v.

51. Ibid.: "S'extima essere morte dele persone xxx^m che dicevo xl^m chi più. Io non ho a iudicare questo. Solo intendo per discritione che 'l danno è stato grandissimo e sì che a nuy tuti si apertene a vivere bene, e stare apparechiati come dice l'evangelio. Maxime ché da docti astronomi s'aspecta in breve una coniunctione de tre pianeti, simile a quella de la gran peste che fu molti anni fa, che incominciando da levante in ponente, spacia dele quatre parte del mondo le tre. Questo però remetto io a Maestro Antonio da Bernadigio e a Mestre Lanfrancho da Bardone più docti de mi."

52. We know that Bardone was still caring for Francesco in 1462, as his name is mentioned in a letter by the Mantuan physician-astrologer Bartolomeo Manfredi to his lord, Ludovico Gonzaga, but very little else about him has emerged so far from my researches. See doc. 14 in Signorini, *Fortuna dell'astrologia*, 369.

53. Majocchi, *Codice diplomatico*, vol. 2, t. 1, 399, and vol. 2, t. 2, 439. For biographical information on Antonio Bernareggi, I have relied substantially on the very informative article by Monica Pedralli, "Il medico ducale Milanese Antonio Bernareggi e i suoi libri," *Aevum* 70 (1996): 307–350.

54. ASMi, SPS 1457, Francesco Sforza to Bianca Maria, Castello di Gaido (possibly near Sondrio), August 1, 1453.

55. On the issue of Francesco's legitimacy to rule over Milan, see Franco Catalano, "La Nuova Signoria: Francesco Sforza," in *Storia di Milano*, vol. 7, *L'età sforzesca dal 1450 al 1500* (Milan: Fondazione Treccani, 1956), 1–224, for an outline of the basic events; and Black, *Absolutism in Renaissance Milan*, for the legal aspects.

56. This included the choice of a wetnurse, for which Agnese wrote that she had collected the opinion of three physicians, Magister Antonio Bernareggi, Magister Dionisio (Reguardati da Norcia), and Magister Guido (da Parati). See ASMi, SPS 1457, Agnese Visconti to Bianca Maria, August 10 [s.a., but most likely 1457]. On November 18, 1457 Bianca was informed by Franceschino Caimi of Elisabetta's indisposition; by November 22 she was well again. For the care of other Sforza children, see also the numerous other letters by Franceschino Caimi (some for the same day at different hours of the day) dated to August and September 1457 in SPS 1457, where it is mentioned that Ludovico had a temperature and the physicians treated him with *cassia* (a purgative). Also, on September 5, Filippo, who had vomited and fallen off the bed, was treated again with *cassia*. Magister Dionisio is the son of the more famous Benedetto Reguardati da Norcia. See Fausto M. de' Reguardati, *Benedetto de' Reguardati da Norcia «medicus tota italia celeberrimus»* (Trieste: Lint, 1977).

57. On this occasion, Agnese had her blood let and was prescribed some medications. ASMi, SPS 1457, Antonio Bernareggi and Dionisio Reguardati, Pavia, September 15 and 29, 1453. See also the letter by a certain "servitrix Donina de Tronamali (?)" to Bianca Maria, Pavia, September 28, 1453; and documents n. 98, 99, 103 (ASMi, Sforzesco, Carteggio Interno 753) transcribed in *Documenti per la storia dell'Università di Pavia*, vol. 1, 110–111, 114–115.

58. ASMi, SPS 1457, Andreotto del Maino to Bianca Maria, Milan, December 10, 1456, and s.d.; Andreotto del Maino, Antonio Bernareggi, and Dionisio Reguardati, Milan, December 17, 1456.

59. ASMi, SPS 1457, Antonio Bernareggi and Cristoforo Soncino to Bianca Maria, Milan, May 25, 1457.

60. ASMi, SPS 1457, Antonio Bernareggi and Cristoforo Soncino to Bianca Maria, Milan, December 5 and 10, 1458; on Elisabetta's indisposition, see also Franceschino Caimi to Bianca Maria, Milan, December 1, 2, 6, 10, and 12, 1458; Antonio Carboni to Bianca Maria, Milan, December 10 and 12, 1458. In the letters dated December 10 it is recounted how Elisabetta suffered "quello accidente che altre volte li vene" and what the physicians indicate as "epilensia." More letters from Franceschino and the doctors may have existed, but they now seem to be lost. Two wetnurses, Giacomina and Margherita, seem to have breastfed Elisabetta, who was born in 1456.

61. ASMi, SPS 1457, Giovanni Andrea Beltrami to Bianca Maria, Milan, May 23, 1459: "Hogi lo illustrissimo signore nostro ha mandato Magistro Gasparo a visitare li prefati fioli de vostra signoria e dice che lo prelibato signore gli ha comandato venga ogni dì a visitarli, e anche domino Magistro Antonio non cessa duoe volte el dì visitarli." The fact that Antonio Bernareggi visited the children twice a day is also reported separately by the courtiers Giovanni Andrea Toscano and Beltramino Pusterla in two letters dated Milan, May 22 and 23, 1459. From other letters it seems clear that both Francesco's mother, Lucia degli Attendoli, and Bianca's mother, Agnese del Maino, were attending to the children's care and living on the same premises. Both women wrote regularly to Bianca with updates on their health and that of the Sforza children, and so did other courtiers and physicians.

62. These documents are transcribed in full in Pedralli, "Il medico ducale," 344–346.

63. Pedralli, "Il medico ducale," 344.

64. ASMi, SPS 1457, Francesco Sforza to Bianca Maria, Milan, July 15, 1460. Pedralli cites this letter secondhand, but dates it, incorrectly, to August 25. See Pedralli, "Il medico ducale," 310. On the almost contractual relationship between doctor and patient in Renaissance Milan, see Nicoud, "Les médecins à la cour de Francesco Sforza." For a broader understanding of this relationship in the early modern period, see Gianna Pomata, *Contracting a Cure: Patients, Healers, and the Law in Early Modern Bologna* (Baltimore-London: The Johns Hopkins University Press, 1994). On Gaspare Venturelli da Pesaro, the sick physician who had served Francesco for over thirty years, see ASMi, *Autografi*, Medici 219 Gaspare, Venturelli da Pesaro, as well as *COM*, vols. 1 (1999), 2 (2000), 3 (2000), and 5 (2003), sub indice.

65. ASMi, SPS 1457, Francesco Sforza to Bianca Maria, Milan, July 16, 1460.

66. From the letters it is not clear where Bianca Maria was residing, possibly in Cremona or Pavia. See ASMi, SPS 1457, Francesco Sforza to Bianca Maria, Milan, July 22, 1460, hora 23.

67. ASMi, SPS 1457, Francesco Sforza to Antonio Bernareggi and Cristoforo Soncino, Milan, July 27, 1460. This letter is transcribed and translated into French in Nicoud, "Les médecins à la cour de Francesco Sforza," 214–215 (discussion at 209–210).

68. Ibid.

69. Possibly Tebaldo Maggi da Sale, who is listed as a promoter for the degree of arts and medicine of numerous Pavian students together with the Lombard physicians Giovanni Ghiringhelli, Biagio Astari, and Lazzaro Datari, all of whom at some time or other held chairs in medicine at the University of Pavia. *Lauree pavesi nella seconda metà del '400*, 3 vols. (Bologna and Milan: Cisalpino, 1995–2008), vol. 1 (1995): (1450–1475) ed. Agostino Sottili, 10–29. He was mentioned caring for Galeazzo Maria Sforza in a letter dated December 4, 1461, and for Francesco himself in a letter dated January 1, 1462. See *COM*, vol. 3 (2000), 359–360; and vol. 4 (2002), 53–54.

70. See Pedralli, "Il medico ducale," 310–312 and documents at 346–347. Other examples are cited at 313 (document at 347); and documents in *COM*, vol. 1 (1999), 302–307, 317, 351; vol. 3 (2000), 195–206; vol. 4 (2002), 504–505.

71. ASMi, SPS 1457, Bianca Maria to Francesco, Castroleone (now Casteleone, near Cremona), October 18, 1458.

72. ASMi, SPS 1457, Francesco to Bianca Maria, Milan, November 3, 1458.

73. Ibid.

74. Another document, dated October 26, 1463 (ASMi, SPS 1457), shows Bianca Maria's persistence in her desire to patronize university lecturers. This time we find her insisting with her husband for the university appointment of the two jurists Ruggero del Conte and Signorollo Amadeo. On the appointment of Ruggero del Conte when he was still "legum scolar," see *Documenti per la storia dell'Università di Pavia*, vol. 2, 314 (doc. 453 dated August 26, 1460). I have not been able to find any information on Signorollo Amadeo.

75. ASMi, *Sforzesco*, Miscellanea 1569. This is probably another copy of the *tacuino* that Bernareggi had sent to Vincenzo della Scalona for Ludovico Gonzaga in January 1461. Bernareggi had sent another *tacuino* to Scalona the previous year. See *COM*, vol. 2 (2000), 58–59, and vol. 3 (2000), 95–96.

76. See Maiocchi, *Codice diplomatico*, sub indice; and Corradi, *Memorie e documenti*, 106. The *rotuli* of the university (much like proofs of payment of university salaries) are not to be taken as a very reliable source of information, however. We know from this and other cases that university professors were often paid in arrears or not paid at all, accumulating credits with the Ducal Chamber. At his death Bernareggi was in credit of two thousand florins "for his salaries." See Pedralli, "Il medico ducale," 315–316. It is possible that records were not kept with the same level of detail under the Sforza.

77. ASMi, SPS 1457, Franceschino Caimi to Francesco, Abbiategrasso, September 30 and October 1, 1464; Antonio Bernareggi to Francesco, Abbiategrasso, October 2, 1464, hora 20a; October 10, 1464, hora 20a. On the "Gonzaga affair," see below.

78. Pedralli, "Il medico ducale."

79. See Chapter 1.

80. In particular, some of these intellectuals rejected Albumasar's much-favored theory of the great conjunctions, which became so popular in the second half of the *Trecento* and the whole of the *Quattrocento* and was at the core of popular astrological prognostications that circulated in large numbers with the advent of printing. On these themes, see especially Zambelli, "Fine del mondo o inizio della propaganda?"; Broecke, *The Limits of Influence;* Grafton, *Cardano's Cosmos.*

81. This is particularly evident in medieval *summae,* such as that of Giorgio Anselmi or that of Guido Bonatti owned by Bernareggi, and is confirmed once more by my analysis of BLO, MSS Canon. Misc. 23 and 24 of the Paduan astrologer Antonio Gazio. See Azzolini, "Refining the Astrologer's Art." *Journal for the History of Astronomy* 42 (2011): 1–25.

82. See Nicoud, "L'expérience de la maladie"; eadem, "Les médicins à la cour"; and particularly Chiara Crisciani's remarks in her "Fatti, teorie, 'narratio' e i malati a corte. Note su empirismo in medicina nel tardo-medioevo," *Quaderni Storici* 3 (2001): 695–718. We can assume that this is equally valid for principles of astrological medicine, which may have sounded even more obscure to the patient and his entourage.

83. See Chapter 4.

84. Little remains of the Visconti archives, which were destroyed almost entirely at the death of Filippo and the instauration of the Ambrosian Republic. For this reason there are serious limits as to what can be known of Filippo's personal views on the matter.

85. The wedding between Isabella d'Este and Francesco Gonzaga was similarly chosen astrologically. The propitious day was established casting an interrogation (*figura*). See Alessandro Luzio, "Isabella d'Este e Francesco Gonzaga promessi sposi," *ASL* 35 (1908), 34–69, esp. 61–62.

86. On the forgery and the relationship between Francesco and the Empire, see Fabio Cusin, "Le aspirazioni straniere sul ducato di Milano e l'investitura imperiale (1450–1454)," *ASL* (1936): 277–369; idem, "L'impero e la successione degli Sforza ai Visconti," *ASL* (1936): 3–116; and idem, "Le relazioni tra l'impero ed il ducato di Milano dalla pace di Lodi alla morte di Francesco Sforza," *ASL* (1938): 3–110.

87. For Francesco's policy of marriage alliances at the time, see also Lubkin, *A Renaissance Court,* 22–24. On the links forged with the realm of Naples see also Chapters 3, 4, and 5.

88. Venice and Milan fought for control over the lands of eastern Lombardy for decades. See Alessandro Colombo, "Abbozzo dell'alleanza tra lo Sforza ed il Gonzaga in previsione di una guerra con Venezia (ottobre–novembre 1450)," *Nuovo Archivio*

Veneto 65 (1907): 143–151. On the fraught relationship between Venice and Milan more generally, see Michael E. Mallett, "Venezia e la politica italiana: 1454–1530," in Alberto Tenenti and Ugo Tucci, eds., *Storia di Venezia,* 12 vols. (Rome: Enciclopedia Italiana Treccani, 1991–2002), vol. 4: *Il Rinascimento: politica e cultura,* ed. Alberto Tenenti and Ugo Tucci (1996), 245–310. For Ludovico Gonzaga's reasons and his relationship with Milan, see Maria-Jose Rodriguez Salgado, "Terracotta and Iron. Mantuan Politics (ca. 1450–ca. 1550)," in Cesare Mozzarelli, Robert Oresko, Leandro Ventura, eds., *La corte di Mantova nell'età di Andrea Mantegna: 1450–1550* (Rome: Bulzoni, 1997): 15–61.

89. See Alessandro Colombo, "Nuovo contributo alla storia del contratto di matrimonio fra Galeazzo Maria Sforza e Susanna Gonzaga," *ASL* 12 (1909): 204–211, esp. 205.

90. The events surrounding these marriage arrangements are recounted in Achille Dina, "Qualche notizia su Dorotea Gonzaga," *ASL* 4 (1887): 562–567, and Luca Beltrami, "L'annullamento del contratto di matrimonio fra Galeazzo Maria Sforza e Dorotea Gonzaga (1463)," *ASL* 6 (1889): 126–132. On Dorotea, see also Isabella Lazzarini, "Gonzaga, Dorotea," *DBI* 57 (2001), 707–708.

91. Beltrami, "L'annullamento," 127.

92. Dina, "Qualche notizia," 564; Maria Nadia Covini, "Introduzione," in *COM,* vol. 6 (2001), 7–31, esp. 9.

93. Barbara seems to have been in regular correspondence with Bianca Maria, on occasion to recommend people for various positions within the territories of the Duchy of Milan, as in the case of Sister Caterina di Bravi, whom Barbara recommended for the job of abbess at the convent of Saint Paul in Parma. See ASMi, SPE Mantova, 394, Barbara to Bianca Maria, Mantua, April 12, 1460. Around the same time, moreover, Barbara was writing to Bianca Maria to inform her of her son Francesco's search for accommodation in Pavia to study at the *Studium* (see letters dated April 13 and 19 and May 5, 1460). For this reason, on May 16, 1460, she wrote again to Bianca Maria thanking her warmly for having sent her doctor Magister Guido Parati (*Guido phisico*) to treat her son, Francesco, who was indisposed; he suffered an *alteratione,* had a temperature, and was coughing. On May 18 she wrote again, moreover, to ask Bianca Maria to intercede with her husband and give Francesco a more suitable room in a building owned by the Duke of Milan in Pavia. Very few letters survive, however, for the month of December, none from Barbara to Bianca Maria or Agnese.

94. "Illustris ac excelsa domina, domina soror honoranda, el non è fin al presente ch'io non habia creduto et creda vostra signoria stare alquanto affannata per lo illustre conte Galeaz. Pur, al presente, non ha quella più a dubitare, però che è prexo tal hordene et facto tanti repari che è quasi imposibile potere accadere veruno errore né mancamento. Però debe vostra preffacta signoria vivere alegramente et de bona

voglia, et tanto più quanto ch'io l'adviso che, per più mia secureza, *ho facto vedere la nativitate sua da molti astrologhi et in specialità da maestro Antonio da Bernaregio,* quale me dice ch'io non debia stare de mala voglia per cosa veruna, però che preffacto conte non è per alcuno pianeta inclinato ad alchuno vitio, anzi più presto constrecto ad vivere bene et virtuoxamente, et sopra l' tuto dice che sarà clemente, piatoso, et magnanimo, senza tema de veruno suo vicino." Emphasis mine. The document is transcribed in Signorini, *Fortuna dell'astrologia,* 70.

95. A wealth of documents attesting Manfredi's position of court astrologer are now available in ibid.

96. The letter is dated Bertana (Mantua), November 29, 1463. The document is transcribed in full in ibid., 370.

97. The letters are dated Bertana (Mantova), December 9, 1463, and December 12, 1463. In the first letter Manfredi specified that Dorotea was born Sunday, December 7, 1449, around the 6th hour and 33 minutes *noctis horologii.* In the second letter Manfredi specified "mando cum questa aligata la natività afigurata de madona Dorothea." Ibid., 371.

98. The letter is dated December 12, 1463. The document is transcribed in full in ibid., 371.

99. The issue of physical disability was treated extensively in the astrological literature. For some remarks on this topic see my "Refining the Astrologer's Art."

100. The letter was dated November 16, 1463, and is discussed in Beltrami, "L'annullamento," 128–129.

101. Dina, "Qualche notizia," 565–567; Beltrami, "L'annullamento," esp. 129–132; and Covini, "Introduzione," 10–11.

102. Giacomo da Palazzo to Ludovico Gonzaga, Milan, January 26, 1464, in *COM,* vol. 6 (2001), 124–126.

103. Vincenzo della Scalona to Barbara, Milan, August 9, 1464, in *COM,* vol. 6 (2001), 436–447. On the epistolary exchange between Bianca Maria and Barbara, see Giuliana Fantoni, "Un carteggio femminile del sec. XV. Bianca Maria Visconti e Barbara di Hohenzollern-Brandeburgo," *Libri e documenti: Archivio storico civico e Biblioteca Trivulziana* 7 (1981): 6–29.

104. Covini, "Introduzione," 13–14.

105. These events are fully documented in *COM,* vol. 6 (2001). Many of the Sforza documents about this *cause célèbre* are also in BNF, MS Italien 1589. See Mazzatinti, "Inventario," 281–282.

106. Dina, "Qualche notizia," 567; Lazzarini, "Gonzaga, Dorotea."

107. These documents are, respectively, nn. 12 and 27 in Signorini, *Fortuna dell'astrologia,* 368–368 and 372–373. Document 11 (at p. 368) includes, among other things, a detailed interpretation of Alfonso of Aragon's nativity.

108. By 1462 the rumors were so insistent that the topic was discussed in all the inns and taverns of Lombardy. To make things worse between the two houses, mem-

bers of the Milanese court blamed the Gonzaga ambassador in Milan, Vincenzo della Scalona, for spreading the rumors. In Mantua, instead, people were blaming Venice. See Isabella Lazzarini, "L'informazione politico-diplomatica nell'età della pace di Lodi: raccolta, selezione, trasmissione. Spunti di ricerca dal carteggio Milano-Mantova nella prima età sforzesca (1450–1466)," *Nuova Rivista Storica* 83 (1999): 247–280, esp. 273–280.

109. Bartolomeo Manfredi to Ludovico Gonzaga, Mantua, November 14, 1461: "[. . .] Mars, dominus eiusdem ascendentis, fuit in Leone tempore eius in domo religionis, quod signum dicitur ascendens Romanorum, ob id signat Romanis impedimentum." As transcribed in Signorini, *Fortuna dell'astrologia*, 368; the document was originally published in Ferdinando Gabotto, *Bartolomeo Manfredi e l'astrologia alla corte di Mantova: ricerche e documenti* (Torino: La Letteratura, 1891), 7.

110. Signorini, *Fortuna dell'astrologia*, 368–369: "Praterea reperi Saturnum in Libra dominum anni in quo natus est ipsumquemet esse principantem et signatorem [possibly a misreading of "significatorem" on Gabotto's part] prefate maiestatis, qui tempore eiusdem nativitatis in mundo et figura obtinuit principatum, unde quod Saturnus cum Libra est significator Christianorum et quia ipse natus est rex, sequitur quod ipse inquiret imperium Christianorum, non causa pietatis et iusticie, sed pocius causa dominandi, eo quod Martis natura est impia et iniusta." Gabotto, *Bartolomeo Manfredi*, 7.

111. Ibid.

112. Ibid.

113. Nancy Bisaha calls Pius II "the greatest crusading pope of the [fifteenth] century." The gathering of Christian princes called by Pius II in Modena, near Mantua, in 1459, had been a failure, and this fact was most certainly known to Manfredi. Nancy Bisaha, *Creating East and West: Renaissance Humanists and the Ottoman Turks* (Philadelphia: University of Pennsylvania Press, 2004), 140.

114. Document transcribed in Signorini, *Fortuna dell'astrologia*, 372.

115. Dina, "Qualche notizia," 566. On the failure of these attempts, see also Lazzarini, "L'informazione politico-diplomatica," 273–279.

116. Signorini, *Fortuna dell'astrologia*, 373.

117. This is BT, MS Triv. 1329, on which more will be said in the next chapter.

3. Astrology Is Destiny

1. Bernardino Corio, *Storia di Milano*, 2 vols. (Turin: UTET, 1978), vol. 2, 1408.

2. See Cecilia M. Ady, *Milan under the Sforza* (London: Methuen & Co., 1907) chap. 5, esp. 94–95; Lubkin, *A Renaissance Court*, 24–26; and more recently Monica Ferrari, *"Per non mancare in tuto del debito mio." L'educazione dei bambini Sforza nel Quattrocento* (Milan: FrancoAngeli, 2000). Much of chap. 2 and the whole of chap. 3 are dedicated to the education received by Galeazzo Maria Sforza.

3. See, especially, Ferrari, *"Per non mancare in tuto al debito mio,"* 32–49.

4. Lubkin, *A Renaissance Court,* 200.

5. Ibid., 196.

6. Galeazzo's passionate love for Lucia Marliani is skillfully recounted in Ibid., 196–202.

7. Helen Ettlinger, "Visibilis et Invisibilis: The Mistress in Italian Renaissance Court Society," *RQ* 47.4 (1994), 770–792. On the specific case of Galeazzo and Lucia, see now Franca Leverotti, "Lucia Marliani e la sua famiglia: il potere di una donna amata," in Letizia Arcangeli and Susanna Peyronel, eds., *Donne di potere nel Rinascimento* (Rome: Viella, 2008), 281–311; and Timothy McCall, "Traffic in Mistresses: Sexualized Bodies and Systems of Exchange in the Early Modern Court," in Allison Levy, ed., *Sex Acts in Early Modern Italy: Practice, Performance, Perversion, Punishment* (Farnham: Ashgate, 2010), 125–136. McCall rightly emphasized that mistresses served as important instruments in cementing the relationship between the ruler and the local patriciate, but that the visibility of favored mistresses could become problematic.

8. Evelyn Welch, "Between Milan and Naples: Ippolita Maria Sforza, Duchess of Calabria," in Abulafia, *The French Descent,* 123–136, esp. 128–129, on one particular instance when her jealousy caused a diplomatic impasse between the two courts.

9. Corio, *Storia di Milano,* vol. 2, 1408–1409; Lubkin, *A Renaissance Court,* 201–202, and Simonetta, *Rinascimento segreto,* 119, n. 33 (for a documented episode with a *scudiere*).

10. For an excellent summary of the thriving astrological culture of Renaissance Ferrara, see Cesare Vasoli, "L'astrologia a Ferrara tra la metà del Quattrocento e la metà del Cinquecento," in *Il Rinascimento nelle corti padane. Società e cultura* (Bari: De Donato, 1977), 469–567. On the astrological frescoes of the Palazzo Schifanoia, see the classic study of Aby Warburg, "Italian Art and International Astrology in the Palazzo Schifanoia, Ferrara," in *Aby Warburg. The Renewal of Pagan Antiquity,* 563–591 [originally published as "Italienische Kunst und internationale Astrologie im Palazzo Schifanoja zu Ferrara," in *L'Italia e l'arte straniera. Atti del X Congresso Internazionale di Storia dell'Arte (1912)* (Rome: Maglione & Strini, 1922), 179–193]. See also the more recent studies of Marco Bertozzi, *La tirannia degli astri: Aby Warburg e l'astrologia di Palazzo Schifanoia* (Bologna: Cappelli, 1985) [republ. Livorno: Sillabe, 1999]; idem, *Lo Zodiaco del Principe. I decani di Schifanoia di Maurizio Bonora* (Ferrara: Tosi, 1992).

11. Lubkin, *A Renaissance Court.*

12. Medieval and Renaissance critiques of astrology, however, often revolved around the issue of free will. For the Renaissance period, see John D. North's nuanced discussion in his "Types of Inconsistency in the Astrology of Ficino and

Others," in Alasdair A. MacDonald, Zweder R. W. M. von Martels, and Jan R. Veenstra, eds., *Christian Humanism: Essays in Honour of Arjo Vanderjagt* (Leiden and Boston: Brill, 2009), 281–302. The issue was already present in late antiquity. See Tim Hegedus, *Early Christianity and Ancient Astrology* (New York: Peter Lang, 2007).

13. BAM, MS S 54 sup., fol. iv: "1444 die 14 Januarii hore noctis sequentis viiii." There is no direct indication that this is the astronomical data needed to construct Galeazzo Maria Sforza's nativity, but the planetary positions, with the exception of Mars (here erroneously given as 12° Leo and retrograde), are in line with those calculated by the astrologer Raffaele Vimercati (on whom more later). The manuscript, therefore, is likely to have been in possession of somebody very close to the Sforza court. On this manuscript, see Jole Agrimi, *Tecnica e scienza nella cultura medievale. Inventario dei manoscritti relativi alla scienza e alla tecnica medievale (secc . XI–XV). Biblioteche di Lombardia* (Florence: La Nuova Italia, 1976), 131–132, and Astrik L. Gabriel, *A Summary Catalogue of Microfilms of One Thousand Scientific Manuscripts in the Ambrosiana Library, Milan* (Notre Dame: Mediaeval Institute, University of Notre Dame, 1968), 350.

14. BT, MS Triv. 1329, fol. 63r: "Explicit liber iudiciorum in nativitate Comitis GaleazMarie Vicecomitis Lugurum futuri ducis dignanter ellecti, quem Raphael de Vicomercato composuit. Finis 1461 die martis secundo mensis Iunii ora octava precise."

15. Cf. the image of Giovanni Bianchini presenting his *Tabulae Caelestium Motuum Novae* to Emperor Frederick III (1452) on the dust jacket of Westman's *The Copernican Question.*

16. The only comparable Italian example that has come to my notice is the sixteenth-century horoscope of Cosimo I de' Medici by the Tuscan astrologer Giuliano Ristori, which is transcribed in full in Raffaella Castagnola, "Un oroscopo per Cosimo I," *Rinascimento* 29 (1989): 125–189 (brief introduction at pp. 125–132, followed by the transcription of the text at pp. 133–189). I wish to thank H. Darrel Rutkin for pointing me to Castagnola's article.

17. Lengthy commentaries on natal charts are relatively rare for the fifteenth century. Horoscopes for individual nativities seem to have been particularly rare in medieval England, but much less so in medieval France. Nonetheless very few include extensive interpretations of the kind given in MS Triv. 1329, which will be examined here. On medieval England, see North, *Horoscopes and History,* 139–142, and Carey, *Courting Disaster;* on France, see Boudet, "Les astrologues et le pouvoir sous le règne de Louis XI" idem, "Un jugement astrologique en français sur l'année 1415," in Jacques Paviot and Jacques Verger, eds., *Guerre, pouvoir et noblesse au Moyen Age. Mélanges en l'honneur de Philippe Contamine* (Paris: Presses de l'université de Paris-Sorbonne, 2000), 111–120; Boudet and Charmasson, "Une consultation

astrologique princière and Boudet and Poulle, "Les jugements astrologiques sur la naissance de Charles VII."

18. Archival sources about him are relatively scarce. He was most certainly *not* the Raffaele Vimercati who was the younger brother of Gaspare Vimercati, a Milanese nobleman who had been instrumental in helping Francesco Sforza gain power in Milan in 1450. A Raffaele Vimercati was elected secretary of the Privy Council by Francesco Sforza on March 1, 1450, but died on September 5 of the following year. See Caterina Santoro, *Gli Uffici del dominio Sforzesco (1450–1500)* (Milan: Fondazione Treccani, 1947), 31. It is possible, however, that he was a member of the same branch of the family and related to Gaspare. This seems to be supported by Raffaele's letter of February 3, 1477, where he recalls the old friendship of his ancestors with the Sforza ("inveterata amicitia de nostri antecesori"). See ASMi, *Autografi,* Medici 219, Raffaele da Vicomercato.

19. See Francesca M. Vaglienti, "Galeazzo Maria Sforza," *DBI* 51 (1998), 398–409.

20. Applying John North's program (*Horoscopes and History,* 198–218) to Vimercati's values, one obtains 2.08 as the hour angle of the sun in minutes and seconds (-31.6 in degrees). The same program also indicates that Vimercati used some astrological tables for 45° latitude. I am very grateful to the late Professor John D. North for helping me verify the astronomical data of Vimercati's chart and for helping me establish that the cusp values of the chart correspond to those of the standard method.

21. The Alphonsine tables seem to have been commonly used in this period. So far I have found no *direct* reference to the active use of an astrolabe when drawing charts. The use of the instrument, however, was taught in the Pavia university curriculum, as I argued in Chapter 1. My thanks go to Professor Jean-Patrice Boudet for kindly checking my data against the Alphonsine values both using Poulle and Gingerich's values [in *Les positiones des planètes au moyen âge: application du calcul électronique aux tables alphonsines* (Paris: Klincksieck, 1968)] and Julio Samso's program Alfin (1990), and to Dorian Greenbaum for supplying the modern data (with Solar Fire).

22. BT, MS Triv. 1329, fol. 21r. This method is rather complex. I summarize here a rough reconstruction of Vimercati's calculations. In *Tetrabiblos* III, 1.2, Ptolemy stressed that "difficulty often arises with regard to the first and most important fact, that is, the fraction of the hour of the birth," and goes on to explain how to "rectify" the chart by taking into consideration the *syzygy*. This is the degree of the new or full moon most recently preceding the birth (astronomically this corresponds, of course, to the conjunction or opposition of the luminaries). Having ascertained the degree at which the conjunction happened (or the opposition), and having chosen the degree of the luminary above the Earth, a professional astrologer should determine what planet ruled it at the time of birth according to one or more of the following criteria:

triplicity, house, exaltation, term, face, or aspect, the so-called planetary dignities. Whatever degree this planet occupies in the sign through which it is passing is also the degree of the ascendant within its sign. But if the degree of the ruling planet is further from the ascendant than from the midheaven (MC), the astrologer should use this degree as that of the midheaven and use that as a starting point in computing the other angles (North, *Horoscopes and History,* 51–52). The latter seems to be the case in our example. The full moon most recently preceding Galeazzo's birth occurred on January 5 at about 23° Cancer. A calculation of the planet with most dignity establishes the Moon as the ruler of the *syzygy* at the time of birth, and as the Moon is at 5°53 Sagittarius (circa 245°), it is indeed further from the ascendant than from the midheaven. For this reason Vimercati placed the midheaven at 5° 53 Virgo and calculated the other angles accordingly, thus placing the new ascendant at 21° Scorpio. For another illustrative study of the importance of rectifying the geniture, see now Hayton, "Expertise ex Stellis," 33–38.

23. The treatise is divided into three parts, each divided into chapters. The first part contains both the calculations for the length of Galeazzo's life and the prognostication regarding how he would die. This first part deals primarily with the body and its infirmities. The second part is dedicated to the soul, and here Vimercati treats also faith and religion. The third part seems to loosely follow the house divisions, treating, in order, father, mother, brothers, wife, sons, servants, fools, friends, secret enemies, overt enemies, minor animals, pilgrimages, riches, reign, and prosperity. See BT, MS Triv. 1329, fol. 20r–v.

24. As Bouché-Leclercq argued in his classic *L'astrologie grecque,* "Le calcul de la durée de la vie, avec indication du genre de mort préfixé par les astres, est le grand oeuvre de l'astrologie, l'opération jugée la plus difficile par ses adeptes, les plus dangereuse et condamnable par ses ennemis." Auguste Bouché-Leclercq, *L'astrologie grecque* (Paris: E. Leroux, 1899), 404. See also my discussion of Cardano's observations below.

25. Ptolemy, *Tetrabiblos,* III. 10 and IV. 10, where he further complicates things by discussing five types of prorogator, that of the horoscope, lot of fortune, the Moon, the Sun, and the midheaven. These are related, respectively, to the body and journeys abroad; property; affections of the soul and marriage; dignities and glory; and finally actions, friendships, and the begetting of children. For the astrological terminology used in the Latin sources (prorogatio, directio, progressio, prorogator, haylag, hyleg), see Broecke, *The Limits of Influence,* 227. For the origin and circulation of these terms in the Middle Ages and the Renaissance, especially in the Arabic sources, see Paul Kunitzsch, *Mittelalterliche astronomisch-astrologische Glossare mit arabischen Fachausdrücken* (Munich: Bayerische Akademie der Wissenschaften, 1977), 35–37.

26. A series of values (minimum, median, maximum) are assigned to each planet. On the "planetary years," see S. Jim Tester, *A History of Western Astrology* (Woodbridge,

Suffolk: Boydell Press, 1987), 86–87, and Otto Neugebauer and H. B. van Hoesen, *Greek Horoscopes* (Philadelphia: American Philosophical Society, 1959), 10–11. For Vimercati's explanation, see BT, MS Triv. 1329, fols. 25v–29r.

27. For our purposes, therefore, it is less relevant to determine which author Vimercati followed than to examine the result obtained and his interpretation of it. Ptolemy's account in the *Tetrabiblos* is extremely lengthy and complex (see Ptolemy, *Tetrabiblos,* III.10). In his widely popular commentary on Ptolemy's *Tetrabiblos,* Haly (Ali Ibn Ridwan, ca. 998–1067) explains, "Sapientes antiqui huius artis concordati sunt quod significatio vite debet accipi a dominio locorum principalium et eorum fortitudine in nativitate. Sunt tamen diversarum opinionum in manieribus quibus sciri potest quantitas vite *quoniam eorunt sunt qui reputant quod quodlibet luminarium quando fuerit in aliquo angulorum in quocumque sit aptum sit esse yleg.* Et generaliter quando multe dignitates coniungentur alicui luminari preponunt illud, et accipiunt ab eo significationem quantitatis vite." Ptolemy, *Liber quadripartiti Ptholemei,* in *Opera astrologica,* fol. 65ra (emphasis mine). Vimercati most likely determined the Moon as the *hyleg* because this luminary is placed on an angle, at 5°54' Sagittarius.

28. Corio, *Storia di Milano,* vol. 2, 1408: "Fu crudele, onde uno sacerdote, essendo dal duca richiesto quanto tempo haveva a dominare, rispuose non attingerebbe a l'undecimo anno, il perché, impregionandolo, li mandò uno piccolo pane, uno bicchiero di guarnaza et una ala di capone, dacendoli dire più altro haverebbe. Con tal cose si mantenne et anche deglutendo il proprio sterco sine a dodici giorni e finalmente morì."

29. MS Triv. 1329, fols. 23v–24v: "Primo per capitulum presens inquiremus alchocoden, qui Iupiter esse, videtur qui lune locum inspicit ubi dominio domus triplicitatis ac termini fungitur. Ideo annorum tuorum dator erit ipse Iupiter, qui, cum sit in angulo maiores, tibi suos contribuet annos, qui iuxta doctorum sententiam dicuntur septuaginta novem; tunc quia retrocedit et erat in fine sue retrogradationis, ut Albumasari placuit, auferenda est pars quinta decenniorum annorum, qua sublata remanebunt anni sexaginta tres dies septuaginta duo, quibus annis aspectus lune ad ipsum alchocoden (qui recipit eam) adet minores ipsius annos qui sunt vigintiquinque. Sed ab eis aufertur quarta pars ex aproximatione caude draconis ad luna per minus duodecim gradibus, et sic remanebunt anni decem octo et menses novem, qui additu annis alchocoden constituent in summa annos octuaginta unum menses undecim, et hii sunt anni quibus tua vita duratura videtur nisi illec infortunia eos breviores efficiant."

30. A calculation of the dignities of *hyleg* (the Moon in 5°54' Sagittarius) would yield the following results: this degree is in Jupiter's mansion (5 points); no planet is in its exaltation; in Jupiter's triplicity by night (3 points); in Jupiter's term according to the Egyptian terms (2 points); in Mercury's face (1 point). This would clearly establish Jupiter as the *alcochoden,* or giver of the years.

31. These values are taken from Bouché-Leclercq, *L'astrologie grecque,* 410. On the "planetary years" see also n. 26 above.

32. BT, MS Triv. 1329, fol. 23r–v.

33. A planet is "received" by another when that planet is located in a sign in which the second planet has an astrological dignity. The condition should be mutual (and if interpreted strictly, should require the planets to exchange the same essential dignity, i.e., either mansion, exaltation, triplicity, term, or face). As Jupiter is in Taurus and the Moon is exalted in Taurus, and as the Moon is in Sagittarius, Jupiter's mansion, then Jupiter is "received" by the Moon. This is, of course, astrological commonplace for any ancient and modern practitioner, and all the sources agree on this. For the planets and their mansions, see Ptolemy, *Tetrabiblos,* I. 17; *Al-Qabisi (Alcabitius): The Introduction to Astrology,* differentia I [13–22], 230–240; and Eade, *The Forgotten Sky,* 59–69.

34. BT, MS Triv. 1329, fol. 29v: "Erit igitur egritudo qua vitam finies de natura Saturni in menbris [sic] pectoris et me[m]bris ventris quasi cum passione genuum quam sequetur mors naturalis, scilicet extinctio caloris innati propter deffectum [sic] humidi radicalis."

35. Ibid., fol. 29r–v: "Ut finem finisque tui modum intelligamus quinque consideranda sunt, scilicet locus casus interfectoris, pars mortis, domus octave, dispositio et qualitas dominorum triplicitatis, quarte domus; et tandem vero qualitas domini octave domus primo namque dicimus quod si locum casus vel directionis interfectoris consideremus et compleveris annos quos promisit alchochodem, non invenimus interfectorem qui tui vitam finire valeat citra annum nonagessimum primum, et tunc interfector cadit in virgine et est saturni quartus aspectus."

36. North, *Horoscopes and History,* 139–142, esp. 139 commenting on English royal horoscopes in the late Middle Ages.

37. Broecke, *The Limits of Influence,* 228. These, however, were not uncommon even earlier, but circulated privately in manuscript form. See the examples discussed in Azzolini, "The Political Uses of Astrology," 136–139.

38. Grafton, *Cardano's Cosmos,* 98–99. See also Anthony Grafton, "Girolamo Cardano and the Tradition of Classical Astrology: The Rothschild Lecture, 1995," *Proceedings of the American Philosophical Society* 142, no. 3 (1998): 323–354, esp. 338–340.

39. Cardano, *OO,* vol. 5, 503: "Harum septem sunt viventium, quatuor olim mortuorum, una quae coepta fuit dum viveret, nunc interim cum ederetur mortui adolescentuli Regis Anglici Edoardi Sexti, quae non solum fortuna quod Rex magnus fuerit, sed virtute, atque casu fortunam superare nedum aequare potest."

40. This horoscope and Cardano's interpretation was first printed in his 1544 edition. The notoriously complex chronology of Cardano's works has been treated exhaustively in Ian Maclean, "A chronology of the composition of Cardano's works," in Girolamo Cardano, *The Libris propriis. The editions of 1544, 1550, 1557, 1562, with supplementary material,* ed. Ian Maclean (Milan: FrancoAngeli, 2004), 43–111.

41. Girolamo Cardano, *OO,* vol. 5, 507: "Ex quo patet, quod non debemus pronuntiare de vitae spatio in debilibus genituris, nisi prius consultis directionibus omnibus aphetarum, processibus, & ingressibus. Et nisi esset quod ego in prognostico quod illis dederam ad haec me reservassem, iure poterant de me conqueri." In the reformed astrology of Cardano, the Greek derivative *apheta* (here used in the plural) is preferred to the Arabic term *hyleg.*

42. Cardano, *OO,* vol. 5, 507: "Solitum aphetas omnes dirigere, si mediam horam adiecissem, poteram & Solem, & Lunam dirigere, quibus directis, ut clarum est, periculum a vestigio apparuisset. Non confisus directioni, statim ad processus & ingressus venissem, tunc res ipsa patuisset, mortis periculum, neque enim tantus sum, ut audeam dicere certam mortem [. . .]."

43. Cardano, *OO,* vol. 5, 507: "Memini me legisse duos exitium praedixisse Principibus, Ascletarionem Domitiano, qui statim praemium tulit verae praedictionis, mortem." The story of Ascletarion is recounted in Suetonius, *The Lives of the Caesars,* transl. G. C. Rolfe (London: William Heinemann/Cambridge, MA: Harvard University Press, 1914), 373–375.

44. Ibid.: "Cum a sacerdote de mortis genere ac de aetate admonitus esset, beneficii loco habuit, illum fame necare."

45. Cardano, *OO,* vol. 5, 482: "Luna ab infortunato Iove opposito, et a Martis quadrato infortunata est, quo ad vitam; ipsa etiam cum cauda mortem decernit violentam, quoniam qui oppugnant in angulo sunt. Caeterum Mars fortis, Iupiter cum draconis capite, Luna in ascendente, regnum inter fratres illi uni adiudicant. [. . .] Cum igitur ascendens ad Martis quadratum et Iovis oppositionem, Luna vero Mercurio per directionem in trino Martis existenti applicuisset, annis perfectis 32."

46. Ann Carmichael's study of the civic records of deaths in Milan in the fifteenth century indicates a Raffaele Vimercati who operated also as doctor in the city around those years (personal communication). No traces of his practice of astrology outside the court have been found so far.

47. On Francesco Medici da Busto, see ASMi, *Autografi,* Medici 213, and his *Taquinus anni 1476 Calculatus in felici urbe Mediolano per artium et medicine doctorem Franciscum de Medicis dictum de Busti,* now in ASMi, *Sforzesco,* Miscellanea 1569. This is an ephemeris, or set of planetary tables, for the year 1476. This may be the same Francesco da Busto mentioned in Pacioli's *De divina proportione.* This Francesco attended the scientific disputation in which both Pacioli and Leonardo took part in February 1498. See Monica Azzolini, "Anatomy of a Dispute: Leonardo, Pacioli, and Scientific Entertainment in Renaissance Milan," *Early Science and Medicine* 9.2 (2004): 115–135, esp. 118, n. 10.

48. On Marco da Bologna, see P. Paolo M. Sevesi O. F. M., "Corrispondenza milanese del B. Marco da Bologna," *Archivum Franciscanum Historicum* 48 (1955): 298–323; Celestino Piana, "Documenti intorno al B. Marco Fantuzzi da Bologna (†

1479)," *Studi Francescani* 25 (1953): 224–235; and idem, "Il beato Marco da Bologna e il suo convento di S. Paolo in Monte nel Quattrocento," *Atti e Memorie della Deputazione di Storia Patria per le Provincie di Romagna* II, n.s. 22 (1971): 85–265. The letter dated February 21, 1472, is republished at pp. 108–109. On his reception among contemporaries, see also Rosa Maria Dessì, "'Quanto di poi abia a bastare il mondo . . .' Tra predicazione e recezione. Apocalittica e penitenza nelle 'reportationes' dei sermoni di Michele Carcano da Milano," *Florensia. Bollettino del Centro Internazionale di Studi Gioachimiti*, 3–4 (1989): 71–90 [republ. as "Entre prédication et réception: le thèmes eschatologiques dans les 'reportationes' des sermons de Michele Carcano de Milan (Florence, 1461–1466)," in *MEFRM* 102 (1990): 457–479.]

49. ASMi, *Sforzesco*, Carteggio Interno 903, Cicco Simonetta to Galeazzo, Milan, February 21, 1472: "Appresso perché luy ha dicto una de queste mattine in Pergolo che haveva havuto aviso che per iudicii facti per astrologi questa cometa che è apparsa novamente minaza peste. [. . .] Et volendo io intendere l'opinione de Magistro Francesco da Busti e Magistro Raphaele da Vimercato li quali vostra Signoria conosce che sonno boni astrologi che iudicio hanno de questa cometa et dele parole che haveva dicto esso frate Marco dell' effecto dessa cometa, me hanno dicto che le parole che esso frate Marco haveva dicto non erano fondate cum alcuna rasone, et che non era da farne caso alcuno. Dicono ben che loro hanno voluto studiare la condicione de questo pianeta et in fine se accordano tutti doi che il prohnostico pur male, como è guerra, la quale non menaza ad queste parte de qua, ma verso lo mezzo dì et di oriente, et precipue pare che debia produrre qualche cattivo effecto verso lo Papa et cristianismo." On the dynamics of gift-giving and scientific patronage, see Mario Biagioli, *Galileo, Courtier. The Practice of Science in the Culture of Absolutism* (Chicago: University of Chicago Press, 1993), esp. Introduction.

50. ibid.: "[. . .] Magistro Raphaele parla molto largamente che non verso la persona vostra né verso lo dominio vostro né queste parte de qua ha ad produrre veruno cattivo effecto, et che vole sostenire questo cum caduno astrologo et me ha dicto che la Signoria vostra sa che ha facto iudicio deli longhi anni che quella ha ad vivere, delle quale cose me è parso mio debito avisarla." This letter is published in its entirety in Gabotto, "Nuove ricerche", 17–18. On the interpretation of this celestial phenomenon in early modern history, see Sara Schechner Genuth, *Comets, Popular Culture, and the Birth of Modern Cosmology* (Princeton, NJ: Princeton University Press, 1997).

51. See ASMi, *Sforzesco*, Miscellanea 1569 (fasc. 1): Pietro Buono Avogario (1r–2v), Gian Battista Piasio (2v), and Giovanni de Bossis Polonio (3r–v). On Avogario, see Thorndike, *HMES*, vol. 4, 422, 432, 460, 463–466, and Cesare Vasoli, "Avogaro (Dell'Avogaro, Avogari, Avogario, Arvogari, Advogaro), Pietro Buono," *DBI* 4 (1962), 709–710. On G. B. Piasio, see Thorndike, *HMES*, vol. 4, 451, 458–459, 703. The Polish astronomer Johannes de Bossis appears for the first time on the university rolls at Bologna in the academic year 1471/72: his treatise on the 1472 comet is still extant

in manuscript. On him, see Curt Bühler, *The University and the Press in Fifteenth-Century Bologna* (Notre Dame, IN: The Medieval Institute, 1958), 22; Thorndike, *HMES*, vol. 4, 422–424; and Bònoli and Piliarvu, *I lettori di astronomia*, 113–114. The apparition of the comet is also recorded in Corio, *Storia di Milano*, vol. 2, 1384.

52. Hayton, "Expertise ex Stellis."

53. ASMi, *Sforzesco*, Carteggio Interno 927, Raffaele Vimercati to Galeazzo, September 2, 1475: "niente dimancho maye non sape exquisire el tempo commodo a fare questa mia offerta, sempre dubitando me reportarne titulo d'esser temperario et imprudente." This letter was first published in Gabotto, "L'astrologia nel Quattrocento," 398–399, with the incorrect date of 1474. As noted earlier, Gabotto's reference to its location is no longer valid.

54. ASMi, *Sforzesco*, Carteggio Interno 927, Raffaele Vimercati to Galeazzo, September 2, 1475: "[. . .] similmente facendo revolutione de dita intronizzatione annuatim potemo intendere se debeno venire gente d'arme nella patria a fargli guerra e da qual parte hanno a venire, e molte altre cose quale numerare serebbe longo dire." For the technique of revolutions, see Chapter 1.

55. Hayton, "Expertise ex Stellis," 30.

56. On Nicolò da Arsago, see ASMi, *Autografi*, Medici 212. Nicolò da Arsago to Galeazzo, Milan, March 2, 1475. In this letter, Arsago talks about the geniture he cast for Galeazzo's son, Gian Galeazzo, and the fact that he was completing one for Galeazzo himself.

57. ASMi, *Sforzesco*, Miscellanea 1569. Giovanni Simonetta to Galeazzo, Milan, November 5, 1475: "Illustrissimo Signore mio, ho recevuto la lettera dela excellentia vostra per la quale me scrive debia dire ad Magistro Raphaele da Vimercato et Magistro Nicolò da Arsago phisici che caduno de loro facia uno iudicio de qualitate temporum et singulorum dierum anni futuri, et che l'uno non sapia dell'altro, et debiano tenere questa cosa secreta. Et per exeguire quanto è dicto ho havuto da me dicti phisici l'uno separato dall'altro et factoli la commissione como me commetté vostra excellentia. Esso Magistro Raphaele ha tolto voluntere el carico de farlo, perché ha caro per fare cosa grata ala excellentia vostra che quella lo adoperi, e domandandoli in quanto tempo lo havera facto, me ha resposto lo farà in uno mese. Dicto Magistro Nicolò similmente dice lo farà voluntere, ma perché vostra signoria scrive se debia fare dicto iudicio de qualitate temporum et singulorum dierum, luy voria essere meglio chiarito da vostra signoria, cioé se dove dice de qualitate singulorum dierium vole dire altro, cioé de electione dierum. Siché vostra signoria lo pò chiarire se li piace de questa parte. Et como serano facti dicti iudicii me li portarano sigillati et io li mandaró ad vostra signoria ala quale me recomando." This letter is published in full in Gabotto, "L'astrologia nel Quattrocento," 399.

58. On Annius's practice of astrology, see now Monica Azzolini, "Annius of Viterbo Astrologer: Predicting the Death of Ferrante of Aragon, King of Naples," *Bruniana & Campanelliana* 14.2 (2008): 619–632.

59. Guido Castelnuovo, "Offices and Officials," in Andrea Gamberini and Isabella Lazzarini, eds., *The Italian Renaissance State* (Cambridge: Cambridge University Press, 2012), 368–384.

60. On the relationship between the University of Pavia and the Sforza court, see Chapter 1. For a striking example of how scientific patronage worked, see Biagioli, *Galileo, Courtier.*

61. See Chapter 5.

62. ASMi, *Autografi,* Medici 219, Raffaele da Vicomercato: "Illustrissimo Signore mio singularissimo, non solamente io desidero con ogni degna subiectione servire ad vostra celsitudine, et con ogni studio cercho de fare piacere ad quela per quanto sapia et possa, ma apresso etiam voria che li fiogli mei havessero qualche luocho che potessero anchora loro insieme demonstrare la fede et devotione ch'aviamo de servigli tutti verso vostra Illustrissima Signoria et del stato suo. Il perché havendo io mio figlio Petropaulo già più tempo usato al scrivere in la corte de vostra Illustrissima Signoria et desiderando de havere qualche firmo luoco in quela, perché meglio se potesse valere con la virtù sua, quando piacesse ad vostra Illustrissima signoria haveria caro et preghola se digna farlo fare coadiutore al consilio suo Secreto con le preminentie solite ad tale officio finch'ello serà sufficiente ad magiore grado, che speriamo sempre potere obtenere da vostra celsitudine per la sua grande munificentia et per la singulare devotione nostra verso quela. Alla quale sempre con dicti mei figlioli continuamente me ricommando. Datum Mediolani die xxiii Nouembris 1474. Servitor M.ʳ Raphael de Vicomercato phisicus etc."

63. It is not clear if Pietro Paolo was later promoted to another job or if he died as I have found no other mention of him in the documents of the chancellery.

64. Vincent Ilardi, "Towards the Tragedia d'Italia: Ferrante and Galeazzo Maria Sforza, friendly enemies and hostile allies," in Abulafia, *The French Descent,* 91.

65. Jacob Burkhardt, *The Civilization of the Renaissance in Italy* (London: Phaidon Press, 1960), 23.

66. David Abulafia, "Ferrante of Naples: The Statecraft of a Renaissance Prince," *History Today* 45.2 (February 1995): 19–25, esp. 19–20.

67. On the long dispute over the Neapolitan lands by the French house of Anjou and the Spanish house of Aragon, see now the article by Alan Ryder, "The Angevin bid for Naples, 1380–1480," in Abulafia, *The French Descent,* 56–69, which traces the history of the kingdom prior to 1459. The early years of Ferrante's reign have received some scholarly attention. For bibliographical references and a useful overview of the current state of the scholarship on Ferrante, see David Abulafia, "The Inception of the Reign of King Ferrante I of Naples: The Events of the Summer 1458 in the Light of the Documentation from Milan," in Abulafia, *The French Descent,* 71–90, esp. 71–72, and notes 1–6, and Elisabetta Scarton, "La congiura dei baroni del 1485–'87 e la sorte dei ribelli," in Francesco Senatore and Francesco Storti, eds., *Poteri, relazioni, guerra nel regno di Ferrante d'Aragona* (Naples: Cliopress, 2011), 213–290. An invaluable source

remains, *Dispacci sforzeschi da Napoli,* 4 vols. (Salerno-Naples-Battipaglia: Carlone-Laveglia&Carlone, 1997–2009), coordinated by Francesco Senatore, of which vols. 1, 2, 4, and 5 have been published so far. For the "politica dell'equilibrio," see also Riccardo Fubini, "Lega italica e 'politica dell'equilibrio' all'avvento di Lorenzo de' Medici al potere," in idem, *Italia Quattrocentesca. Politica e diplomazia nell'età di Lorenzo il Magnifico* (Milan: FrancoAngeli, 1994), 185–219, and in English in *Journal of Modern History* 67, suppl. (1995): 166–199.

68. The situation finally resolved when Calixtus died in August 1458 and Aeneas Sylvius Piccolomini was elected Pope as Pius II. Abulafia, "The inception," 81–88. For the tense relationship between Ferrante and the papacy, see also Chapter 4.

69. For a broad (albeit now quite dated) overview of Francesco's political maneuvering to maintain peace in the Italian peninsula, see the essays by Franco Catalano, "La pace di Lodi e la Lega Italica," and "La politica Italiana dello Sforza," in *Storia di Milano,* vol. 7 (1956), 3–81, 82–172. Despite the relatively quiet years between 1454–1494, destabilizing political maneuvers were still attempted, often by enlisting political exiles as sources of disruption. See Christine Shaw, *The Politics of Exile in Renaissance Italy* (Cambridge: Cambridge University Press, 2000), 2, and *passim*. On the often strained relationship between Ferrante and the Neapolitan barons, together with Abulafia, *The French Descent,* see also Henri Francois Delaborde, *L'expédition de Charles VIII en Italie: histoire diplomatique et militaire* (Paris: Firmin-Didot, 1888), 189–190; and Shaw, *The Politics of Exile,* 24–26, 135–136, 171, 182, 193–194, 237.

70. The correspondence between Ferrante and Francesco often had an intimate, familiar tone that played upon images of brotherhood and fatherhood between the two men. See Abulafia, "The inception," 80, 82–83.

71. Alfonso's marriage with Ippolita was celebrated in 1465, while that of Sforza Maria and Eleonora was dissolved in 1472. Ilardi, "Towards the Tragedia d'Italia," 94. On the dissolution of this second marriage, see also Nicola Ferorelli, "Il ducato di Bari sotto Sforza Maria Sforza e Ludovico il Moro," *ASL* 41 (1914): 389–433. Eleonora later married Ercole I d'Este and had two daughters, Isabella and Beatrice.

72. Abulafia, "The inception," 88–89. In the Franco-Milanese alliance, Louis XI abandoned any claims over Genoa, Milan, and Naples in exchange for support in his continuing struggle against his own feudal barons, who were increasingly dissatisfied with Louis's internal policies. See Ilardi, "Towards the Tragedia d'Italia," 93. On the claims of the House of Orléans over Milan, see also A. Mary F. Robinson, "The Claim of the House of Orleans to Milan," *English Historical Review* 3.9 (1988): 34–62, and eadem, "The Claim of the House of Orleans to Milan (continued)," *English Historical Review* 3.10 (1988): 270–291.

73. See Ilardi, "Towards the Tragedia d'Italia," 91–122. For an overview of the relationship between Milan and the Kingdom of Naples, see Gigliola Soldi Rondinini, "Milano, il Regno di Napoli e gli Aragonesi (secoli XIV–XV)," in *Gli Sforza a*

Milano e in Lombardia e i loro rapporti con gli Stati italiani ed europei (1450–1535). Convegno internazionale, Milano 18–21 maggio 1981 (Milan: Cisalpino-Goliardica, 1992), 229–290 [repr. in eadem, *Saggi di storia e storiografia visconteo-sforzesca* (Bologna: Cappelli, 1984), 83–129], and Paul M. Dover, "Royal Diplomacy in Renaissance Italy: Ferrante d'Aragona (1458–1494) and his Ambassadors," *Mediterranean Studies* 14 (2005), 57–94. For the diplomacy of the period, see Alberto Aubert, *La crisi degli antichi Stati Italiani (1492–1521)* (Florence: Le Lettere, 2003), esp. chap. 1.

74. Ilardi, "Towards the Tragedia d'Italia," 94.

75. On the rift that developed, see Lubkin, *A Renaissance Court,* 39–48, 53–54, 62–65; and Ilardi, "Towards the Tragedia d'Italia," 94–96.

76. Paolo Margaroli, "Bianca Maria e Galeazzo Maria Sforza nelle ultime lettere di Antonio da Trezzo (1467–1469)," *ASL* 111 (1985), 327–377.

77. This is minutely detailed in Ilardi, "Towards the Tragedia d'Italia," and other essays in Abulafia's volume. On the patterns of alliances that developed after the Italian League (and that effectively undermined it), see Fubini, "Lega Italica e 'politica dell'equilibrio'," esp. 206–219.

78. Ilardi, "Towards the Tragedia d'Italia," 106. On Venice's battle against the Turks for the colony of Negroponte, see the recent article by Margaret Meserve, "News from Negroponte: Politics, Popular Opinion, and Information Exchange in the First Decade of the Italian Press," *RQ* 59 (2006): 440–480. For Galeazzo's machinations against Venice at this time, see Pietro Magistretti, "Galeazzo Maria Sforza e la caduta di Negroponte," *ASL* 1 (1884): 79–120 (fasc. I), and 337–356 (fasc. II). See also Kenneth M. Setton, *The Papacy and the Levant (1204–1571),* 4 vols. (Philadelphia: American Philosophical Society, 1978), vol. 2, 271–313.

79. ASMi, SPS 1464, ca. 1455, Galeazzo to the Marquis of Monferrato, the Duchess of Savoy, Antonio Appiani, Sforza Bettini (Milan's resident ambassador in France), and the bishop of Novara: "Ell' è stato rasonato et praticato più mesi sono fra el Serenissimo Signore Re Ferrando e nuy per mezo de nostri ambaxatori de contrahere certo matrimonio, cioé dare per legitima sposa et mugliere la inclyta Madonna Isabella figliola del Illustrissimo Signore Duca de Calabria et de la Illustrissima Madonna Hyppolita Duchessa sua consorte lo inclyto Zoanne Galeazzo nostro primogenito Conte de Pavia. Tandem deo duce ad dì xiiii del presente in Napoli per mezo del nostro ambaxiatore che habiamo li è stato concluso dicto parentato. Unde a contentamento vostro l'habiamo voluto significare, rendendone certo ne recevereti piacere insieme cum nuy." On the same page, a similar note is drafted in Latin and addressed to the King of France and the members of the Consiglio Segreto (the Privy Council). See also, Carlo Canetta, "Le sponsalie di Casa Sforza con Casa d'Aragona (Giugno-Ottobre 1455)," *ASL* 9 (1882): 136–144, and idem, "Le 'sponsalie' di Casa Sforza con Casa d'Aragona," *ASL* 10 (1883): 769–782. Far from being a genuine attempt to pursue the common interests of the two houses, this marriage was possibly

a desperate maneuver to prevent the double marriage proposed by Louis XI to the King of Naples: his son Charles would marry Isabella and his daughter Anne, Isabella's brother, Ferdinand. See Ilardi, "Towards the Tragedia d'Italia," 115.

80. Ibid., 118–120.

81. See ASMi, SPE, Napoli 227, Ippolita to Galeazzo, Castello Capuano, November 12, 14, 16, 18, and 19, 1475, in which she sends medical bulletins "hour by hour." The letters of November 12, 16, 18, and 19 have been published by Ferdinando Gabotto in *Lettere inedite di Joviano Pontano in nome de' reali di Napoli* (Bologna: Romagnoli Dall'Acqua, 1893), 64–73, together with other letters on this subject (74–89). Ippolita waited eight days before writing to her brother about Alfonso's illness. Since the seventh day from the inception of an illness is deemed to be the first *dies iudicativus* in Galen's theory of the critical days, she must have waited to receive a reliable pronouncement from the physicians. On the Galenic theory of the critical days, see Dell'Anna, *Dies critici*. On the application of these theories at the Sforza court, see also Chapter 4.

82. Welch, "Between Milan and Naples," 132–136.

83. ASMi, SPE, Napoli 227, Ippolita to Galeazzo, Castello Capuano, on November 14 and 16, 1475. Other letters were sent to Galeazzo by Antonello Petrucci, *regio segretario,* on November 15 and 27. When Petrucci wrote on November 15 he still feared greatly for the life of the king and openly asked Galeazzo to pledge his support for Alfonso, his children, and his state "per rispecto de la strictissima parentela fra le vostre Illustrissime Signorie." By November 27 Petrucci could inform Galeazzo that his letter reassuring the king and his son of his unconditional support pleased the king enormously and "li ha dato grandissimo juvamento ala sua convalescentia." He also confirmed that Ferrante was feeling better and was no longer in danger of dying.

84. In response to his sister's letter, and the letter of Petrucci dated November 15, Galeazzo sent three separate letters of reassurance on November 21, one to his sister, one to her husband, Alfonso, and one to Petrucci. In his letter to Alfonso he stated, "Grandissima molestia et affanno ho preso del accidente de vostra Signoria ma molto magiore de la Maestà del Re vostro padre per essere in periculoso termine pregando Dio che lo conserve in vita et prosperità. Ma se pure al divino consiglio piacesse disponerne altramente ve dico che prendiate bon coraggio et conservative l'animo constante et forte che in tale bisogno farò per la S. V. quanto lei saperà domandare et mettarò tutte le mie facultà et bisognando etiam la propria persona per conservatione et stabilimento del vostro stato et ponerò omne mia cosa per voi nel tavoliero senza alcuno reservo, perché el vostro bene è mio proprio." The same day, Galeazzo wrote to Roberto Sanseverino asking him to prepare his troops for intervention in case of Ferrante's death. In following letters dated November 25 and 26 he reassured Ippolita of his support and indicated that his troops in Romagna and Lombardy had already received instructions to march toward Naples should the need arise.

85. Ilardi, "Towards the Tragedia d'Italia," 119. On Trezzo see also Francesco Senatore, *«Uno mundo de carta». Forme e strutture della diplomazia sforzesca* (Naples: Liguori Editore, 1998).

86. Sacramoro's letters contain a wealth of information regarding the situation on the ground, the attitude of the barons, and the opinions that his pledge of support had generated at court. See ASMi, SPE, Napoli 227, Sacramoro to Galeazzo, Gaeta, November 29, 1475; Carinola (Caleno), November 30, December 1; Naples, December 4 (two letters), December 5, December 7. Sacramoro Sacramoro, better known as Sacramoro da Rimini, was elected bishop of Piacenza in October of the same year, only to leave the appointment on January 15, 1476, to become bishop of Parma. He held the post until 1482, the year of his death. See Konrad Eubel, *Hierarchia catholica medii aevi sive Summorum pontificum, S. R. E. cardinalium, ecclesiarum antistitum series e documentis tabularii praesertim Vaticani collecta, digesta edita*, 8 vols. (Monasterii: sumptibus et typis librariae regensbergianae, 1898–1979), vol. 2, 235 (Parma), 239 (Piacenza); Gianluca Battioni, "La diocesi parmense durante l'episcopato di Sacramoro da Rimini (1476–1482)," in Giorgio Chittolini, ed., *Gli Sforza, la Chiesa lombarda, la corte di Roma. Strutture e pratiche beneficiarie nel Ducato di Milano (1450–1535)* (Napoli: Liguori Editore, 1989), 115–213; idem, "Indagini su una famiglia di 'officiali' fra tardo medioevo e prima età moderna: I Sacramoro da Rimini (fine secolo XIV–inizio secolo XVII), *Società e Storia* 52 (1991): 271–295; and Franca Leverotti, *Diplomazia e Governo dello Stato: I 'famigli cavalcanti' di Francesco Sforza (1450–1466)* (Pisa: GISEM-ETS Editrice, 1992), esp. 205–207.

87. ASMi, SPE, Napoli 227, Antonello Petrucci to Galeazzo, Carinola (Caleno), November 27, 1475.

88. ASMi, *Sforzesco,* Miscellanea 1569, Annius of Viterbo to Galeazzo, November 24, 1475. The interrogation, with an explanation of its astrological meaning and a transcription of the text is included in Azzolini, "Annius of Viterbo Astrologer," 630–632.

89. ASMi, *Sforzesco,* Miscellanea 1569, Annius of Viterbo to Galeazzo, November 24, 1475: "Si autem diem sui decubitus haberem, audacissime in unam partem me conferrem. Vale felix ex cellula nostra die 24 Novembris hore 23 et si potes scire diem decubitum mitte."

90. Recent osteoarcheological research on Ferrante's body revealed that Ferrante died of bowel cancer, most likely caused by a genetic predisposition and the large amount of red meat in his diet. See Gino Fornaciari, Antonio Marchetti, Silvia Pellegrini, and Rosalba Cirrani, "K-ras Mutation in the Tumor of King Ferrante I of Aragon (1431–1494) and Environmental Mutagens at the Aragonese Court of Naples," *International Journal of Osteoarcheology* 9 (1999): 302–306.

91. See Chapter 5. As research into the Milanese *carteggio* continues, other interrogations of this kind may well emerge. The difficulty of locating such information is partly due to the nature of these sources. Although in Annius's case we have the

celestial figure and the astrologer's interpretation, other interrogations appear as parts of longer letters and contain only the interpretation itself. In addition, the results of these interrogations could be mentioned by astrologers or clients in their correspondence. Even today, there is no certainty as to where these types of documents might be found in the *mare magnum* of the Carteggio Sforzesco.

92. A *iudicium* of 1473 is in ASMi, *Sforzesco, Miscellanea* 1569. A letter from Galeazzo to Cicco dated November 4, 1475, instructed him to speak with Annius and exhort him to send the 1475 prognostication. This letter is quoted in Gabotto, "L'astrologia nel Quattrocento," 401. The personal prognostication for 1473, which addressed the duke's health, the wars that would be relevant to the duchy, political success, and the future of his progeny ("primo de sanitate aut egritudinibus quibus succurris possit causis ad plenum cognitis. Secundum de bellica expeditione et imperio acquirendo. Tertio de processo dominii. Quartum de successione heredum quorum"), is accompanied by a letter where Annius explicitly declares that he made the *iudicium* on the basis of Galeazzo's nativity. The letter and the parchment with the *iudicium* are in ASMi, *Diplomatico,* Diplomi e Dispacci Sovrani, Milano, 6 (Genoa, January 10, 1473). This very iudicium "in carta capretina" is mentioned also in a letter from Genoa by Giovanni Pallavicino da Scipione dated January 13, 1473. The two letters are transcribed in Edoardo Fumagalli, "Aneddoti della vita di Annio da Viterbo O.P. (III)," *Archivum Fratrum Praedicatorum* 52 (1982): 197–218, esp. 205–206; Pallavicino's letter is transcribed also in Paola Mattiangeli, "Annio astrologo e alchimista," in *Annio da Viterbo. Documenti e ricerche* (Rome: Consiglio Nazionale delle Ricerche, 1981), 269–275, esp. 271 (n. 58).

93. ASMi, *Sforzesco,* Miscellanea 1569.

94. Ibid.

95. On Annius's anti-Sforza campaigns, see Edoardo Fumagalli, "Aneddoti della vita di Annio da Viterbo O.P. (I–II)," *Archivum Fratrum Praedicatorum* 50 (1980): 167–199. On Ferrante as the leader of a new crusade, see Roberto Weiss, "Traccia per una biografia di Annio da Viterbo," *Italia medioevale e umanistica* 5 (1962): 425–441, esp. 429; and Annius of Viterbo, *De futuris Christianorum triumphis in Saracenos, seu Glossa super Apocalypsin* [*sic*] (Genoa: Battista Cavallo, 1480), sig. f8r.

96. See Fumagalli, "Aneddoti (I–II)."

97. The conspiracy was treated extensively in Bortolo Belotti, *Il dramma di Gerolamo Olgiati* (Milan: s.a. [1929]), later republished as *Storia di una congiura (Olgiati)* (Milan, 1965). More recently, the wider national and international implications of the conspiracy have been explored in Vincent Ilardi, "The Assassination of Galeazzo Maria Sforza and the Reaction of Italian Diplomacy," in Lauro Martines, ed., *Violence and Civil Disorder in Italian Cities 1200–1500* (Berkeley-Los Angeles-London: University of California Press, 1972), 72–103; Riccardo Fubini, "L'assassinio di Galeazzo Maria Sforza nelle sue circostanze politiche," in Lorenzo de' Medici, *Lettere,* 13 vols. (Florence: Giunti Barbèra, 1977–2010), vol. 2, ed. Riccardo Fubini

(1977), 247–250 and 523–535; and idem, "Osservazioni e documenti sulla crisi del ducato di Milano nel 1477 e sulla riforma del consiglio segreto ducale di Bona Sforza," in Sergio Bertelli and Gloria Ramakus, eds., *Essays Presented to Myron P. Gilmore*, 2 vols. (Florence: La Nuova Italia, 1978), vol. 1, 47–103 with its rich documentary appendix. Francesca M. Vaglienti's article "Anatomia di una congiura. Sulle tracce dell'assassinio del duca Galeazzo Maria Sforza tra scienza e storia," *Atti dell'Istituto Lombardo. Accademia di Scienze e Lettere di Milano* 136 (2002): 237–273, does not add any new information to earlier accounts.

98. Since the publication of three major articles on Italian court astrology by Ferdinando Gabotto at the end of the nineteenth century, little systematic effort has been made to study comprehensively the role and function of astrology at Italian Renaissance courts. While Gabotto made reference to a wealth of material scattered in Italian archives that revealed the ubiquitous presence of astrology in fifteenth- and sixteenth-century diplomatic courtly correspondence, political historians have generally considered astrology as mere superstition and thus have given it only anecdotal value. See Gabotto, "L'astrologia nel Quattrocento"; idem, "Bartolomeo Manfredi"; and idem, *Nuove ricerche*.

99. On the political circumstances that led to the French invasion, see Abulafia, *The French Descent*, and the dated, but still extremely useful volume by Delaborde, *L'expédition de Charles VIII*.

100. Lubkin, *A Renaissance Court*, 206–207 and passim.

101. "Havendone instantissimamente li Ill.mi S.ri Duca de Barri et messer Ludovico nostri fratelli rechiesto che li vogliamo lassare andare a vedere il mondo [. . .]." Letter of Galeazzo Maria to the Milanese ambassador in France, Francesco Pietrasanta, November 28, 1476, as quoted in Achille Dina, "Ludovico il Moro prima della sua venuta al governo," *ASL* 3 (1886): 737–776, esp. 766. See also Fubini, "L'assassinio di Galeazzo Maria Sforza," 531. Fubini rightly stresses the fact that Galeazzo's brothers, Sforza Maria and Ludovico, contested from the start Galeazzo's succession to power as the eldest son. On Galeazzo's death, see also Lubkin, *A Renaissance Court*, 239–241, 257.

102. See Dina, "Ludovico il Moro," 764–766; Fubini, "L'assassinio di Galeazzo Maria Sforza," 534; and Shaw, *The Politics of Exile*, 7. Cf. Lubkin, *A Renaissance Court*, 239 n. 124.

103. Corio, *Storia di Milano*, vol. 2, 1399: "Il proximo giorno, dedicato al martire, nel quale fece uno acerbo freddo, il duca se mise la corazina, quale cavò dicendo parebbe troppo grosso."

104. Since Babylonian times celestial portents of this sort, and especially comets, were believed to presage murders, conspiracies, and changes of government. See Genuth, *Comets, Popular Culture, and the Birth of Modern Cosmology*, 20–26. The *Cronica gestorum in partibus Lombardie et reliquis Italie* contains numerous examples of celestial portents of this kind. See the *Cronica gestorum in partibus Lombardie et*

reliquis Italie, ed. Giuliano Bonazzi, *RIS²,* t. 22, pt. III (Città di Castello: S. Lapi, 1904). For instance a comet was reported to have forecast the woes of Italy in Parma on November 20, 1480: "Vigesimo secundo novembris 1480 circa horam vigesimam quartam apparuit in medio platee magne communis Parme ignis ad modum stelle cadentis, qui finivit super tecto palacij potestatis, absque ulla lexione, qui a multis est visus et multis prebuit terrorem." This was accompanied by lamentations that announced the conjunction of Jupiter and Saturn of 1484 (at 84–85). Another comet was believed to have foretold the death of Ludovico Maria Sforza's opponent, Pier Maria Rossi, in 1482: "Apparuit his temporibus una nocte ingens stella in aiere versus montes Parmenses cum maxima cauda ignea, que prognosticabat mortem illustris domini Petrimarie Rubei de proximo futuram" (at 114).

105. Corio, *Storia di Milano,* vol. 2, 1398.

106. Ibid., 1399–1400.

107. *Cronica gestorum,* 4: "Quod quidem sic gestum toti Lombardie terrorem indixit et vix poterat credi; nam princeps ipse, *qui timebat talli morte perire et a multis astrologis et sanctis viris prenosticatum ei fuerat,* maxima sibi preparaverat custodias armatorum; sed nihil profuit, quin ad tallem exitum perveniret" (emphasis mine).

108. ASMi, SPE, Napoli 218, Antonio da Trezzo to Galeazzo, Naples, January 15, 1470, where Antonio writes: "Illustrissimo Signore mio, per altre mie ho avisato la Illustrissima S. V. de quanto havea seguito in exequutione [sic] de le littere vostre le quale me comandavano operasse cum la Maestà del Signor Re che la Excellentia V. *non fosse in alcuno modo nominata in alcuno iudicia per quelli doctori et astronomi che li sogliono fare ogni anno.* Et come mis[s]er Silvestro quale è tenuto homo singularissimo et doctissimo in questa scientia era restato contento de non nominare essa V. Excellentia et me havea nominato uno mis[s]er Augello de Benevento che molto se delecta fare talli iudicia etc., doppoi io mandai messo proprio a Benevento al prefato mis[s]er Augello al quale scripse quanto bisognava circa questo, dal quale ho havuto la resposta che mando inclusa, per la quale la celsitudine V. serà chiara che esso non ve nominarà in suo iudicia quale ordinariamente fa ogni anno. Se potrò intendere quod in regno hoc sia alcuno che faci iudicia provederò in modo che serà satisfacto alla voluntà de V. Excellentia." The letter of reply of Augello da Benevento is not included in the same *filza.*

109. Galeazzo to Sacramoro da Rimini, Pavia, July 14, 1474: "Voi sapete de quanta importantia sia quando in li populi si divulga qualche sinistra opinione de li mali eminenti et che possono intervenire, como fanno spesse volte questi astrologi temerarii e legeri, li quali mettendo suo studio in divinare et fantasticare delle cose occulte, reservate solo in arbitrio de Idio, stultamente predicono la morte di principi, guerre e carestie, e descendono usque ad individua in signare ad chi credono debia toccare la sorte; et licet li homini de bon sentimento et gravi poco prestino fede a simile prenosticatione, tamen el vulgo pur li presta orecchie et istano cun li animi suspesi, et fanse spesso penseri che generano scandalo in li stati et principati. Nostro parere

seria che la Santità de nostro Signore faci excommunicare tutti et singuli astrologi et mathematici, li quali presumeranno in li loro iudicii nominare nè specificare alcuno principe nè signore, nè farne alcuna mentione tacita vel expressa, ma solamente li sia permesso dire de le cose universale, perché le particularitate possono mettere turbatione et sonno pericolose, et la religione e la fede christiana prohibisce queste tale superstitione." The letter is quoted in Gabotto, "L'astrologia nel Quattrocento," 404. In this and following instances I have occasionally modified Gabotto's transcriptions to follow more modern criteria. So far I have been unable to locate the document in the Archivio di Stato, Milan.

110. Ibid. The Ferrarese astrologer mentioned in this letter is probably Pietro Buono Avogario.

111. ASMi, SPE, Romagna 183, Giovanni Simonetta to Gerardo Cerruti, Cassano, July 15, 1474: "Gerardo, volemo che dextramente sapii da quelli astrologi li, de li quali ne hai mandato el iudicio de questo anno, si epsi in particolarità hanno veduto qualche pronosticatione de noi et si implicite per loro iudicii ne hanno voluto sentire alcuna cosa che pertenghi ad noi et al nostro stato." Quoted in Gabotto, "L'astrologia nel Quattrocento," 403. On Cerruti's role as Sforza ambassador, see Tommaso Duranti, *Il carteggio di Gerardo Cerruti, oratore sforzesco a Bologna (1470–1474)*, 2 vols. (Bologna: CLUEB, 2007).

112. ASMi, SPE, Romagna 183, Giovanni Simonetta to Gerardo Cerruti, Cassano, July 15, 1474: "Et fate dire el tutto in scripto et bene expresso, pregandoli che qui innanzi non voglino più fare mentione di facti nostri per alcuno modo, et si pur ne faranno mentione, la facino in bene et comodo nostro, et non in contrario."

113. Galeazzo to Giovanni Battista (Biancoli) da Cotignola, Cassano, July 17, 1474: "Messer Zoan Baptista, voi andarete da lo Illustrissimo Duca Hercule et per vostra parte li exponerete che avendo Magistro Petro Bono astrologo de Signoria Sua divulgato el suo iudicio de questo anno per tutta Italia, è acapitato in le mano di questi nostri physici peritissimi de astrologia, li quali, lecto epso iudicio et examinato bene le particularità che sonno in quello, ne hanno dechiarato (quantunché extimano che in epso iudicio non habia voluto fare mentione de noi, tam perché simile cose capitano per mano del vulgo, facilmente se poteria pensare) ch'el dicto Magistro Petro Bono apertamente inferisce che noi in questo anno havemo ad incurrere in pericolo della vita et anche del stato, como per le parole tolte dalli capitoli de epso iudicio et notate qua soto se pò comprendere." Quoted in Gabotto, "L'astrologia nel Quattrocento," 405. So far I have not been able to locate this letter in the archives.

114. Ibid.: "Et benché crediamo questo sia proceduto senza consentimento della signoria sua, perché como quella che ne ama non lo averia potuto, tamen volemo la pregate che voglia admonir dicto astrologo che da qui inante, nec explicite, nec implicite, parle de noi; ita che mai in li soi iudicii se possa fare coniectura sopra li facti nostri; che, quando lui faci el contrario, et nunc volemo ne excusiate cum la signoria sua, che ad dicto astrologo demonstreremo quanto ne despiaca che lui voglie pronosticare

de noi [. . .] et certificate la signoria sua che quando noi volessimo usare simile arte, havemo chi anche saperia astrologare non cum manco doctrina et rasone che se facino li altri, ma ne pare che sia officio alieno da reale principe."

115. *Iudicium Magistri Petri Boni Ferrariensi Mcccclxxiiii,* in ASMi, *Sforzesco, Miscellanea* 1569. Another *iudicium,* for the year 1471, is also preserved in the same location. The same *fondo* also houses Avogario's prognostication for the comet of 1472. See note 51 above.

116. "Magistro Petro Bono, voi astrologati e fati iudicio d'altri e non sapeti astrologare né fare iudicio de periculi vostri imminente, perché il Duca di Milano ha mandato lì per farve tagliare a pezi e tuttavia ne manda de li altri per fare questo, che, non potendolo uno, venghi facto all'altro' e azò credati ve dicha el vero, se fate ponere mente ad le bollete et ad le porte, trovareti che tra li altri ve capitarà uno Zorzo Albanese di piccola statura et homo scuro in faza, et l'altro Iohanne de Lucoli, grande, rubicondo, cum li capelli longhi di colore castano, et va uno poco zoppo. State advertente che non ve parlo senza casone." Quoted in Gabotto, "L'astrologia nel Quattrocento," 407. I have not been able to locate this letter.

117. "Unus vel rex dominus periculum magnum patietur ex parte animalium brutorum, vel ex parte inimicorum occultorum," as quoted in Gabotto, "L'astrologia nel Quattrocento," 407. We still possess Manfredi's *iudicium* for 1474 (together with those of 1469, 1470, 1471, and 1473), but Marsilio's *iudicium* seems to be lost. His 1473 *iudicium* is still preserved, however. See ASMi, *Sforzesco, Miscellanea* 1569. On Manfredi, see also Thorndike, *HMES,* vol. 4, 451, 459–461, 480, 575, 578.

118. ASMi, SPE, Romagna 183. Galeazzo to Giovanni Battista da Cotignola, Cassano, July 17, 1474: "Et de nostra parte precarete le loro Magnificentie che admoniscano epsi astrologi che da qui inante mai più in li iudicii loro implicite nec explicite facino mentione né parola de noi, aut di nostri amici, confederati et colligati che se possa coniecturare sopra li facti nostri. Perché, quando facessero el contrario, ex nunc volemo essere excusati, che gli faremo demonstrare che ad noi despiacerà sommamente el iudicare loro, et forse anche così ne li faremo pentire." Quoted in Gabotto, "L'astrologia nel Quattrocento," 406.

119. Excerpts from this letter are quoted in Ibid., 407–408. Gabotto does not give a clear indication as to its location.

120. ASMi, *Sforzesco, Miscellanea* 1569, *Iudicium huius anni 1474 domine Johannes Artoni ducatus aulici,* fol. 2v. I have not been able to read the original prediction glued underneath.

121. ASMi, SPE, Romagna 183. Giovanni II Bentivoglio to Galeazzo, Bologna, July 26, 1474; and Giovanni Battista da Cotignola to Galeazzo Maria Sforza, Bologna, July 1474.

122. Ibid. Marsilio da Bologna to Galeazzo, s.d.

123. Thorndike, *HMES,* vol. 4, 460; Gabotto, "L'astrologia nel Quattrocento,"

403. On Cola's role in the conspiracy and the support and patronage he had received from Galeazzo himself (including his appointment as professor of rhetoric at the University of Pavia), see Lubkin, *A Renaissance Court*, 241, 361, n. 130. The edition of Ptolemy's *Geographia* was published in Bologna (1477, misdated 1462) and saw the collaboration of Pietro Buono Avogario, Girolamo Manfredi, and Galeotto Marzio, with Cola Montano. For this edition, see the facsimile *Cosmographia: Bologna, 1477,* with introd. by R. A. Skelton (Amsterdam, 1963).

124. See ASMi, *Autografi,* Medici 212, Pietro Buono Avogario to Ludovico, Ferrara, July 7, 1494, where he pleads for Ludovico to intervene in the restitution of the goods of a dear friend of his accused of murder and reminds him of his services to him as a skilled astrologer; and Pietro Buono Avogario to Ludovico, Ferrara, December 5, 1496.

125. Pietro Buono Avogario's geniture of Ludovico Avogario in BLO, MS Canon. Misc. 24, fol. 11r.

126. Two Ferrarese chroniclers reported the Avogario's prediction in relation to the duke's death: Zambotti, *Diario ferrarese,* 29, and Ferrarini, *Memoriale estense,* 58.

127. BLO, MS Canon. Misc. 24, fols. 16r (nativity), 118r (revolution and profection), 118v (nativity). On this geniture collection and the Sforza horoscopes herewith contained, see my "Refining the Astrologer's Art."

128. BLO, MS Canon. Misc. 24, fols. 16r: "Mortus est autem in ecclesia dum haberet, quia Saturnus qui malam sibi mortem promiserat erat in nona retrogradus, qui est domus itinerum et religionis." Saturn's placement is 27°21' Gemini, but for his purposes, Gazio considers it close enough to the cusp to be in the ninth. In classic and medieval astrology, in most cases it is possible to accept a leeway of up to five degrees. In his most comprehensive nativity in fol. 118v, however, Gazio states that Saturn is in the eighth house. On the demarcation of the cusps, see Eade, *The Forgotten Sky,* 46–47.

129. BLO, MS Canon. Misc. 24, fol. 118r: "Vide quem aperte hec revolutio mortem duci Galeatio demonstravit et conminata. [. . .] Mars in sixta in domo servorum promittit malum et interfectionem a servo. Animadverte quoque Saturnus retrogradus in duodecima in domo inimicorum malum inimicis et inimicorum [esse?] portendentem."

130. Ibid.: "Vide Martem in octava in signo humano mortem ab homine denotantem."

131. Ibid., fol. 118v: "Fuit autem vulneratus in ventre ex eo quod Mars retrogradus fuit in Virgine, quae ventre habet." On *melothesia,* the theory that relates the heavenly bodies and the zodiacal signs to different parts of the human body, see the Introduction.

132. Ibid.: "Fuit autem luxuriosus valde ex eo quod Venus in signo calido et solis iuncta."

4. The Star-Crossed Duke

1. This seems to have been the case also in Naples and the rest of Italy. The etiology, symptoms, prognosis, and treatment outlined in the letters sent by Ippolita Sforza and other Neapolitan courtiers mentioned in the previous chapter bear striking similarity to the letters discussed in this chapter. For a comprehensive overview of the extremely rich and varied classical and medieval literature on the critical days, see Dell'Anna, *Dies critici.*

2. Arsenic was one of the main ingredients used for embalming and preservation. Therefore, without knowing how Gian Galeazzo's body was prepared for burial, it may be impossible to determine if he was poisoned or not. On the highly controversial results of the "Operazione Medici," which plans to exhume the remains of forty-nine members of the Medici family, see Francesco Mari, Aldo Polettini, Donatella Lippi, and Elisabetta Bertol, "The Mysterious Death of Francesco I de' Medici and Bianca Cappello: An Arsenic Murder?," *British Medical Journal* 333 (December 23, 2006): 1299–1301, and the rapid responses to the paper posted at http://www.bmj .com/cgi/eletters/333/7582/1299 (last accessed: September 25, 2011). On the practice of embalming and the use of arsenic, see Silvia Marinozzi and Gino Fornaciari, *Le mummie e l'arte medica: per una storia dell'imbalsamazione artificiale dei corpi umani nell'evo moderno* (Rome: Università la Sapienza, 2005). Similar research has been carried out on the earthly remains of Pandolfo III Malatesta and Ferrante of Aragon. See Rosalba Cirrani, Laura Giusti and Gino Fornaciari, "Prostatic Hyperplasia in the Mummy of an Italian Renaissance Prince," *Prostate* 45.4 (2000): 320–322; Gino Fornaciari, Rosalba Cirrani, and Luca Ventura, "Paleoandrology and prostatic hyperplasia in Italian mummies (XV–XIX century)," *Medicina nei secoli* 13.2 (2001): 269–284; Rosalba Cirrani, Valentina Giuffra, and Gino Fornaciari, "Ergonomic pathology of Pandolfo III Malatesta," *Medicina nei secoli* 15.3 (2003): 581–594; as well as Gino Fornaciari, "Malignant tumor in the mummy of Ferrante Ist of Aragon, King of Naples (1431–1494)," *Medicina nei Secoli* 6.1 (1994): 139–146, and Gino Fornaciari, Antonio Marchetti, Silvia Pellegrini, Rosalba Cirrani, "K-ras Mutation in the Tumor of King Ferrante I of Aragon (1431–1494) and Environmental Mutagens at the Aragonese Court of Naples," *International Journal of Osteoarcheology* 9 (1999): 302–306.

3. Francesco's wife, Bianca Maria, had been expressly excluded from the succession to the duchy and, at the death of the last Visconti, other states laid claim to Lombardy, first and foremost Alfonso of Aragon and the House of Orlèans. For the validity of these claims (or lack thereof), see Gary Ianziti, *Humanistic Historiography under the Sforzas: Politics and Propaganda in Fifteenth-Century Milan* (Oxford: Clarendon Press, 1988), 21–22, and Lubkin, *A Renaissance Court,* 22–24.

4. On all legal aspects of Sforza rule, see Black, *Absolutism in Renaissance Milan.*

5. On the guidelines of Galeazzo's physician, Cristoforo Soncino, on how to live a healthy life, see Ferrari, *"Per non manchare in tuto del debito mio."* Soncino's *Ordine della vita del conte Galeazzo Maria* is fully transcribed at pp. 58–80; a detailed analysis of the text and its reception and consumption at court is provided at pp. 32–57.

6. Lubkin, *A Renaissance Court,* 25.

7. Ibid., 48.

8. For these events, see ibid., 48–54.

9. The second marriage ceremony was celebrated after the consummation of the marriage in Vigevano. Ibid., 53.

10. Ibid.

11. On the importance of issues of dynastic lineage, fertility, and the process of reproduction in the later Middle Ages and the Renaissance, see now Park, *Secrets of Women,* chap. 3. We do not know if the day and time of the marriage and its consummation were determined astrologically. On this astrological practice, see Chapter 5.

12. Lubkin, *A Renaissance Court,* 79–80.

13. Lubkin reports how dynastic considerations dictated not only the choice of the name, but also the place of birth. Lubkin, *A Renaissance Court,* 79–80. On Gian Galeazzo Visconti, first Duke of Milan, see Chapter 2.

14. Battista Bendodoni to Galeazzo Maria, Tortona, June 21, 1468, ASMi, SPS 1464, doc. 1. "Illustrissime princeps excellentissime dux, domine mi colendissime etc. Havendo nuperime veduto et inteso per lettere de vostra celsitudine a mi directe del prospero et foelice parto de la inclitissima et excellentissima Madonna Duchessa dignissima consorte vostra, non tanto per essere scaricata senza danno et periculo de la persona, quanto per havere parturito uno bellissimo figliolo ne ho pigliato quello piacere et gaudio per reverentia de la vostra Sublimità che meritamente io debo come affecionatissimo et devotissimo servitore de quella. Cussì per questa mia a quale demonstratione de ciò mi è parso debito a congratularmene cum la prefata vostra signoria pregando quello che'l tuto rege et move che cussì come gli ha prestato de essere nato da tale gloriosissimi padre et madre quali sono le vostre excellentie et in tanto excellentissimo et triumphantissimo stato quale è quello de la vostra excellentia cussí per la sua natura li presti quilli gloriosi et foelicissimi successi li quali siano secondo el desiderio et appetito de la vostra prelibata celsitudine: *a perpetua exultatione et contento de le vostre sublimità et ad eterna memoria et laude del nome vostro et de la excellentissima casa dei Visconti."* Emphasis mine. On the high value that Galeazzo placed on his Visconti lineage, see also Lubkin, *A Renaissance Court,* 198.

15. ASMi, SPS 1464, Galeazzo to the Marquis of Monferrato, the Duchess of Savoy, Antonio Appiani, Sforza Bettini, and the Bishop of Novara: "Ell'è stato rasonato et praticato più mesi sono fra el Serenissimo Signore Re Ferrando e nuy, per mezo de nostri ambaxatori, de contrahere certo matrimonio, cioè dare per legitima sposa et mugliere la inclyta Madonna Isabella, figliola del Illustrissimo Signore Duca de Calabria et de la Illustrissima Madonna Hyppolita Duchessa sua consorte, lo

inclyto Zoanne Galeazzo nostro primogenito Conte de Pavia. Tandem, deo duce, ad dì xiiii del presente in Napoli per mezo del nostro ambaxiatore che habiamo lì, è stato concluso dicto parentato. Unde a contentamento vostro l'habiamo voluto significare, rendendone certo ne recevereti piacere insieme cum nuy." On the same page, a similar note is drafted in Latin and addressed to the King of France and the members of the Consiglio Segreto (the Privy Council). On the Aragonese of Naples, see Jerry H. Bentley, *Politics and Culture in Renaissance Naples* (Princeton, NJ: Princeton University Press, 1987); on Alfonso II, see also, George L. Hersey, *Alfonso II and the Artistic Renewal of Naples, 1485-1495* (New Haven and London: Yale University Press, 1969).

16. Francesco Sforza had applied the same matrimonial strategy with Galeazzo himself, pursuing the union of the houses of the Sforza and Savoy to establish firmer ties with the French crown and thus neutralize any possible Orleanist intervention against the Sforza's interests.

17. On the risks associated with large consumptions of meats and wine, see the reflections of Gino Fornaciari, "Renaissance Mummies in Italy," *Medicina nei Secoli* 11.1 (1999): 85-105; and Fornaciari et al., "K-ras Mutation in the Tumor of King Ferrante I." On gout and meat consumption at the Medici court, see Emiliano Panconesi and Lorenzo Marri Malacrida, *Lorenzo il Magnifico in salute e in malattia,* with an introduction by Eugenio Garin (Florence: Alberto Bruschi, 1992). On dietary habits in the Renaissance, see also Ken Albala, *Eating Right in the Renaissance* (Berkeley and London: University of California Press, 2002).

18. ASMa, AG 1623, Zaccaria Saggi to Barbara Gonzaga, Milan, October 8, 1469. Quoted in Lubkin, *A Renaissance Court,* 8.

19. Niccolò Machiavelli, *Istorie Fiorentine,* ed. Alessandro Montevecchi, in Machiavelli, *Opere,* 4 vols. (Turin: UTET, 1984-1999), vol. 2 (1986), Book VII, chap. 28, 682-683.

20. ASMi, SPS 1464, Galeazzo to Boldrino Crivelli, Vigevano, November 8, 1473: "Volemo che subito alla recevuta de queste te ritrove con tutti quelli monesteri de donne et homini chi sonno nostri devoti, alli quali facemo elimosina, et gli dighi che facciano oratione per il Conte de Pavia nostro primogenito a ciò che lo reducano in bona sanità." Boldrino Crivelli was *ufficiale delle vettovaglie* at court from at least 1477 until 1489, when he was dismissed, with three other *ufficiali,* under fraud allegations. See Caterina Santoro, *I Registri delle lettere ducali del periodo Sforzesco* (Milan: Castello Sforzesco, 1961), 172 (n. 157), 181 (n. 208), 241 (n. 23).

21. See subsequent letters in ASMi, SPS 1464.

22. See Chapter 3.

23. See Fubini, "L'assassinio di Galeazzo Maria Sforza"; and idem, "Osservazioni e documenti sulla crisi del ducato."

24. See ASMi, SPS 1464, Bona of Savoy and Gian Galeazzo Maria Sforza to the Governor of Asti, Milan, May 28, 1477 and following documents. See also Corio,

Storia di Milano, vol. 2, 1410–1416. Apparently Ascanio had asked the support of the Republic of Venice to launch an assault against Bona and Gian Galeazzo, but Venice had turned him down and remained in the League with Milan and Florence (Bona to Filippo Sacramoro, Milan, June 14, 1477). On Roberto Sanseverino, see Machiavelli, *Istorie Fiorentine,* Book VII, chap. 13, 717–718, and Niccolò Machiavelli, *Il Principe,* ed. Rinaldo Rinaldi, in Machiavelli, *Opere,* vol. 1 (1999), pt. 1, chap. 12, 248. See also Emilio Motta, "I Sanseverino feudatari di Lugano e Balerna, 1434–1484," *Periodico della società storica comense* 2 (1882): 155–310, and Marco Pellegrini, "Ascanio Maria Sforza: La creazione di un cardinale 'di famiglia'," in Giorgio Chittolini, ed., *Gli Sforza, la chiesa lombarda, e la corte di Roma* (Naples: Liguori, 1989), 215–298. On these events see also Carlo de' Rosmini, *Dell'Istoria di Milano,* 4 vols. (Milan: Tip. Manini e Rivolta, 1820), vol. 4, 163–167; idem, *Dell'istoria intorno alle militari imprese e alla vita di Gian-Jacopo Trivulzio detto il Magno,* 2 vols. (Milan: Tip. Gio. Giuseppe Destefanis, 1815), vol. 2, 16–24; and, much more recently, Shaw, *The Politics of Exile,* 8–9, 80, 86.

25. ASMi, SPS 1464, Ludovico to Bona of Savoy, Pisa, May 20, 1478: "[. . .] certo la contritione mia è tale che merita perdono: e così supplico vostra excellentia che seguendo lo comandamento del nostro Signore Dio mi voglia remettere e perdonare ogni manchamento e offesa havesse havuta da mi non perhò volontariamente."

26. In fact, Corio suggested that a romantic relationship had developed between Bona and Antonio and that this dangerously impaired her judgment. See Corio, *Storia di Milano,* vol. 2, 1422–1423.

27. ASMi, SPS 1464, Bona and Gian Galeazzo to Filippo Sacramoro, Milan, September 14, 1479. This letter is countersigned by the new secretary, Bartolomeo Calco.

28. ASMi, SPS 1464, Bona and Gian Galeazzo to Roberto Sanseverino, September 10, 1479: "Magnifice etc., havendo nuy cognossuto expressamente la iactura et danni seguiti al stato et subditi nostri et quasi ad tutta Italia esser proceduti per mancamento et perversità de Cicho Symoneta et de Zoanne suo fratello et de Orpheo de Ricavo, n'è parso provederli opportunamente per salute et quiete del stato et populi nostri, et hogi li haveno facti destenire tutti tre, mediante el quale remedio et molti altri tutti li nostri subditi se retrovano tanto alegri et di bona voglia et in tanta tranquillità che'l non poteressimo exprimere [. . .]." As is well known, Cicco Simonetta became the scapegoat and was accused of having turned Bona against her brothers-in-law unjustly. For this reason, Cicco, his brother Giovanni, and the Florentine Orfeo Cenni da Ricavo, were imprisoned and Cicco was later executed. See also ASMi, SPS 1464, Gian Galeazzo to Filippo Sacramoro (Milanese ambassador in Florence), September 14, 1479; and Medici, *Lettere,* vol. 4 (1981): (1479–1480), ed. Nicolai Rubinstein, letter 427, 206–209. Lorenzo exerted his influence at the Sforza court and Orfeo was exiled to Tuscany, where he died on January 5, 1482.

29. Once again, Corio insinuated that her decision was motivated by Antonio's mysterious departure from the court of Milan en route to Venice "with a great sum of money and pearls." Corio, *Storia di Milano*, vol. 2, 1429–1430.

30. ASMi, SPS 1464: "Instrumentum tutelle [sic] et administrationis Illustrissimi domini Joannis GaleazMarie Ducis Mediolani etc. decrete Illustrissimo domino Lodovico Marie Duci Bari eius Patruo."

31. Pietro Bembo, *History of Venice*, ed. and trans. by Robert W. Ulery, Jr. (Cambridge, MA.: Harvard University Press, 2007), 80–81.

32. Ibid.

33. ASMi, SPS 1464, Gian Galeazzo to Ascanio Sforza, Milan, September 28, 1483: "Come per altre ve scripsimo essendoci sopravenuta hogi el *quarto giorno* certa *alteratione* cum la febre freda et calda in vero tutto heri et questa nocte passata fin ad le nove hore ci dede grande ambastia et passione in tanto che ne pariva pur troppo alla *compressione* [sic] et età nostra de adolescentia. Nondimeno circa le nove hore commenzò ad fare meglioramento et cusì per tutto hogi siamo stati assay bene. <Per la tal> cosa speramo in la divina clementia reuscirne ad bono porto et presto remanere libero. Et però ad vostra consolatione vi ne havemo voluto dare noticia ad ciò se fusse stato scripto altramente ne possiati remanere cum l'animo reposito, et de ciò ne communicarete cum lo Illustrissimo Signor Duca de Calabria." Emphasis mine.

34. Danielle Jacquart, "De *crasis* à *complexio:* Note sur le vocabulaire du tempérament en Latin médiéval," in eadem, *La science médicale occidentale entre deux renaissances* (Aldershot: Ashgate, 1997), chap. VI, 71–76 (originally published in Guy Sabbah, ed., *Textes médicaux latins antiques,* Saint-Etienne: Université de Saint-Étienne, 1984, 71–76). Galen's theory is best elucidated in his treatise *De temperamentis.* For the Greek and Latin version, see Kühn, vol. 1, 509–694; for its English translation, see Galen, *Selected Works,* trans. with an introduction and notes by P. N. Singer (Oxford: Oxford University Press, 1997), 202–289.

35. Ptolemy's *Tetrabiblos* (I, 2) offers the following explanation for the phenomenon: "A very few considerations would make it apparent to all that a certain power emanating from the eternal ethereal substance is dispersed through and permeates the whole region about the earth, which throughout is subject to change, since, of the primary sublunary elements, fire and air are encompassed and changed by the motions of the ether, and in turn encompass and change all else, earth and water and the plants and animals therein."

36. See Ptolemy, *Tetrabiblos,* I, 5: "[. . .] because two of the four humours are fertile and active, the hot and the moist (for all things are brought together and increased by them), and two are destructive and passive, the dry and the cold, through which all things, again, are separated and destroyed, the ancients accepted two of the planets, Jupiter and Venus, together with the Moon, as beneficent because of their tempered nature and because they abound in the hot and the moist, and Saturn and Mars as producing the effects of the opposite nature, one because of his excessive cold

and the other for his excessive dryness." Also, Ptolemy, *Tetrabiblos,* III, 10–12. See also Galen, *De diebus decretoris,* 3.6 (Kühn, vol. 9, 911–912): "Si enim ad planetas temperatos steterit, quos jam nominant, salutares faustos ac bonos dies producere, si ad intemperatos, graves molestosque." On the influence of the Moon in different signs of the zodiac, see Galen, *De diebus decretoris,* 3.5–6 (Kühn, vol. 9, 908–913). For an example of the key role played by these concepts in medieval medical astrology, see Paganica, *Compendium medicinalis astrologiae.*

37. The literature on the plague is immense and cannot be successfully summed up here. For recent discussions of this issue of its celestial origins, see Danielle Jacquart, "Theory, Everyday Pratice, and Three Fifteenth-Century Physicians," in eadem, *La science médicale occidentale,* chap. XIII, 142–153 (orig. publ. in Michael McVaugh and Nancy Siraisi, eds., *Renaissance Medical Learning: The Evolution of a Tradition* (Osiris, vol. 6) (Philadelphia: History of Science Society, 1990: 140–160); and Smoller, "Of Earthquakes, Hail, Frogs, and Geography." For an equivalent explanation of the spread of syphilis, see Hayton, "Joseph Grünpeck's Astrological Explanation of the French Disease."

38. This discussion is *par force* very general and does not do justice to the rich medieval and Renaissance literature on the topic. For an elegant summary of these concepts, see Nancy Siraisi, *Medieval and Early Renaissance Medicine: An Introduction to Knowledge and Practice* (Chicago: University of Chicago Press, 1990), 97–106, 115–123.

39. Dell'Anna provides an extensive bibliography of medieval and Renaissance sources that included a discussion of this theory. These spanned from about the tenth to the nineteenth century. See Dell'Anna, *Dies critici,* vol. 1, esp. 9–36; Maclean, *Logic, Signs, and Nature,* 305–306. For a recent translation of Galen's text from the Arabic, see Glen M. Cooper, *Galen, De diebus decretoriis, from Greek into Arabic: A Critical Edition, with Translation and Commentary, of Hunayn ibn Ishaq, Kitab ayyam al-buhran* (Farnham: Ashgate, 2011). Cooper rightly stresses the general dearth of studies on this influential Galenic text.

40. Galen makes a distinction between the *mensis medicinalis* and the *mensis lunaris.* See Dell'Anna, *Dies critici,* chap. 1; Pennuto, "The Debate on Critical Days;" and Galen, *De diebus decretoriis,* 3.9 (Kühn, vol. 9, 928–933). Galen's treatment of this theory was occasionally critiqued by medieval and Renaissance physicians and astrologers. For some examples, see Pennuto, "The Debate on Critical Days," and Grafton and Siraisi, "Between Election and My Hopes," esp. 87–89.

41. Dell'Anna, *Dies critici,* esp. vol. 1, 83–153; Y. Tzvi Langermann, "The Astral Connections of Critical Days. Some Late Antique Sources Preserved in Hebrew and Arabic," in Akasoy, Burnett, and Yoeli-Tlalim, *Astro-Medicine,* 99–118.

42. Albumasar, *Introductorium in astronomiam Albumasaris abalachi octo continens libros partiales* (Augsburg: Erhard Ratdolt, 1489; in the translation of Hermann of Carinthia), Book I, chap. 1, sig. a5r: "Medicus quidem elementorum alterationibus

operam dat. Astrologus stellarum motus sequitur elementarie ad alterationis causas." See, also Lemay, vol. 8, 8.

43. Ibid, Book I, chap. 4, sig. b2r:" Si enim in arte sua quam profitentur noti essent, nec astrologie summam ope ignorarent, quod Ypocras attestans in libro quodam. [. . .] Nec enim astrologia parvam in medicina obtinet partem, qua sentencia phisicorum artificio aditos instruit temporum alterationes motusque naturarum sidereos cursus consequi, ut precipuum sit medicis astrologie fore participes quatinus artis sue fundamentum et principium recognoscant, cui quantum astrologia prestet perpendi potest. Cum enim previderit astrologus cui medendum sit, et quare ac quantum, demum medicus utiliter accedit [. . .] unde tam Ypocrati quam Galieno quam ceteris fere omnibus philosophis compertum, astrologiam plane phisice ducatum obtinere, ut qui astrologiam damnet phisicam necessario destruit." See Lemay, vol. 8, 20 (with variants).

44. If the cycle were protracted beyond 120 days, the fever would be declared chronic; otherwise it was considered acute. For a treatment of these complex aspects of the theory, see Dell'Anna, *Dies critici,* vol. 1. 66–76. See also Hippocrates, *Aphorisms,* 2.24 in Hippocrates, *Nature of Man. Regimen in Health. Humours. Aphorisms. Regimen 1–3. Dreams,* transl. W. H. S. Jones (London: W. Heinemann/Cambridge, MA: Harvard University Press, 1931), Book 4, 115; and Galen, *De diebus decretoriis* 2.3 (Kühn, vol. 9, 848–852).

45. Dell'Anna, *Dies critici,* vol. 1, esp. 46–48, 69–70. Opinions differed as to how to count these days and determine the outcome of an illness. Together with Dell'Anna, see also Siraisi, *Medieval and Early Renaissance Medicine,* 135–136; and Pennuto, "The Debate on Critical Days."

46. Recent historiography has stressed how practitioners and patients shared the same Galenic worldview based on humoral theory and that this knowledge was not limited to learned physicians. See Gianna Pomata, *Contracting a Cure: Patients, Healers, and the Law in Early Modern Bologna* (Baltimore and London: The Johns Hopkins University Press, 1998), and John Henderson, *The Renaissance Hospital: Healing the Body and Healing the Soul* (New Haven and London: Yale University Press, 2006).

47. ASMi, SPS 1464, Ludovico to Ascanio, Milan, September 28, 1483. The expulsion of substances from the lower orifices was supposed to restore the natural humoral balance in the body. For this reason patients were often treated with powerful laxatives.

48. ASMi, SPS 1464, Ludovico to Ascanio, Milan, September 19, 1483: "Reverendissime et Illustrissime Domine Frater Cordialissime, circa le xvii hore sopragiunse el fredo ad questo Illustrissimo Signore et poy el caldo el quale gli dura ancora ad quest'hora 24; el *parocismo* è stato asay minore de quello de l'alter heri et benché la *materia peccante* sia de natura che dimonstra dovere protrahere el male qualchi dì, nondimeno sperano li medici redurre la excellentia sua ad bon porto. Non se è

ordinato ancora darli medicina per expectare quello possi fare la natura da se la quale fin qua se è aiutata asay bene et etiam per lassare passare la presente *coniunctione* in la quale trovandosi la luna con marte faria furere la medicina." This letter is cited also in Carlo Magenta, *I Visconti e gli Sforza nel castello di Pavia, e loro attinenze con la certosa e la storia cittadina,* 2 vols. (Milan: Hoepli, 1883), vol. 2, 535 (n. 1). On astrological medicine at the Sforza court, see also my "Reading Health in the Stars," 183–205.

49. Dell'Anna, *Dies critici,* esp. vol. 1, 60–61, 83–86.

50. Lemay, vol. 5, 306 (Book VII, chap. 6, in the trans. of John of Seville): "Debilitas quoque planetarum est ut sint coniuncti malis in uno signo, aut in oppositione eorum, vel in quarto aspectu, aut in trino sive sextili aspectu eorum et fuerit inter eos et malos minus termino planete. Et ut sint in terminis malorum aut in domibus eorum." See also Charles Burnett, Keiji Yamamoto, and Michio Yano, eds., *Abū 'Masar. The Abbreviation of the Introduction to Astrology* (Leiden and New York: Brill, 1994), 55: "The misfortune of the planets is if they are in conjunction with the malefics or in their opposition or in their quartile or their trine or their sextile, or between them and the body of the malefic or its rays there is less than the term of the planet, or if they are in the terms of the malefics or in their houses [. . .]." For Ptolemy's scheme of injuries and diseases that befell human beings on the basis of their planets at birth or at the start of a disease, see Ptolemy, *Tetrabiblos,* III, 12.

51. One of the most popular medieval texts to propound these theories is the ps.-Albertus Magnus *Liber aggregationis, seu liber secretorum de virtutibus herbarum, lapidum et animalium quorundam,* which had wide manuscript circulation in the Middle Ages and was printed in the mid–1470s both in Italy and Germany. A Strasbourg edition by the press of Henricus Ariminensis (Georg Reyser?) appeared between 1474 and 1479 (GW 619), a Ferrarese and a Bolognese edition were published around the same time by the presses of Severino Ferrarese (around 1477; GW 630), and Johannes Schriber (around 1478; GW 361). Numerous other editions appeared in the 1480s and 1490s in Italy, Germany, England, France, and the Low Countries. For the Latin text with French translation, see Isabelle Draelants, *Le «Liber de virtutibus. Herbarum lapidum et animalium (liber aggregationis)». Un texte à succès attribué à Albert Le Grand* (Florence: SISMEL-Edizioni del Galluzzo, 2008); for its English translation, see *The Book of Secrets of Albertus Magnus of the Virtues of Herbs, Stones and Certain Beasts: Also, A book of the Marvels of the World,* ed. Michael R. Best and Frank H. Brightman (Oxford: Clarendon Press, 1973). Many of the same principles of celestial influence are also illustrated in Marsilio Ficino's work of medical astrology *De vita comparanda.* See Ficino, *Three Books on Life.*

52. BL, MS Arundel 88, fol. 37r. The short passage in a note entitled "Electio pro pharmacis recipiendis" reads: "Quoad aspectus et coniunctiones planetarum cum luna aliqua vitanda sunt, aliqua eligenda. Coniunctio etim Solis, Martis, seu Saturni vitanda est, et etiam coniunctio Iovis cum humores retineat." This section echoes

those rubrics that are occasionally present in annual prognostications, thus suggesting a common origin in university teaching. One example is in the *iudicium* for 1464 of a certain "Magistro Martino da Cracovia Polonium" (possibly Martinus Rex from Zurawica, professor of mathematics in Cracow, Prague, and Bologna) for the cardinal legate of Bologna Angelo Dei. This *iudicium* contains a rubric entitled "Tempora electa pro sumendis farmaciis." See BAR, MS 102, fols. 64v–66r. On BL, MS Arundel 88 and Gian Battista Boerio, see the Introduction and Chapter 1.

53. Astrologically, the Moon is *combusta* when it is roughly within eight degrees of the Sun and receives its rays. See Eade, *The Forgotten Sky*, 84. Cf. Albumasar, *Introductorium in astronomiam,* Book VII, chap. 2, sig. g6r; and Lemay, vol. 5, 280 (where Albumasar states that it is combust within six degrees).

54. ASMi, SPS 1464, Gian Galeazzo to Ascanio, Milan, October 2, 1483: "Respondendo ad quanto me scrive la signoria vostra per le soe del penultimo del passato condolendose dela *alteratione* nostra dela febre dicemo che siamo certissimi che sua signoria insieme con quello Illustrissimo signore Duca di Calabria nostro patre ne habino havuto displacentia et in vero al principio la ce dede grande affano et ambastia. Pur da domenica in qua semo stati assai meglio et infine tutto martedì continuamente facessemo meglioramento et se trovamo in tutto mondi de febre. Da heri in qua ne havimo pur sentita uno pocho del che credemo sia stato casone *la combustione dela luna.* Pur per gratia del nostro Signore Dio circa le xxii hore hogi ne fumo remasti necti et speramo in la sua clementia che ne remaneremo presto liberi perché la è stata molto più legere che non fu in li principii." Emphasis mine.

55. Albumasar, *Introductorium in astronomiam,* Book VII, chap. 6, sig. h2r: "Impedimenta Lune seu corruptiones singulares undecim sunt. Prima est eclipsis validissima in signo radicali seu in trigono eius aut tetragono eius. Secunda sub radiis. Tercia inter ipsam et oppositionem Solis minus 12 gradibus. [. . .]." Cf. Lemay, vol. 5, 307 (in the trans. of John of Seville): "Impedimentum quoque Lune fit XI modis. Uno, cum fuerit obscurata et eo gravius si obscuretur in signo in quo fuerat in radice nativitatis hominis, aut in eius triplicitate, seu quarto aspectu. Secundo, ut sit sub radiis Solis, sintque inter ipsam et corpus eius XII gradus ante vel retro." And Albumasar's *Ysagoga minor,* "The corruption of the Moon is in eleven ways. One of them is if it is eclipsed, and the strongest form of this is if it is eclipsed in the sign in which it was in the base nativity of a man or in its trine or its quartile aspect. The second is if it is under the rays of the Sun." Burnett, Yamamoto, and Yano, *Abû 'Masar. The Abbreviation of the Introduction to Astrology,* 59.

56. I have been unable to locate this letter in the archives.

57. ASMi SPS 1464, Ludovico to Ascanio, Milan, October 5, 1483: "Reverendissime et Illustrissime Domine Frater Cordialissime, heri sera scripse ala Signoria vostra Reverendissima del melioramento de questo Illustrissimo signore nostro *havendo bene operato la medicina;* per questo l'aviso como questa nocte l'è stato molto bene e *senza veruna excrescentia* et questa matina se è sentuto meglio et continuando

fin quest' hora xxiii. Dicono li medici che dale 16 hore in qua l'hano trovato *penitus mondo* et havere hogi havuto la meliore giornata che l'habia anchora havuto stando sua excellentia alegra et di bona voglia et pigliando deli piaceri commo se non havesse havuto male alchuno in modo se spera debia presto redurse a bona et optima convalescentia." Emphasis mine.

58. O'Boyle, *Medieval Prognosis and Astrology*, 18–20.

59. BL, MS Arundel 88, fols. 70v–86v. On this manuscript and this text see also discussion in Chapter 1. The text in MS Arundel 88 presents interesting variants when compared to that edited by O'Boyle. While this is not the place for a detailed analysis, it seems possible to speculate that these differences could be due to its oral delivery as part of a lecture series rather than to another textual tradition in addition to the two indicated in O'Boyle's study.

60. On the doctrine of signs in Renaissance medical theory and practice, see Maclean, *Logic, Signs, and Nature in the Renaissance*, 276–332, esp. 303–06.

61. The contract was already signed in 1472, as courtly documents attest when announcing the marriage to the leaders of Italian and European states, including Bona, the Marquis of Monferrat, and the King of France. See ASMi, SPS 1464 (Latin and Italian version of the same message).

62. For a contemporary account of the event, see Tristano Calco, *Nuptiae Mediolanensium Ducum, sive Ioannis Galeacii cum Isabella Aragona, Ferdinandi Neapolitanorum Regis*, in *Tristani Chalci Mediolanensis Historiographi, Residua e Bibliotheca Patricii Nobilissimi Lucii Hadriani Cottae* (Milan: Giovanni Battista and Giulio Cesare Malatesta, 1644), 59–85. On the music celebrations and the festivities, see also Paul A. Merkley and Lora L. M. Merkley, *Music and Patronage in the Sforza Court* (Turnhout: Brepols and Pietro Antonio Locatelli Foundation, 1999).

63. "Il duca in tal forma restò afaturato che gran tempo stette che non puotte con la bella sposa compire li amorosi intenti." Corio, *Storia di Milano*, vol. 2, 1475. Similarly, Borgia speaks of a "magico veneficio," while Sanuto reports that, "celebrate le sponsalicie, tamen esso Duca per uno tempo stette che non poté usar con lei, o fusse ligato acciò che non generasse, o che se fusse, pur alla fine, volente Deo, la ingravedò." See BMV, MS Lat. 3506, Girolamo Borgia's *Historiae de bellis italicis ab anno 1494 ad 1541*, Book III, fol. 44v; and Marino Sanuto, *La spedizione di Carlo VIII in Italia*, ed. Riccardo Fulin (Venice: Tipografia del Commercio di Marco Visentini, 1883), 30–31.

64. *Maleficium* was the term most often used to refer to sexual impotence caused by magic; in some cases maleficium could be performed using astrological talismans. See Catherine Rider, *Magic and Impotence in the Middle Ages* (Oxford: Oxford University Press, 2006), esp. 76–80 and 195–205; on astrology and sexuality, see also Helen Lemay, "The Stars and Human Sexuality: Some Medieval Scientific Views," *Isis* 71.1 (1980): 127–137; on astrology and homosexuality, see the recent articles by H. Darrel Rutkin and P. G. Maxwell-Stuart in Kenneth Borris and George S. Russeau, eds., *The Sciences of Homosexuality* (London: Routledge, 2007).

65. Weill-Parot, Les "images astrologiques."

66. BAV, MS Vat. Lat. 4080, fols. 42r–43r. We do not know, however, if Ansel-mi's other work, his *Opus de magia* [*sic*] *disciplina,* circulated also in Pavia. Nicolas Weill-Parot ranks this among the most significant fifteenth-century works on talismans. On Anselmi, see now Weill-Parot, Les "images astrologiques," 622–636. On Guainieri, see Rider, *Magic and Impotence,* 198–205.

67. BNF, MS Italien 1524, fols. 142r–v (for male potency and impotence), 146v (against female sterility), 152r (against male impotence), 155r (to find out who is sterile if a couple does not procreate), 157v (if you want to make a man and a woman copulate). I wish to thank Professor Jean-Patrice Boudet for generously sharing with me his knowledge of this manuscript.

68. While more research is needed in this area, we know that the astrologer-physician Girolamo Manfredi constructed astrological talismans for Ludovico and Barbara Gonzaga. See Signorini, *Fortuna dell'astrologia,* for a golden talisman in the shape of a lion (368, doc. 12).

69. On the context of this secret negotiation, see Marco Pellegrini, *Ascanio Sforza: La parabola politica di un cardinale-principe del Rinascimento,* 2 vols. (Rome: Istituto storico italiano per il Medio Evo, 2002), vol. 1, 218 and 228. It seems that in 1487 Ludovico's offer was declined. Later, rumors circulated that Ferrante offered Isabella in marriage to Emperor Maximilian I, possibly with the sole intention of forcing Ludovico's hand and speeding up the marriage between Gian Galeazzo and Isabella. See ASMo, *Materie,* Medicina, Giacomo Trotti to Ercole d'Este, Milan, September 10, 1487. Ferrante's renewed proposal to Ludovico is reported also in BMV, MS Lat. 3506, Borgia, *Historiae,* Book 1, fol. 4v. I would like to thank Elena Valeri for providing reference to these and other passages in Borgia's manuscript cited in this chapter. On Girolamo Borgia, see Elena Valeri, *Italia dilacerata. Girolamo Borgia nella cultura storica del Rinascimento* (Milan: FrancoAngeli, 2007).

70. ASF, MAP 50, Pietro Filippo Pandolfini to Lorenzo de' Medici, Milan, January 5, 1489/1490, doc. 186, fol. 193r.

71. ASMo, Ambasciatori, Milano 6, Giacomo Trotti to Ercole d'Este. The letters cover the period 1489–1491. From January until April of 1489, Trotti reported regularly to his duke about Gian Galeazzo's lack of sexual prowess, including the fact that when in bed Gian Galeazzo eschewed Isabella's kisses, preferring instead to tell her stories about birds and horses, to the dismay of the Neapolitan women that accompanied Isabella (Giacomo Trotti to Ercole, Milan, February 11, 1489).

72. ASF, MAP 50, Pietro Filippo Pandolfini to Lorenzo de' Medici, doc. 190, fol. 197r: "Il signor Ludovico mi afferma che il Duca non può fare, etc. Ma quando pure per opera di costoro se operasse lui lo facessi una volta, dice essere chiaro che poco fructo farà. Perché il Duca subito tornerà al suo naturale [stato] et in somma conclude faccino costoro quello voglino, il Duca non è per aver figlioli, et su questo fondamento ha fatto il disegno vi scripsi a xxviiii del passato."

73. Ibid.

74. The letter dated December 28, 1489 (ASF, MAP, 50, doc. 181) and subsequent letters (for instance, doc. 185, dated January 2, 1489/1490) suggests that Ludovico was already actively plotting against the Aragonese of Naples by trying to put himself between Innocent VIII and Ferrante of Aragon. Ludovico had attempted a similar double game already during the Barons' War and he was to do the same in the years immediately preceding the French descent. See Pellegrini, *Ascanio Sforza,* esp. vol. 1, 169–186 and vol. 2, 446–452, 463–478. On the relationship between Ferrante and Ludovico, see also Aubert, *La crisi degli stati Italiani,* 21 and notes.

75. ASF, MAP 50, Pietro Filippo Pandolfini to Lorenzo de' Medici, Milan, January 14, 1489, doc. 190, fol. 197r: "Et cum la diligentia di queste matrone per questi del Re si pensa ogni via per fare pruova se il Duca può consumare il matrimonio, in fino ad confortare la Duchessa che aiuti il Duca [. . .]." On the same issues, see also Pandolfini to Lorenzo de' Medici, docs. 191, fol. 198r (January 20, 1498/1490); 193 (on the *matrone;* January 22, 1498/1490); 195 (with Ferrante suggesting that Ludovico take Isabella as wife and make himself Duke of Milan; s.n., probably January 25, 1489/1490); 196, fol. 193 bis-r (again with the proposal that Ludovico take Isabella as wife and make himself Duke of Milan; January 27, 1489/1490). See also Delaborde, *L'expédition de Charles VIII en Italie,* 218.

76. ASF, MAP 50, Pandolfini to Lorenzo de' Medici, doc. 190, fol. 197r: "Il signor Ludovico fa al continuo tante careze & dimostrationi verso la Duchessa che chi non sapessi il secreto direbbe che lui gli volesse bene e fussi inclinato a torla."

77. ASMi, SPS 1464, Antonio to Bartolomeo Calco, Vigevano, April 26, 1490: "Magnifice patre, nudius quidem tertius advenienentem coniugem Isabellam magno honore ac leticia Dux noster excepit. Mutata sua & affinium Principum in Gallicanum habitum veste, Antonio etiam Vicecomite tabellarii in more ornato, deducta in cubiculum communis quidem fuit thorus cum viro, sed quem mox frebicula invasit impedivitque omnem veneris voluptatem. *Hac vero nocte certissimis apparentibus signis magno effectu exercuerunt.* Hacque de eam [sic] Belpratus confestim tabellarium summa celleritate cum hoc nuncio Neapolim evolare iusit, et Ludovicus pater magna leticia affectus cum utroque nepotum [sic] cungratulatus diligentissime est." Emphasis mine. Vincenzo Belprato was the Neapolitan ambassador in Milan.

78. ASMi, SPS 1464, Agostino to Bartolomeo Calco, Pavia, July 15, 1490: "Princeps noster non levi febricula tentatur, quae nudius tertius cepit hora circiter secundam ante occasum solis; eadem heri paulo maturius incepta in lucem usque tenuit magna primo vi frigoris tum caloris. Repetiit hodie quoque sed paulo levior. Itaque tertianam duplam censent medici. Alligatas literas ad te mitto ut distribuantur quibus debent."

79. Unfortunately it is not clear to whom this letter was addressed: the back of the letter seems to indicate it was sent to a bishop or archbishop. See ASMi, SPS 1464. A Dionisio Confalonieri appears as *coadiutor* in the *Cancelleria Segreta* of Milan.

According to Caterina Santoro (at p. 59) his mandate antedates 1489. He is present in a list of the *Cancelleria Segreta* for the year 1489. He also appears to have been *cancellario ducale*, with appointment dating prior to 1495. Santoro also indicates that numerous letters of 1495 in the Registro delle Missive (ASMi, RM 199) are signed by him. He is certainly occupying this role in 1499. See Santoro, *Gli Uffici del dominio sforzesco*, 57.

80. ASMi, SPS 1466, letters dated between August 2 and September 14, 1491.

81. ASMi, SPS 1466, Dionisio Confalonieri to Bartolomeo Calco, Pavia, September 24, 1491. See also SPS 1466, Dionisio to Bartolomeo Calco, Pavia, September 26, 1491: "Magnifice eques et prestantissime Domine et benefactor mi observandissime, hieri lo Illustrissimo signore nostro stete bene et hogi fin alle 20 hore, ne la qual hora gli è venuto il fredo che li è durato circa ad una hora et meza, et sopragionto il caldo andando in decrescimento in modo alle 2 hore è stato mondo. Per haver anticipato et essere stato meno il fredo et caldo de l'altro *parocismo* si spera che l'excellentia sua sarà presto libera. Alla qual ho facto l' ambassata sua che molto li è stata grata"; and letter dated September 30: "Magnifice eques et prestantissime domine et benefactor mi observandissimo, rimando alla magnificenza vostra l' inclusi ordeni et litere sottoscripte de mane del nostro Illustrissimo signore. Ne più presto le ho possuto far expedire per essere stato la excellentia sua alterata. Pur essendo riducto ad pristine valitudine come è quando hogi che li dì suspecto staghi bene come fece hieri, et l' altro non m'è parso differire più la expeditione per essere el predetto signore questa matina levato molto di bona voglia."

82. The correspondence consulted is not complete. Judging from internal evidence, it is almost certain that part of the doctors' correspondence has been lost. The letters still in the archives, however, seem to show that at the time of the illness that eventually killed Gian Galeazzo, up to three to four letters a day were sent to report on his health.

83. ASMi, SPS 1464, Dionisio Confalonieri to Ludovico il Moro, Mortara, April 10, 1494: "Dopoy l'excellentia sua è andato dal canto della duchessa la sera: *quando va ad dormire fano colatione insolite, con mandar ad tore da bevere. Et per questo stando* l'excellentia sua sobria, credo non haverà altro male." Emphasis mine.

84. O'Boyle, *Medieval Prognosis and Astrology*, 14: "Patiens obediens debet esse in omnibus medico ita ut non erret super se ipsum, error enim bonam crisim retardat, malam autem accelerat," and at p. 89: "[. . .] Assistentes solliciti debent esse, et administrare cibum et potum horis congruis et secundum quod medicus precipit, et cooperire secundum quod oportuerit."

85. ASMi, SPS 1464, Isabella to Ludovico, Pavia, July 19, 1494.

86. These letters are dated between July and October 1494 and, unless stated otherwise, they are in ASMi, SPS 1464.

87. Dionisio to Ludovico, Pavia, July 21, 1494: "L'Excellentia vostra per littere di medici che li consignerà Cesare Birago intenderà come al predetto signore è fallito el

parocismo, e stano in speranza debia avere pocho male. Tuttavolta io non li tacerò che l'excellentia sua ha il stomacho guasto e debile per modo che dala pistata in fora magna pocho altro, come da epso Cesare potrà intendere che è stato presente al disinare, ultra che dice non potersi sostenere suso le gambe."

88. Ibid: "Quisti medici rasonando con mi attribuiscono che'l male suo sia causato ob minimum cohitum." For an exhaustive discussion, see Mikkeli, *Hygiene in the Early Modern Tradition.* On the regimen sanitatis literature, with great attention to Milan, see Nicoud, *Les régimes de santé au Moyen Âge,* and eadem, "Les pratiques diététiques à la cour de Francesco Sforza."

89. Dionisio to Ludovico, Pavia, July 20, 1494: "De pocho inanzi lo Ilustrissimo signor Duca habia desinato, li è stato dicto, rasonando come si fa, che hier sira li fu adaquato el vino, et si ne è alterato con Maestro Gabriele et col credenzero, in modo che, beuto la pistata et magnato quattro cugali di pannata, non ha voluto né magnare, né bevere altro stando alquanto sopra di sé et infra pocho si è adormentato."

90. Dionisio to Lodovico, Pavia, September 8, 1494; and Gabriele Pirovano to Ludovico, Pavia, September 9, 1494. Dionisio similarly reported on September 11 that Gian Galeazzo had spontaneously confessed his sins once again.

91. Gabriele Pirovano to Ludovico, Pavia, September 11, 1494. *Terzana* and *quartana* are fevers that come at regular intervals, have different symptoms, and release different type of putrid matter (depending on the humor in the body mostly affected). They were often the result of malarial diseases. Bloodletting and purging were often used to treat these fevers. On these and other types of fevers, see Galen, *De febrium differentiis* (Kühn 1964–1965, VII, 273–405; with a brief definition of tertian and quartan fevers at 298–299), Avicenna, *Canon,* Book 4, Fen I (esp. chaps. XXXIV–XXXIX for tertian fevers, and chaps. LX–LXV for quartan fevers); Michele Savonarola, *Practica canonica de febribus* (Venice: Giunta, 1552). On fevers, see also William F. Bynum and Vivian Nutton, eds., *Theories of Fever from Antiquity to the Enlightenment,* Medical History Suppl. N. 1 (London: The Wellcome Institute, 1981), esp. the essay by Iain M. Lonie, "Fever Pathology in the Sixteenth-Century: Tradition and Innovation," 19–44; Dell'Anna, *Dies critici,* vol. 1, esp. 155–390; and Siraisi, *Medieval and Renaissance Medicine,* 130–131.

92. Gabriele Pirovano to Ludovico, Pavia, September 13, 1494; Gian Galeazzo to Ludovico, Pavia, September 13, 1494. Bembo's account of the meeting, where he states that Gian Galeazzo was also present, is therefore factually inaccurate. See Bembo, *History of Venice,* 98–99.

93. Dionisio to Ludovico, Pavia, September 14 and 16, 1494. Rhubarb pills are also prescribed as laxatives, as indicated in another letter by Pirovano dated September 18, 1494.

94. See for instance the letter sent on September 22 by Dionisio, where it is said that the duke confessed to him that "omne volta montava li doleva il stomacho, et il palmo dele mane, in modo non poteva tenere le redene dela bria, et smontato se

moveva la collera. La qualcosa credo proceda per havere il stomacho talmente debile che forsi non digerisse come bisogneria per magnare assai bene come fa, et si dole de non possersi refare, come anche nel volto dimonstra per essere pallido et magro."

95. Jacopo Pusterla (*castellano*) to Ludovico, Pavia, September 28, 1494: "Ho parlato con Maestro Lazaro de Prasentia [sic] quali dici il male esser grande; non di manco spera però che se redurà ad bona convalescentia et de quanto succederà darò aviso ala prelibata vostra excellentia ala quale de continuo me ricomando."

96. Gabriele Pirovano to Ludovico, Pavia, October 12, 1494: "Illustrissimo et excellentissimo signore mio colendissimo, continuando lo adviso del stare del Signore duca secondo comadano le lettere della excellenza vestra et è mio debito. Havendole dato questa matina *secundo la hora electa* tre pillule e la manna, essendoli convenientissima medicina, et hano operato tanto laudabilmente, quanto desideravenno, col collere grosse putride cum che se causa il male et assitate [?] rose et gialde, che è cosa certo assay miranda come se ascondasi le materie diverse ne lo corpo humano, et non è poy maraveglia se sono varie febre et egritudine diverse. Sua Signoria è stata obediente a recevere dicta medicina, ma non è stato fato obediente al disinare. Imperó ha voluto disinare ad tavola cum la duchessa et ha passato la debita proportione del bevere; pui li havemo facto fare la penitentia, che non li havemo dato cosa alcuna in fino a cena, ne la quale ha tolto la pistata e ha bevuto pocho più di quello gli fu limitato in questo. Hore tre si è misso ad dormire et starà assai bene che il Dio il driza in meglio." Considering Pirovano's expertise in astrology and the general context, it is quite clear that the time chosen was determined astrologically.

97. Dionisio to Ludovico, Pavia, October 14, 1494: "Illustrissimo et excellentissimo Signore mio singularissimo, l'excellentia vostra vederà quanto li hano scripto questi medici del stare del Illustrissimo Signore Duca. Et per il desordine hano compreheso ha facto ne stano de malissima voglia, conoscendo el periculo che porria accadere. [. . .] mi e parso habia la lingua molto dopia, et perché li medici non voleno se li daghi fino una zania de vino et aqua quando magna, omne volta domanda de voler lavare la bocha per volerli inganare ad questo modo." The letter is cited in Magenta, *I Visconti e gli Sforza*, vol. 2, 536 (n. 1). According to Magenta, a *zania* is about one-fifth of a liter.

98. Dionisio to Ludovico, Pavia, October 18, 1494.

99. The ducal physicians to Ludovico, Pavia, October 19, 1494: "Hogi di giorno è andato sey volte de materie diverse et le ultime due egestione hano qualche dispositione de materia epaticha. Et havendo compreheso per epse egestione che ha magnato brugne, pere, et pome crude per tenerne continuamente da capo del lecto con monstrare de odorarle, ne è constato essere vero ne ha magnato. La qual cosa ha conturbati molto per conoscere si è misso in manifesto periculo et che uno simile desordine lo porria ridure ad termine che quanti rimedii li potessimo fare non li gioveariano." The letter is cited also in Magenta, *I Visconti e gli Sforza*, vol. 2, 536 (n. 1). A letter from Dionisio reports similar preoccupations.

100. Dionisio to Ludovico, Pavia, October 20, 1494.

101. ASMi, SPS 1464, Nicolò Cusano, Gabriele Pirovano, Lazzaro Dactilo, and Pietro Antonio Marliani to Ludovico, Pavia, October 20, 1494. The letter is cited in full in Magenta, *I Visconti e gli Sforza*, vol. 1, 461. Gabriele Pirovano appears as lecturer in medicine (*Ad lecturam medicinae*) at Pavia in the year 1481; during the 1480s and 1490s he was assigned to the care of Gian Galeazzo Sforza. Lazzaro Dactilo (o Dataro), who was *promotore* of degrees in arts and medicine during the Sforza period, taught *Ad lecturam logicae ordinariae de mane, et sophistariae* in 1467; *Ad lecturam medicinae de nonis* in 1472; *Ad lecturam medicinae ordinariae de mane* in 1480 and again in 1496 and 1508. Less involved with the court, he was nonetheless a strong presence in Pavia. Pietro Antonio Marliani was the son of the famous ducal physician Giovanni Marliani. On these figures, see Corradi, *Memorie e documenti,* vol. 1, 118–120. Pirovano is the author of a *Defensio astronomiae* seemingly written in response to Pico's *Disputationes*.

102. ASMi, SPS 1464, Nicolò Cusano and Gabriele Pirovano to Ludovico, Pavia, October 20, 21 hour (3:30 p.m.), 1494: "Illustrissime et excellentissime domine nostre colendissime, dopo lo adviso de hore 17 lo Illustrissimo duca se resvegliò circa hore 18 et ad noy non c'è parso redonato alla virtute secundo che speravamo dovesse fare dopo tale refectione et sompno [sic] ma ne pariva quodammodo opressa più la virtute et multo debile et havendoli dato uno pocho de stilato fra pocho li sopravene uno salto et moto tremulo nel stomaco cum una gurgitatide di ventositate et strepito di aqua che sensibilmente se sentiva movere et li faceva dolore et suffugatione che comunicava al fidago et ala milsa cum dolore maggiore nel fidago al tochare; pur cessava et pariva descendesse tale rugito ad le intestine, cosse che sono di grande timore in medicina, et che haveriemo poche o nulle medicine. Continuaremo maxime stando etiam lo *influxu horribile di celo et per ecclipse et per directione di la nativitate* cum se vostra excellentia per il passato perché etiam in medicina solus casus virtutis est per se signum malum noy continuaremo li remedii apportuni et necessarii quanto poteramo et non li mancharemo di previssione ali casi futuri et provedere ad quello poria accadere continue et sopravenire di novo facelmente in tale caso che poy accadendo non saria in nostra libertate la reductione ma seria in manifesto periculo di manchare subito che dio non voglia." Emphasis mine.

103. ASMi, SPS 1469, October 22, 1494, Ludovico communicates to the ducal general Galeazzo Sforza that he had been appointed Duke of Milan by the Privy Council, by the finance ministers, and with "grande alegrezza de tucto questo populo." On his election, this too determined astrologically, see Chapter 5.

104. The term "directione" has quite clearly a technical meaning in this context. The astrological technique of medieval directions is still very poorly understood by modern historians of astrology. The inspection of Gian Galeazzo's nativity and that of the chart of the eclipse preceding his death seems to suggest that in the year when Gian Galeazzo died, the Sun, which was the *hyleg* (giver of life) of the natal chart,

was "directed" so as to be in square to Saturn, who was the lord of the illness, and in its "fall" in Aries. This was indeed an inauspicious situation. On these terms, see Eade, *The Forgotten Sky*. I wish to thank Robert Hand and Dorian Greenbaum for explaining this technique to me.

105. As noted, internal evidence indicates that some of the medical bulletins are now lost, and this might help explain why medical astrology is for the most part absent from the later correspondence.

106. Francesca M. Vaglienti, "Gian Galeazzo Maria Sforza, duca di Milano," *DBI* 54 (2000), 391–397.

107. BLO, MS Canon. Misc. 24, fol. 120r. The two charts are on fols. 20r (nativity) and 19v (eclipse): "Is miser a patruo ut dicitur primum subiugatus deinde maleficiis, infectus, insipiens vel amens effectus est et currente 1495 obiit ut creditur morte biothanate. Nam patruus eius Ludovicus Ducatum subiit quamvis filium cui regnum pervenibat dictus Joannis relinquisset."

108. J. L. Heiberg, *Beiträge zur Geschichte Georg Valla's und seiner Bibliothek* (Leipzig: Otto Harrassowitz, 1896), 426–427, Giorgio Valla to Gian Giacomo Trivulzio, Venice, September 30, 1494: "Ceterum reor tibi innotuisse, quod nuper palam factum est Johannem Galeacium Mediolani ducem vita defunctum esse, et statim Ludovicum Sfortiam cum tunica aurata ducalique capitis tegmine prodiisse in vulgus et a multitudine vociferatum ducis nomine. An exciderit tibi, quod duodecimo iam anno significavi, nescio. Repetam, si forte minus meministi. Nam cum Johannes Marlianus Ducis Johannis Galeacii forte mederetur aegritudini, protuli ipsi medico fore, ut tali aegritudine e vita decederet. [Scilicet veneno, ut palam fertur]. Id medicum sollicitum egit meque nescente Ludovico significavit; qui statim e cubiculariis suis unum ad se me misit accitum, cumque me in arcem contulissem, significatumque Ludovico esset in senatu, advenisse, cum legatos quosdam missos fecisset, omnes alios amovit arbitros. Restitimus ambo; de duce quid opinarer, percunctatus est; respondi fore, ut tali langore esset e vita decessurus, futurusque dux Mediolani ipse Ludovicus; qui respondit 'erit, ut volam.' Iteravi ego 'ita fore videtur.' Inde abeunti mandavit, ea apud me essent. Ego vero soli tibi omnem rem aperui. Tu respondisti: bonus profecto est Ludovicus, neque unquam crediderim, etiam si possit, se ducem effecturum. Nunc, quid sit, tenes. Post, quae in aliis litteris exposimus, ut opinor, intuebere: non diu fore censeo, cum prorsus Sfortiarum nomen delebitur. Vale, et cum has legeris, te etiam atque etiam oro, ut eas igni immergas."

109. Ibid., 12.

110. Lubkin, *A Renaissance Court,* 192–193. Rosmini, *Vita di Gian-Jacopo Trivulzio,* vol. 1, 7.

111. On Trivulzio's thriving career under the French, see Letizia Arcangeli, "Gian Giacomo Trivulzio marchese di Vigevano e il governo francese nello stato di Milano," in eadem, *Gentiluomini di Lombardia. Ricerche sull'aristocrazia padana nel Rinascimento* (Milan: Unicopli, 2003), 3–70.

112. Rosmini, *Vita di Gian-Jacopo Trivulzio*, vol. 1, 212 and vol. 2, 203.

113. Heiberg, *Beiträge zur Geschichte Georg Valla's*, 368–369.

114. Paolo Giovio, *Le vite de i dodeci Visconti, e di Sforza prencipi di Milano . . . tradotte per M. Lodovico Domenichi* (Venice: Gabriel Giolito de' Ferrari, 1558), 216: "[. . .] Carlo Ottavo Re di Francia andando all'acquisto del Regno di Napoli per l'antica ragione della heredità Angioina, passate l'Alpi se ne venne a Pavia, per visitare Giovan Galeazzo, il quale di là a due giorni haveva a morire. Il quale poi che fu morto, et non senza sospetto di veleno, Lodovico suo zio prese l'insegne, fu gridato Duca et Prencipe di Milano."

115. Pietro Giustiniani, *Rerum Venetarum ab urbe condita historia* (Venice: Comino Tridino da Monferrato, 1560), 346: "Inde Carolus sex millibus equitum septus, Ticinum proficiescitur: ubi Ioannes Maria Galeatii Sforciae Mediolani Ducis filius gravi morbo pressus iacebat: qui non multo post non sine veneni a Ludovico Patruo insidiose dati suspicione e vita decessit. Ipseque ducatum per violentum nepotis obitum tyrannice invasit."

116. Guicciardini, *The History of Italy*, 53.

117. Ibid., 54.

118. According to Corio, Ludovico was willing to offer any sum of money (*impetrationem eam pecuniae summam quantacunque*) to receive the investiture. On the investiture, see Corio, *Storia di Milano*, vol. 2, 1507. On the background to the investiture and what this entailed, see now Pellegrini, *Ascanio Sforza*, vol. 1, 432–433 and vol. 2, 463–471, 526, 539, 561–562. On Ludovico's complex political maneuvers at the time, see also Paolo Negri, "Milano, Ferrara e l'Impero durante l'impresa di Carlo VIII in Italia," *ASL* 44 (1917): 423–571, and Chapter 5.

119. Corio, *Storia di Milano*, vol. 2, 1508–1509, 1524–1525. The same amount is quoted by Giovio, who probably relied on Corio. See Giovio, *Le vite de i dodeci Visconti, e di Sforza principi di Milano*, 215. On the legal issues surrounding this investiture, see Black, *Absolutism in Renaissance Milan*.

120. Ludovico to the Bishop of Brixen, December 1494, as quoted in Magenta, *I Visconti e gli Sforza*, vol. 2, 469–470: "Factum est ab R. D. V. quod summa ipsius bonitas et nostra singularis erga eam affectio postulabat cum se honoris nostri defensorem prebuerit contra illos, qui ad calunniam nostram iactabant illustrissimum nepotem nostrum non morti naturali defecisse, nam preter quod se ostendit amicitie bene memorem, suscepit defensionem innocentie nostre et veritatem tutata est. Qua in re etsi nihil ab ea inexpectatum habeamus, gratias tamen agimus quod et ipsa veri amici officium implevit, et ut idem faciat germanus suus apud serenissimum regem, si opus fuerit, litteris suis monuerit. Ut autem D.V. possit liberius loqui contra hos detractores narrabimus quemadmodum hec mors sucesserit et vobis et aliis inopinata videri potuit." See also Ludovico's letter to Maximilian I in Giuseppe Canestrini, "Lettera di Ludovico il Moro all'Imperatore Massimiliano (30 Settembre 1495)," *ASI*, Appendice 3 (1846): 113–124, esp. 120; and his letter to Alexander VI, quoted in

Arturo Segrè, "Ludovico Sforza, detto il Moro, e la Repubblica di Venezia (dall'autunno 1494 alla primavera 1495)," *ASL*, 18 (1902), 249–317, and 20 (1903), 33–109, 368–443, esp. 252–253.

121. Ludovico to the Bishop of Bressanone, December 1494, as quoted in Magenta, *I Visconti e gli Sforza*, vol. 2, 469–470: "Egrotavit plusquam mensem illustrissimus nepos noster in opido Papie: astabat illustrissima eius coniunx, curabant eum tres medici qui eius naturam usque ab incunabilis noverant; hii morbum eius levissimum et sanabilem affirmabant: id propter de egritudine nihil scribendum videbatur. Nos dum egrotabat propter christianissimi Regis in Italiam adventum in magnis occupationibus detinebamur et quandoque Alexandrie, quandoque Annoni, quandoque alibi morabamur et ab Papia fere semper abfuimus; non tamen defuimus quin in eo visendo, cum data est opportunitas, et in solicitando medicos ut eum diligenter curarent parentis officium peregerimus. Christianissimus rex cum venit Papiam accessit ad eum, patuit et accessus aliis proceribus Galic, qui se verum fateri volunt negare nequeunt quin morte naturali defecerit. Preterea novit ne R. D. V. nos eiusmodi nature esse ut cupiditate dominandi vellemus cum perditione anime nostre per totum orbem infames fieri? Non profecto si totius orbis imperium assequi possemus. Nihil enim ab natura nostra alienius fuit quam non modo cogitare in morte illustrissimi nepotis nostri, quem semper paterna charitate complexi sumus, sed nec in morte eorum quos scimus odio capitali nos prosequi."

122. Ibid.: "Processit egritudo, nec medici salutem eius desperarunt nisi per unum diem ante quam e vita migraret. Placentie cum christianissimo rege eramus, ibi gravitas egritudinis nunciata fuit, et, dum properabamus Papiam ut ea omnia ageremus que salutem recuperare possent, in itinere nunciatum est eum obiisse, quod summo nos affecit dolore."

123. Ibid.: "Potest itaque et debet R. D. V. in defensione honoris nostri audacter pergere cum amicum innocentem et veritatem ipsam sit defensura, quod, ut faciat, eum rogamus et obsecramus. Nec minus etiam defensionem nostram suscipiat rogamus contra hos qui adventum christianissimi regis in Italiam in sinistram partem accipiunt. Nihil enim egimus nec agimus, inscio serenissimo rege, cuius voluntati precipuum est nobis optemperare."

124. ASMi, SPS 1464, Dionisio to Ludovico, Pavia, October 7, 1494. Dionisio reported how: "Facto andare da canto ogniuno, me ha dicto, in secreto, se credeva che l'Excellentia Vostra li volesse bene. Li ho risposto maravegliarmi dell'Excellentia Sua et le demonstratione passate et presente li debono essere bon testimonio, con altre parole che per la veritate li ho dicto. La qualcosa credo sii proceduta per relatione de qualche bone lingue. Dopoy me ha dicto se l'Excellentia Vostra dimonstra haver displacentia del male suo, li ho risposto quello mi è parso et che per dimonstratione ho compreso nel tempo che sono stato lì." Quoted also in Magenta, *I Visconti e gli Sforza*, 537 (n. 1). For other historians who commented on Gian Galeazzo's death, see also Segrè, "Ludovico Sforza," 251–252.

125. Guicciardini, *The History of Italy,* 53–54. It is interesting to notice that in 1485 Theodoro Guainieri (or Guarneri) da Pavia, son of Antonio, was the personal physician of Ascanio Sforza. Apparently, Ascanio later dispatched him to France as an informant in 1488. See Pellegrini, *Ascanio Maria Sforza,* vol. 1, 175, 283. The relationship between the two men would explain why Pier Martire d'Anghiera corresponded extensively with both. See Pier Martire d'Anghiera, *Opus epistolarum [Petri Martyris Anglerii Mediolanensis, Protonotarii Apostolici . . .* (Amsterdam: Daniel Elzevier, 1670). The story of Teodoro's examination of Gian Galeazzo and the fact that he recognized the signs of poisoning is recounted also in BMV, MS Lat. 3506, Borgia, *Historiae,* Book I, fol. 9r.

126. Guicciardini, *The History of Italy,* 55.

127. "Ecce Maurus Sfortia vipereo insigni suum praeferens ingenium, exitium Alfonso paratum prospiciens, tanquam metu liberatus et patentiorem peccandi campum nactus, quo liberius tyrannide frueretur (ut superius memini) Jo. Galeatium fratris filium juvenem innocentem verum mediolanensium principem veneno tolli curavit. Et cuius iniuriam defendere ac salutem curare debuit, eum nefario extinxit-poculo, cum citra invidiam et odium regnare potuisset et Imperii habenas pro libidine moderari si adolescentis tutelam uti coeperat administrare perseverasset. Id cum cives advenaeque passim atque impune omnes praedicarent: nemo erat qui hominis impietatem tantam non quovis etiam maledicentiae genere damnaret facinus videlicet inauditum, atrox, miserabile et christiano [sic] potissimum in homine detestandum, atque eo quidem praecipue in homine qui pater, qui rector, qui rei omnis principatusque universi temperator esset, atque administrator, idem pro voluntate suaque ex sententia cuncta regeret. Quis tunc facinus illud violentum non ruiturum atque in Tyrannum ipsum et filios casurum est auguratus?" Borgia, *Historiae,* Book I, fol. 14r–v.

128. Bembo, *History of Venice,* vol. 1, 98–101: "This traveling there and back was the last journey Gian Galeazzo ever made. Having taken to his bed with an apparent case of diarrhea (*profluvio ventris*)—though really, it was believed, through drinking poison given him by his uncle (*veneni a patruo dati haustu*)—he died the following month. Such is the hold that ambition and the corrupt desire to rule have on the hearts and minds of men. [. . .] Ludovico [. . .] returned to Milan the next day on hearing of his nephew's death. As he walked through the city dressed in ducal clothes, he permitted himself to be hailed as duke, though in reality he craved it more than anything else—nor was this action any the less deliberate for the fact that Gian Galeazzo had left behind at his death two children, Francesco and Maria." See also Sigismondo Tizio, BNCF, MS II, V, 140, vol. 6, 346 (erroneously listed under 1495): "Mortuo autem Joanne iam extra adolescentia constituto Galeaz Ducis Mediolani filio, quem variis *ex causis veneno extinxisse Ludovicus Patruum homines suspicabantur;* Ducatum enim Ludovicus semper a Ferdinando Neapolis Rege, et Joanne ipso nepote a die interfecti Galeaz fratris gubernaverat. Monetae insuper et aureae, et argenteae

tametsi ex uno latere Joannis nomen onstenderent, in altero tamen ita scriptus erat, LUDOVICO PATRUO GUBERNANTE. Mors itaque Joannis Ludovicus omnibus exosum reddiderat." Emphasis mine. On Simone Del Pozzo's testimony, see Felice Fossati, "Ludovico Sforza avvelenatore del nipote? (Testimonianza di Simone Del Pozzo)," *ASL* 2, serie 4 (1904): 162–171, esp. 169.

129. Magenta claims that, "The letters of the doctors and of the *famigliari* of Ludovico are such that murder can be excluded," but, as suggested in this chapter, it would be fairer to say that the evidence is inconclusive. See Magenta, *I Visconti e gli Sforza*, vol. 1, 533–537.

5. The Viper and the Eagle

1. Much of the information on Bonatti comes either from himself or from the second edition of the history of Florence written by the Florentine Filippo Villani. See Talbot R. Selby, "Filippo Villani and his Vita of Guido Bonatti," *Renaissance News* 11.4 (1958): 243–248. For an accessible account of Bonatti's life, see now Benjamin N. Dykes, "Introduction," in Guido Bonatti, *Book of Astronomy*, trans. Benjamin N. Dykes, 2 vols. (Golden Valley, Minnesota: The Cazimi Press, 2007). For discussion of his real and alleged clients, see xxxvi–xliv and index.

2. Cardano, *OO*, vol. 5.4, 104. This passage is translated in Grafton, *Cardano's Cosmos*, 144. The passage reveals feeble echoes of Villani's narrative about the kind of services that Bonatti provided to his patron Guido da Montefeltro, albeit put in a negative light. See Filippo Villani, *De origine civitatis Florentie et de eiudsem famosis civibus*, ed. Giuliano Tanturli (Padua: Antenore, 1997), 459–460.

3. See APD, 401, doc. 11 (Pavia, June 1, 1491, Gian Galeazzo Maria Sforza nominates Ambrogio Varesi *senator*), and doc. 17 (Milan, November 11, 1493 "diploma" of concession of the fief of Rosate to Ambrogio Varesi), and document in ASMi, *Autografi*, Medici 219, Varesi, dated 1551. I wish to thank Countess Carola Pisani-Dossi for allowing me to consult the material preserved in her family archive. On Ludovico's serious illness in August 1487, Corio recounts how: "[I] Genovesi al vigesimo terzo de agosto dodeci ambasciatori mandarono a Milano per la conferma-tione de li loro capituli con il duca, ma per essere Ludovico Sforza vexato da gravis-sima infermitate, solo Luca Grimaldo in loco de tutti costituirono per suplire a la legatione e gli altri doppo septe giorni, grandemente essendo honorati dal principe, ritornarono a Genua, dove per suo duce crearono Giovanne Galeazo Maria Sforza, duca illustrissimo de Milano. [. . .] e tanto Ludovico se redusse in extremo che quasi come morto fu deliberato meterlo fuor dil castello, ma doppo varii appareri fu rete-nuto." See Corio, *Storia di Milano*, vol. 2, 1470–1471. On Ludovico's illness, see also Alberto M. Cuomo, *Ambrogio Varese: un rosatese alla corte di Ludovico il Moro* (Rosate: Amministrazione comunale, 1987), 18–19.

4. APD, 401, doc. 2 (dated Milan, November 26, 1480): "Illustrissimus Dominus Lodovicus Maria Sfortia de Angleriae Patruus et Gubernator generalis noster carissimus superioribus diebus assignavit egregio Doctori Magistro Ambrosio de Rosate Phisico nostro dilecto Ducatos centum auri [. . .] pro honorantia ipsius officii, quod quidem preadictus Illuster Dominus Ludovicus ideo fecit ne singularis doctrina, integritas, ac sincera fides eiusdem Magistri Ambrosii omni ex parte destituta et aliquo honesto premio vacua videretur." See also Cuomo, *Ambrogio Varese,* 14–15, and 185 for a transcription of the document. Cuomo's study includes the transcription of some of the letters preserved in the ASMi and some of the most significant documents in APD. Since the publication of his book, however, the Pisani-Dossi Archive has been re-catalogued and Cuomo's references have been superseded by the new numeration indicated here.

5. Cuomo, *Ambrogio Varese,* 144.

6. Cardano was certainly one of the harshest critics, but already in the fifteenth century the Mantuan astrologer Bartolomeo Manfredi expressed his preference for genethlialogy over the use of interrogations. See Signorini, *Fortuna dell'astrologia,* 374 (doc. 35).

7. Grafton, *Cardano's Cosmos,* 96–108.

8. On Cardano's mission to reform astrology by returning to the classics, and particularly to the lesson of Ptolemy, see Grafton, *Cardano's Cosmos,* esp. 127–155. As far as we know Varesi never taught astrology at the *Studium* of Pavia, only medicine. See Cuomo, *Ambrogio Varese,* 139–145.

9. On marriage in the Renaissance, see Michela Di Giorgio and Christiane Klapisch-Zuber, eds., *Storia del matrimonio,* 3 vols. (Rome: Laterza, 1996).

10. See Chapter 4.

11. On Maximilian I's patronage of astrologers and his use of astrology for political propaganda see Hayton, *Astrologers and Astrology in Vienna;* and Mentgen, *Astrologie und Öffentlichkeit,* 242–244. On Maximilian's reign see also Larry Silver, *Marketing Maximilian: The Visual Ideology of a Holy Roman Emperor* (Princeton, NJ: Princeton University Press, 2008).

12. Julia Cartwright, *Beatrice d'Este, Duchess of Milan, 1475–1497: A Study of the Renaissance* (London: J. M. Dent, 1912), 6–7. On Beatrice, see now also Luisa Giordano, ed., *Beatrice d'Este, 1475–1497* (Pisa: ETS, 2008).

13. Alessandro Luzio and Rodolfo Renier, "Delle relazioni di Isabella d'Este Gonzaga con Ludovico e Beatrice Sforza," *ASL* 17 (1890): 74–119, 346–399, 619–674, esp. 76–77; Dina, "Isabella d'Aragona," 276; and Cartwright, *Beatrice D'Este,* 8–9.

14. Giulio Porro, "Nozze di Beatrice d'Este e di Anna Sforza. Documenti copiati dagli originali esistenti nell'Archivio di Stato di Milano," *ASL* 9 (1882): 483–534, esp. 484–487.

15. Ibid., 492.

16. Now in ASMi, SPS 1470, *Instrumentum ratificationis sponsalium inter Illustrissimum dominum Ludovicum Mariam Sforciam Vicecomitem et Illustrissimam dominam Beatricem filiam Illustrissimi dominis Ducis Ferrarie.* See also Cuomo, *Ambrogio Varese*, 60.

17. *Instructio particularis Francisci Casati ducali secretari ituri Ferrariam* (Milan, April 12, 1490): "non ne è parso impertinente consultare li astronomi nostri da li quali ne è confortato el decimo octavo de Julio proximo per fortunato et prospero alla coniunctione nostra con lei." The document is quoted in full in Porro, "Nozze di Beatrice," 491. Presently it could not be located in ASMi, SPS 1470, as indicated in Guido Lopez, *Festa di nozze per Ludovico il Moro: Fasti nuziali e intrighi di potere alla corte degli Sforza, tra Milano, Vigevano e Ferrara* (Milan: Mursia, 2008), 39. On it, see also Cuomo, *Ambrogio Varese*, 61.

18. At the time Ercole and Eleonora became deeply concerned that the marriage would never happen. It was well known that Ludovico Sforza had a long-time lover, Cecilia Gallerani, of whom he was very fond, and that he had considered making her his legitimate wife at some point. Her beauty and virtues are immortalized in Giovanni Bellincioni's sonnets and in Leonardo da Vinci's portrait, *The Lady with the Ermine,* now at Cracow. See Cartwright, *Beatrice d'Este,* 52–55.

19. ASMo, Ambasciatori, Milano 6, Giacomo Trotti to Ercole d'Este, Milan, November 16, 1490: "Dapoi intrando in ragionamento, io li dixi chel me dicesse ciò chel voleva pur chel non ragionasse de differir et dillatare la cosa perché seria cum troppo gran vostro charico & suo & anche pocha vostra satisfatione."

20. Ibid. "che li la sposaria & conduria dapoi a Parma a chavallo, non chel se volesse accompagnare in quello loco *per salvare l'ora & il puncto dela astrologia.*" Emphasis mine.

21. ASMo, Ambasciatori, Milano 6, Giacomo Trotti to Ercole d'Este, Vigevano, November 21, 1490.

22. Ibid.: "Pregandome che volando scriva a vostra celsitudine li debba fare intendere se sel dormirà in nave et in aqua & come & chi allogiarà in terra de masculi & femme per ordinare le cosse sue per camino & viaggio al meglio che potrà [. . .]. De la qual comitiva & del tuto el voria la lista & esser chiarito de predictis punctualmente cum questa condictione: che non passi li xvi de zenaro che l'excellenza de Madama nostra cum la spoxa siano a Pavia, *perché a li 18 la possa sposar & coniungerse cum epsa,* perché non li essendo a dicto termine, potría poi passare parechi giorni et forsi anche qualche mese chel non potría ni sposare ni acompagnarse, e che spoxato & acompagnato chel sia la condurà a Milano." Emphasis mine. The letter is transcribed, with a few imperfections, in Lopez, *Festa di nozze,* 46–48. To indicate its importance, the pressing issue of the date—the party had to reach Pavia before the sixteenth—was repeated once more in a letter dated November 22.

23. ASMo, Ambasciatori, Milano 6, Giacomo Trotti to Ercole d'Este, Pavia, January 17, 1491: "Domane ale xvi hore se spoxa & benedise la Duchessa de Bari

molto cerimoniosamente per puncto de astrologia, che è il die de Marti, & la nocte seguente se acompagna cum epsa lo Illustrissimo signor Ludovico." From the context it is possible to infer that the "nocte seguente" was the same night of the marriage. The outcome of that night was not immediately known, but in a letter dated to the following day, Trotti indicated that the young bride has been "well instructed and admonished [on what to do]."

24. Cartwright, *Beatrice d'Este,* 8. Porro, "Nozze di Beatrice," 483.

25. ASMo, Ambasciatori, Milano 6, Giacomo Trotti to Ercole d'Este, Vigevano, November 30, 1490: "[Ludovico] ha messo quisti suoi astrologi insieme per elegere uno giorno che sia foelice & bono per spoxare la predetta Madama Anna. [. . .] Messer Giovan Francesco Ghilino suo secretario [. . .] me ha dicto *che li astrologi li hanno refferto collegialmente che ali xviiii de zenaro avanti le xvii hore seria bono & foelice die & hora* ma che non è possibile perché acompagnandose sua excellentia ali xviii a Pavia esser ali xviiii a Milano a fare le sposaglie, perché ali xx el partirà da Pavia cum tuta la comitiva & venirà a Binascho che è al mezo dela via da Pavia a Milano che sono x miglia et ali xxi serà a Milano, et conoscendo questa imposibilità li *predicti astrologi li hanno facto intendere che li xxiii de zenaro avanti le xvi hore un pochetto serà bono & foelice die & hora ma che chiaramente & determinatamente non possono respondere non havendo la nativitate del predicto illustrissimo Domino Alfonsio, se ben hanno quela de Madonna Anna.* Per la qual cosa il predicto signor Ludovico desideraria che subito per vostra signoria li fusse dato notitia de epsa nativitate punctualmente e quanto più presto meglio." Emphasis mine. See also Lopez, *Festa di nozze,* 49–50.

26. ASMo, Ambasciatori, Milano 6, Giacomo Trotti to Ercole d'Este, Pavia, January 17 and 18, 1491. To attend it were all the major players of Milan's political life, with the exception of Duke Gian Galeazzo and his wife, Isabella. See also Luzio and Ranier, "Delle relazioni di Isabella Gonzaga d'Este con Ludovico e Beatrice Sforza," 83, n. 1; and Tristano Calco, *Nuptiae Mediolanensium et Estensium Principium scilicet Ludovici Mariae cum Beatrice, Alphonsi Estensis sorore, vicissimque Alphonsi cum Anna, Ludovici nepote,* now reproduced and translated in Lopez, *Festa di Nozze,* 118–142, esp. 124–129.

27. For these arrangements, and others related to the marriage celebrations, see documents cited in Porro, "Nozze di Beatrice," 492–534. On the decorations, see also Calco, *Nuptiae Mediolanensium et Estensium Principium.* On the significance of these and other decorative undertakings, see also Welch, *Art and Authority,* esp. chaps. 7 and 8. For the lavish celebrations and the week-long festivities that followed the marriages of Ludovico to Beatrice, and of Anna to Alfonso, see also Cartwright, *Beatrice d'Este,* 65–73.

28. Calco, *Nuptiae Mediolanensium et Estensium Principium.* On the genre of the *prose epithalamium,* or marriage oration, in the Renaissance, see Anthony d'Elia, *The Renaissance of Marriage in Fifteenth-Century Italy* (Cambridge, MA: Harvard University Press, 2005), esp. chaps. 2 and 3.

29. ASMo, Ambasciatori, Milano 6, Giacomo Trotti to Ercole d'Este, Milan, January 21, 1491: "dapoi disinare, venissimo da Pavia a Binascho, dove allogiassimo la nocte, et questa matina che è Venerdì ali xxi del presente, avanti le xiii hore, fussimo tutti a chavallo, et per uno teribelissimo & crudelissimo fredo, per puncto de astrologia se ne venissimo a Sancto Eustorgio [. . .]." Quoted also in Lopez, *Feste di nozze*, 79–80, and Cuomo, *Ambrogio Varese*, 62. The 13th hour of the twentieth would have corresponded roughly to 6 a.m. of January 21. On that day in Milan the Sun would have risen shortly before 8 a.m.

30. Calco, *Nuptiae Mediolanensium et Estensium Principium*, in Lopez, *Festa di Nozze*, 124–125. Binasco is about 15 km south-west of Milan, a distance that, on a good day, could be reached in less than four hours by foot (probably two or three hours on a horse, depending on the pace). The 5th hour of the day would correspond roughly to 1 p.m. on January 21. We can presume, therefore, that the contingent woke up when it was still dark (around 6 a.m.), got ready, and set off to reach the city walls before midday. The small discrepancy between the two sources can be easily explained by assuming that the contingent may have taken a few breaks along the way and that Ludovico may have given himself some margin of maneuver to reach the city with ease.

31. Calco in Lopez, *Festa di Nozze*, 122–123 and 137 (where Calco connects the comet with the favorable birth of Francesco II Sforza, son of Gian Galeazzo and Isabella). Depending on a large number of factors such as speed, direction, shape, color, and position in the sky, comets could herald positive or negative events. See Schechner, *Comets, Popular Culture*, chap. 3. The comet was visible in the sky over Nurenberg on January 17, as recorded by Regiomontanus's assistant Bernard Walther. See Johannes Regiomontanus, *Opera Collectanea. Faksimiledrucke von neun Schriften Regiomontans und einer von ihm gedruckten Schrift seines Lehrers Purbach. Zusammengestellt und mit einer Einleitung hrsg. von Felix Schmeidler* (Osnabrück: Otto Zeller, 1972), 682.

32. Attilio Portioli, "Nascita di Massimiliano Sforza," *ASL* 9 (1882): 325–334.

33. ASMa, AG 1630, fol. 143r, Teodora Angeli to Isabella d'Este, Milan, February 4, 1493: "poi da questa se usciva < . . . > salla tucta parata de bellissimi racci, dove dà udientia lo signore cum lo consi<glio>, et dove è da uno di capi in dicta salla almegio lo astrolagio [sic] de magistro Ambroxio, senza quello non si fa niente." This letter is cited in Portioli, "Nascita di Massimiliano Sforza," 329. The author, however, mistakenly reads "astrolagio" as "astrologio," and omits the important "de," thus missing the important reference to Varesi's astrolabe, which sat in the middle of the room where Ludovico met with members of the Privy Council.

34. Herlihy and Klapisch-Zuber have calculated that in Renaissance Florence one in five married women died of childbirth complications. David Herlihy and Christiane Klapisch-Zuber, *Tuscans and Their Families: A Study of the Florentine Catasto of 1427* (New Haven and London: Yale University Press, 1978), 276–277.

[originally published as *Les Toscans et leurs familles. Une étude du Catasto florentin de 1427* (Paris: Fondation Nationale des Sciences Politiques, 1978)].

35. ASMa, AG 1630, fol. 152r, Teodora Angeli to Isabella d'Este, Milan, February 23, 1493: "La illustrissima sorella s'è levata del parto et il mercorì che fu il primo de quadragesima, per puncto d'astrologia, che senza ciò far non si polle, ad hore xviii et uno terzo tucte due le duchesse inpaiolate et mo fora de paiolle, insieme cum la excellentia de madama et madonna Anna se andò a Madonna Sancta Maria dale Gratie ad referire gratie et laude deli loro parti bene discaricati." The letter is quoted in Portioli, "Nascita di Massimiliano," 331–332. Cf. ASMa, AG 1630, fol. 148r, (probably) Bernardino de' Prosperi to Isabella d'Este, Milan, February 20, 1493, where it is said, "Hozi dappoi mangiar a hore xviii e tri quarti queste illustrissime duchesse sono andate cum la compagnia de madama, de madonna Biancha et madonna Anna a presentarse ala chiesa de Santa Maria dele Gratie [. . .]." In November 1493 Varesi himself sent some remedies for a healthy pregnancy to Isabella through a midwife that Isabella had sent expressly to Milan. Cf. ASMa, AG 1630, fol. 170r, Ambrogio Varesi to Isabella d'Este, Milan, November 6, 1493.

36. ASMa, AG 1630, fol. 148r, (probably) Bernardino de' Prosperi to Isabella d'Este, Milan, February 20, 1493: "Et la partita nostra è ordinata, consultata però prima cum Magistro Ambroso per questo Illustrissimo Signore secundo la usanza, etc. Luni che viene che serà li 4 de marzo Madama cum pocha brigata andarà a stare a Cusago tri giorni [. . .] et la se starà sina ali xiii, poi se torna qui et ali xvii andemo a Pavia. Dui giorni se starà a Pavia, dui a Peasenza, et altretanti a Cremona." Beatrice and her own retinue had previously made a short trip to Cusago on February 25, only to return to Milan shortly after. This trip was also organized astrologically. See ASMa, AG 1630, fol. 152v, Teodora Angeli to Isabella d'Este, Milan, February 23, 1493: "Passando domane che serà luni a xxv, ad ore 15, per astrologia se partimo de qua et andiamo a Cuxago et là stiamo dui dì fermi; poi per st[r]ologia ne partimo [. . .]."

37. ASMa, AG 1630, fol. 183v, Bernardino de' Prosperi to Isabella d'Este, Vigevano, March 6, 1493, fol. 183v: "Magistro Ambroso de Roxate ha mandato a dire al Signore Ludovico che sua mogliere gli ha a fare un altro figliolo prima passa un anno. Non scio mo' se vostra signoria vorà patire che vostra sorella ne faci dui prima che vui uno. Sicché, patrona mia, stati de bona voglia che anchora vui cum la gratia de Dio non vi mancharà questa contenteza perché non seti già pegio accompagnata de uno marito che lei. Il predetto signore ha invidato madama al parto sperando nel prenostico del suo astrologo."

38. ASMa, AG 2991 (Isabella's Copialettere), fol. 51r, doc. 173, Isabella d'Este to Francesco Gonzaga, Ferrara, May 12, 1493. A puzzled Isabella reported: "non scio dove se causi questa subita partita, avisando vostra signoria chel vuole intrare qua per puncto de astrologia ale tredice hore, et a Parma se firmarà per aspectarli la moglie [Beatrice]." The document is cited without proper archival reference in Luzio and Renier, "Delle relazioni di Isabella d'Este," 368, n. 2.

39. Isabella's husband, Francesco, had recently been appointed capitain-general of the Venetian forces, and Isabella had been invited by the Doge to visit the city. Cartwright, *Beatrice d'Este*, 174.

40. ASMa, AG 1232, fol. 599r, Bernardino de' Prosperi to Isabella d'Este, Ferrara, May 24, 1493: "domatina a hore x e meza è ordinato de partirse, che è puncto preso per astrologia." The document is cited without archival reference in Luzio and Renier, "Delle relazioni di Isabella d'Este," 374.

41. Cartwright, *Beatrice d'Este*, 186–204, for a detailed account of this visit. For a detailed account of their arrival , see ASMi, SPS 1470, Beatrice d'Este to Ludovico, Venice, May 27, 1493, translated in Cartwright, *Beatrice d'Este*, 189–193.

42. Ibid., 186.

43. Segrè, "Ludovico Sforza," 257, 262, and 266. On these events, see also Pellegrini, *Ascanio Sforza*, 423–437, esp. 423–426; and Aubert, *La crisi degli antichi stati Italiani*, 7–37, esp. 20–26.

44. ASMi, SPS 1470, Beatrice d'Este to Ludovico, Venice, May 30, 1493. The letter is translated in Cartwright, *Beatrice d'Este*, 200–202 (the date, however, is incorrectly indicated as May 31, 1493).

45. ASMi, SPS 1470, Luigi Marliani to Ludovico, Venice, May 30, 1493: "L'aviso che la causa de la sua debile et minima indispositione è stata una descessa de humiditate alquanto accutta, quale ha causato nel declutire qualche et pocho dolore et l'origine de questo è nasciuto da divversi movimenti, mutatione de ayere, de vestimente, et da la privatione del dormire consueto de la sua signoria, né de questo me ne maraviglio perché la qualità de questo ayere essendo situata questa città in lochi palustri, è causa propinqua ad produre simile effecto."

46. ASMi, SPS 1470, Ludovico il Moro to Luigi Marliani, s.d. (but before May 26): "Magistro Luigi, non ne poteria essere stato più grato quello che ne haveti scripto del ben stare de nostra mogliere perché cosa non possemo sentire che ne piacia più che como intendere che la sia sana acioché habi ad retornare cum quella sanità cum la quale se partì da noi & perché havevamo inteso che el tempo del partire vostro era Sabato che fo heri, havendo noi veduto la nota de Magistro Ambrosio che diceva che era el Lunedì alli 3, che serà domane, si ne meravigliavamo molto che non havesseno servar l'ordine dato da Magistro Ambrosio. Ma havendo depoi inteso che ve partireti domane ne siamo restati satisfacti & benché noi havessemo dicto che el Sabato fosse el tempo, l'havevamo dicto persuadendosi che così havessi dicto Magistro Ambrosio. Per questo mo havereti havere advertentia al partire da Ferrara peroché el sii cum la bona hora [. . .]." The rest of the letter goes over the same content in different words.

47. See other letters between Beatrice and Ludovico in ASMi, SPS 1470 (where Ambrogio's updates on Ercole's health occur frequently), as well as letters of Bartolomeo Calco and others to Ludovico in ASMi, SPS 1468. Nicolò Cusano seems to have joined Varesi in caring for Ludovico's children at the birth of Ludovico's

second child, Sforza Sforza (later known as Francesco II Sforza). More updates on the children's health are included in ASMi, *Autografi,* Medici 219, Ambrogio Varesi.

48. ASMi, *Autografi,* Medici 219, Ambrogio Varesi, Varesi to Ludovico Sforza, Abbiategrasso, March 16, 1494: "per essere domane la quadratura de Saturno piacendo ala Illustrissima Signoria Vostra demoraremo qua demane er possodomane saremo li pasate le 18 hore." Varesi was possibly comparing the chart for March 17, 1494 with Ercole's nativity.

49. It took over two years before Erasmo Brasca, Ludovico's envoy in Paris, could inform him of the success of his mission. See Delaborde, *L'expédition de Charles VIII,* 220–224.

50. Ibid. 227–228.

51. Ibid., 228. On these negotiations, see also Pietro Filippo Pandolfini to Lorenzo de' Medici, ASF, MAP 50, letters dated December 1491–January 1492. Pandolfini seems rather critical of Ludovico and remarks on his excessive self-confidence about the league with the King of France.

52. Cartwright, *Beatrice d'Este,* 119–120. Delaborde, *L'expédition de Charles VIII,* 236–250.

53. "In el quale essendo immenso el piacere quale habiamo de essere con questi dui vinculi restituiti al loco nostro del amore consueto con quella christianissima casa, parimente la Maestà regia ne ha recevere et sentire contenteza per havere facto evidentia che con queste due firmissime cathene se ha colligato el stato nostro ad tutti li soi propositi, el quale voi con tutte le facultà, gente d'arme et persone proprie nostra et de lo Illustrissimo Signore nostro barba li offerereti et affirmareti essere prompto in omne suo bono piacere quanto le altre parte del regno [. . .] direti che in le cose de Italia dovi siamo per la parte nostra quanto alcuno altro, non essere stato quale più ne possa ne voglia per la existimatione et dignità de sua Maestà [. . .]", in ASMi, SPE, Francia 549, *Instructio M.corum Comitis Caiace, D. Hieronymi Tottaville, D. Vicecomitis et Comitis Caroli Bezoiosi, et Agustini Chalci secretarii proficientim ad Christianissimum Francorum Regem* (Milan, February 21, 1492). See also Delaborde, *L'expédition de Charles VIII,* 237.

54. Cartwright, *Beatrice d'Este,* 120.

55. ASMi, SPE Francia 549, Giovanni Francesco d'Aragona di Sanseverino (Count of Caiazzo), Girolamo Tuttavilla, Galeazzo Visconti and Carlo Belgioioso to Ludovico, Paris, March 29, 1492.

56. ASMi, SPE Francia 549, Agostino Calco to Ludovico, Paris, March 29, 1492.

57. "direti che javendo de resente havuto dal Re de Ingelterra le lettere quale ve havemo date, ce è parso col mezo vostro farne participatione a sua Maestà, prima perché epsa cognosca che essendo venuti cum lei ad queste redintegratione de la consueta benevolentia havemo procedere cum omne sincerità et non tenerli celata cosa [. . .] Anci nostro fermo proposito premonere l'amicitia sua a tutte le altre del mundo [. . .]." ASMi, SPE Francia 549, *Instructio secretior Magnifici Comitis Caiacie de his*

quae Christianissimo Francorum Regi dicturis est seorsum ab aliis (Milan, February 21, 1492). See also, *Instructio particularis Magnifice Comitis Caiace* and another *Instructio* (whose title is not fully legible) in the same location (both dated Milan, February 22, 1492). See also Delaborde, *L'expédition de Charles VIII*, 238.

58. *Instructio particularis.* The Count of Caiazzo mentions Ferrante's machinations against Ludovico in a letter dated Lyon, March 9, 1492. Both documents are in ASMi, SPE Francia 549. See also Delaborde, *L'expédition de Charles VIII*, 239.

59. ASMi, SPE, Francia 549. See, for instance, the letter sent by the Count of Caiazzo, Tuttavilla, and Carlo de Longhi to Ludovico from Turin dated February 27, 1492. They later complained that, at their entry into Lyon, they had not been received by any dignitaries.

60. ASMi, SPE Francia 549. In his letter dated Paris, March 11, 1492, Erasmo Brasca said clearly that the king was "not only short of cash, but in great debt." Other letters from Brasca make clear that the king's favorites had requested remuneration in exchange for exerting their influence on the king to confirm the fief of Genoa and the league. See also Delaborde, *L'expédition de Charles VIII*, 220–224, 239–240.

61. Delaborde, *L'expédition de Charles VIII*, 248 on the document, preserved in French in BNF, MS Latin 10133, fol. 478r, and in Italian in ASMi, SPE, *Trattati*. See also Abel Desjardins, *Négociations diplomatiques de la France avec la Toscane*, 5 vols. (Paris: Imprimerie imperiale, 1859–1875), vol. 1, 545–556.

62. On Charles VIII's ambiguous position and the pressures exerted on him by other courtiers, see Delaborde, *L'expédition de Charles VIII*, 257–258.

63. On these issues, see Black, *Absolutism in Renaissance Milan.*

64. For the history of France in this period, see the concise narrative of David Potter, *A History of France, 1460–1560: The Emergence of a Nation-State* (Basingstoke: Macmillan, 1995), 251–260; and Robert J. Knecht, *The Rise and Fall of Renaissance France, 1483–1610* (Oxford: Blackwell, 2001; first ed. 1996), 22–45.

65. On this alliance and Ascanio's delicate role in it, see Pellegrini, *Ascanio Sforza*, vol. 1, 423–437, esp. 423–426.

66. Ibid., 426.

67. A large number of fifteenth- and sixteenth-century historians blamed Ludovico for the invasion of Italy and the start of the Italian Wars. This included the Milanese Corio, who spoke of "an unquenchable fire" set up by Ludovico Sforza—a fire that not only destroyed the Sforza dynasty, but "almost the whole of Italy" (Corio, *Storia di Milano*, vol. 2, 1480–1481). At times, the responsibility was shared with Ercole d'Este and Alexander VI. See Segrè, "Ludovico Sforza," 254–255; and Michael Mallett, "Personalities and Pressures: Italian Involvement in the French Invasion of 1494," in Abulafia, *The French Descent*, 151–163.

68. One of the first to challenge the old interpretation was Arturo Segrè, who, however, was sometimes unnecessarily apologetic toward Ludovico. Segrè, "Ludovico Sforza."

69. Ibid., 257. Mallet, "Personalities and Pressures," 152–153.

70. Pellegrini, *Ascanio Sforza*, vol. 1, 428–430. Mallet, "Personalities and Pressures."

71. Pellegrini, *Ascanio Sforza*, vol. 1, 431. Delaborde, *L'expédition de Charles VIII,* 230, 246.

72. Cartwright, *Beatrice d'Este*, 196–197.

73. ASMi, *Autografi*, Medici 219, Ambrogio Varesi to Ludovico, Milan, October 8, 1493, h. 15: "Chel sya in via me il manifesta il Sole signore del 10ª in lo ascendente, Venere significatrice dela domanda in lo asendente, la Luna in la xiª, la separatione dela luna da Mercurio, quale cosse indubitamente denotano il meso [sic] essere partito et essere in viagio, et così che la lettera essere siglata, et perché Mercurio se parte da Venere fortuna del cielo, in dita lettera gli sariano cose piaciarano ala Illustrissima et excellentissima Signoria Vostra."

74. Ibid. The letter, which is extremely confused and makes constant reference to other letters we no longer possess, is transcribed in full in Cuomo, *Ambrogio Varese,* 67–68. Interestingly, Varesi used the revolution of Ludovico Sforza's nativity to predict these events.

75. ASMi, *Autografi*, Medici 219, Ambrogio Varesi to Ludovico, Milan, October 8, 1493, h. 15: "et per salvare l'altra resposta, quale fu che forse tardariano insina 25 del presente et che ali tri dì del mese proximo per la revolutione dela nativitade dela Illustrissima Signoria Vostra et dela tutella, quale così demostravano credo questo sarà poy la venuta deli altri oratori forse; overo che circha quelli dì si celebrarà poy lo efeto deli oratori."

76. Ibid.: "perché in cello non posso, né se pò vedere così particularmente chomo le cose de qua soto."

77. Cartwright, *Beatrice d'Este*, 209.

78. Much like medicine, astrology was founded on theoretical principles, but its practice was conjectural and its certainly was an issue of debate within the discipline. For the example of medicine, partly *scientia,* partly *ars,* see Siraisi, *Taddeo Alderotti,* 118–139; Michael McVaugh, "The Nature and Limits of Certitude at Early Fourteenth-Century Montpellier," in McVaugh and Siraisi, *Renaissance Medical Learning,* 62–84, esp. 67–75; Katharine Park, "Natural Particulars: Medical Epistemology, Practice, and the Literature of Healing Springs," in Anthony Grafton and Nancy Siraisi, eds., *Natural Particulars: Nature and the Disciplines in Renaissance Europe* (Cambridge, MA: The MIT Press, 1999), 347–367.

79. The story of the negotiations and of the marriage with Bianca Maria is concisely recounted in Hermann Wiesflecker, *Maximilian I: das Reich, Österreich und Europa an der Wende zur Neuzeit,* 5 vols. (Munich: Oldenbourg, 1971–1986), vol. 1, 363–372. Maximilian ruled as *de facto* emperor until 1508, when he was formally invested with the title in Trent. On Maximilian's reign, see also Manfred Hollegger, *Maximilian I (1459–1519). Herrschen und Mensch einer Zeitenwende* (Stuttgard: W. Kohlhammer, 2005).

80. Cartwright, *Beatrice d'Este*, 211. A very lengthy and detailed letter from Beatrice to Isabella is translated in full in Ibid., 211–16. The ceremony and the journey to Innsbruck are described in detail also in Corio, *Storia di Milano*, vol. 2, 1528–1529.

81. Wiesflecker, *Maximilian I*, vol. 1, 363–368.

82. Frederick III wanted to consummate the marriage on the day of the conjunction of Venus with the Sun. The other important conjuction seems to have been that of Venus with Mercury, which had also occured on the marriage day of a number of Habsburg rulers. See Helmut Grössing and Franz Graf-Stuhlhofer, "Versuch einer Deutung der Rolle der Astrologie in den persönlichen und politischen Entscheidungen einiger Habsburger des Spätmittelalters," *Anzeiger der Österreichichischen Akademie der Wissenschaften*, phil.-hist. Kl., 117 (1980): 267–283, esp. 275–283.

83. Gregory Harwell, "Ruled by Mars: Astrology and the Habsburg-Sforza Alliance of 1494," unpublished paper, Renaissance Society of America, Cambridge (UK), 2005. Harwell makes a reasonably convincing case in favor of this hypothesis. I wish to thank Gregory Harwell for kindly sending me his unpublished paper. Other evidence presented here confirms that Maximilian had planned his marriage ceremony astrologically.

84. Grössing and Stuhlhofer, "Versuch einer Deutung der Rolle der Astrologie," 282.

85. The conjunction between Venus and the Sun occurs every 9.6 months, while that of Venus and Mercury, despite being much more frequent, happens at irregular intervals. Grössing and Stuhlhofer, "Versuch einer Deutung der Rolle der Astrologie," 276.

86. The horoscope is in Regiomontanus, *Opera collectanea*, 1–33. See also Johannes Schöner, *De iudiciis nativitatum libri tres* (Nuremberg: Johannes Montanus and Ulricus Neuber, 1545). Grössing believes it unlikely that the young Regiomontanus, then only fourteen years old, was entrusted with the task. The horoscope was more probably cast by his teacher Peuerbach or another court astrologer, from whom Regiomontanus must have copied it. See Grössing and Stuhlhofer, "Versuch einer Deutung der Rolle der Astrologie," 270–273. On this horoscope, see also Hayton, *Astrologers and Astrology in Vienna*, 32–35.

87. Cardano, *OO*, vol. 5.7, 501.

88. Schöner, *De iudiciis nativitatum*, fol. 8v: "Principem itaque nativitatis huius propositae Martem esse censeo, Venerem sibi coniungendo. Nec Solem ab hac partecipatione secludendum puto, quod in perfecta ac amicabili cum Marte commixtione reperiatur, cum in dictis locis quoque 16 dignitates habeat."

89. This may be a veiled allusion to Maximilian's dynastic marriages and the political alliances that they brought with them. Ibid., fol. 60r: "Multos natus habebit amicos & bonos fautores, a quibus utilitatem & commodum sentiet, arguit

primus dominus triplicitatis undecimae domus, fortis in septima in sua exaltatione."

90. Ibid., fol. 62r: "Multos habebit natus inimicos occultos. Id accipitur a Saturno Almuten significatorum inimicorum, domino ascendentis quadrato aspectu premixto. Idem affirmatur ex eo quod idem est dominus ascendentis, & duodecimae."

91. According to Schöner, the fact that the luminaries of the two charts came together was a favorable factor for a successful marriage: "Cum luminaria in utraque nativitate tam viri quam mulieris adinvicem convenient tam ex parte signorum quam ex parte ascendentium, aut ex trino vel sextili mutuo sibi associabuntur luminaria [. . .], erit perpetuus amor inter eos" (fol. 51r).

92. "Saturnus etenim geniturae dominus in octavo, [. . .] plurima decernit mala." See Cardano, *OO*, vol. 5.7, 463–464.

93. This is briefly analyzed in Harwell's paper. Giasone de Mayno, *Epithalamion in nuptiis Maximiliani et Blancae Mariae* (Milan: Leonardo Pachel, after April 8, 1494). I have consulted BAV, Inc. IV. 354 (9).

94. See Cuomo, *Ambrogio Varese*, 70–71.

95. Del Mayno, *Epithalamion*, sig. a2r–v.

96. Ibid., sig. a7r–v: "Non ignoras quam altissima sit Divus Ludovicus prudentia, quam immensa sapientia, quo divino ingenii acumine excellat, quanta animi magnitudine praestet, quantam rei militaris scientiam teneat, quanta auctoritate polleat, quantum gratia."

97. Ibid., sig. a7v: "Non ignoras quod non solum Mediolanense imperium moderatur, verum etiam caeterorum Italiae principum & regum vota pro arbitrio disponit. Non ignoras Divum Ludovicum esse rerum Italicarum arbitrum. Quod in eius manu sit pacis aut belli [. . .] Divus Ludovicus Italicae pacis non solum auctor verum etiam conservator [. . .]."

98. Indeed, all Renaissance marriage orations placed much emphasis on the political value of such unions and often praised the rulers of the two houses as much as the bride and groom. This included drawing parallels with classical figures and inventing elaborate genealogies. See D'Elia, *The Renaissance of Marriage*, 54–75.

99. Del Mayno, *Epithalamion*, sig. a7v: "Non solent planetae, nisi geminentur, magnum aliquid moliri. Sed ubi coniunguntur, ut Saturno Iupiter, aut Sol Mercurio, magnos in terris effectus portendunt. Per Blancham augustam vipera aquilae tonantis aliti colligatur, & Ludovicus tanquam Mercurius Iovi, hoc est, Maximiliano terrarum domino coniunctus est."

100. This is the case in both Cardano's chart and that of the "ultramontanus" and of Pietro Buono Avogario included in BLO, MS Canon. Misc. 24.

101. Tristano Calco, *Residua*, 109–110, as quoted in Cuomo, *Ambrogio Varese*, 70.

102. Harwell, "Ruled by Mars."

103. For this astronomical data, see Johannes Regiomontanus, *Ephemerides 1475–1506* ([Nuremberg]: Johannes Regiomontanus, 1474) (digital edition BSB:

http://daten.digitale-sammlungen.de/0003/bsb00031043/images/index.html?fip=193
.174.98.30&id=00031043&seite=558; accessed October 14, 2010).

104. Corio, *Storia di Milano,* vol. 2, 1529. Corio, however, does not mention astrology but adduces religious motives instead.

105. Sanuto, *La spedizione,* 353.

106. Ibid., 352. In a letter dated November 25, 1495, the astrologer Giovan Maria de Albinis (or de Albricis) wrote to Duke Ercole reminding him that, years before, he had predicted to him that Ludovico would have elected himself Duke of Milan. For this letter, see Gabotto, "Nuove ricerche," 7. As the letter refers to Ercole's planned trip to San Jacopo of Galizia (Santiago de Compostela), the prediction can be safely dated to 1487. See Corio, *Storia di Milano,* vol. 2, 1468.

107. Pellegrini, *Ascanio Sforza,* vol. 1, 127–135.

108. Ibid., 161–163.

109. Ibid., 167, 169–214; 338–339. Mallet, "Personalities and Pressures," 153–155.

110. Pellegrini, *Ascanio Sforza,* vol. 1, 338–339.

111. See my "Annius of Viterbo Astrologer," as well as Chapter 3.

112. ASMi, *Autografi,* Medici 219, Ambrogio Varesi da Rosate to Ludovico, Milan, July 20, 1492. For its transcription and other considerations on this document, see my "Reading Health in the Stars," 183–184, n. 1. For a discussion of this type of interrogation, see Azzolini: "The Political Uses of Astrology," 139–140.

113. This data was derived from Arquato's annual prognostication for the year 1492. See Antonio Arquato, *Astrorum fata* (Ferrara: Lorenzo de Rubeis da Valenza, after September 21, 1491), sig. a2r (ISTC: ia01082200).

114. Azzolini, "Reading Health in the Stars," 184, n. 2.

115. Pellegrini, *Ascanio Sforza,* vol. 1, 371–373.

116. ASMi, SPE, Roma 106, Ludovico to Ascanio, Vigevano, July 28, 1492. The letter is quoted in the original in Pellegrini, *Ascanio Sforza,* vol. 1, 376. I have not yet found the geniture of Ascanio Sforza. His birthday is indicated as March 3, 1455.

117. Pellegrini, *Ascanio Sforza,* vol. 1, 380–381.

118. Stefano Infessura, *Diario della città di Roma* (Rome: Forzani, 1890), 282, originally quoted in Pellegrini, *Ascanio Sforza,* vol. 1, 385. The vice-chancellery was the most prestigious office, second only to the papacy.

119. Ibid., vol. 1, 387. On Alexander's pontificate, see now Maria Chiabò, Silvia Maddalo, Massimo Miglio, and Anna Maria Oliva, eds., *Roma di fronte all'Europa al tempo di Alessandro VI. Atti del convegno, Città del Vaticano-Roma, 1–4 dicembre 1999,* 3 vols. (Rome: Roma nel Rinascimento, 2001).

120. Infessura, *Diario,* 280–281.

121. Pellegrini, *Ascanio Sforza,* vol. 1, 428. This alliance was finally signed by Ferrante's son, Alfonso, in April 1494. See ASMi, SPE, Napoli 252, Antonio Stanga to Ludovico Sforza, Naples, April 6 and 8, 1494.

122. Pellegrini, *Ascanio Sforza,* vol. 1, 433–435. Mallet, "Personalities and Pressures," 157–158. A good starting point on the Italian Wars with reference to Milan remains Francesco Catalano, "La fine della signoria sforzesca," in *Storia di Milano,* vol. 7 (1956): 431–508. For a broad overview, see Marco Pellegrini, *Le guerre d'Italia, 1494–1530* (Bologna: Il Mulino, 2009). For recent specialized studies, see Danielle Boillet and Marie-Françoise Piéjus, eds., *Les guerres d'Italie. Histoire, pratique, représentations* (Paris: Université de Paris III Sorbonne Nouvelle, 2002); Jean-Louis Fournel and Jean-Claude Zancarini, eds., *Les guerres d'Italie. Des batailles pour l'Europe (1494–1559)* (Paris: Gallimard, 2003); and Giuseppe Galasso and Carlos José Hernando Sánchez, eds., *El reino de Nápoles y la monarquía de España: Entre agregación y conquista (1458–1535)* (Rome: Real Academia de España en Roma, 2004).

123. Cartwright, *Beatrice d'Este,* 223.

124. ASF, MAP 50, Piero Alamanni to Piero de' Medici, docs. 239 (April 2, 1494), 242 and 247.

125. ASF, MAP 50, Piero Alamanni to Piero de' Medici, doc. 242: "e poi lunghamente mi ragionò delle cose di Francia e della Magna, cose però che tutte tendono a sua gloria e honor, dicendo 'vedete ibasciatore se io ho a trattar cose grandi che Maximiliano ha messo le poste da casa sua insino a me, & simile ha facto il re di Francia [. . .] Ma seguitò in dire che Maximiliano et Re de Francia pareva certassino a ghara chi più lo potessi honorare et desiderare." On Ludovico's *hubris,* see also Delaborde, *L'expédition de Charles VIII,* 276; and Gigliola Soldi Rondinini, "Ludovico il Moro nella storiografia coeva," in *Milano nell'età di Ludovico il Moro. Atti del convegno internazionale, 28 Febbraio–4 Marzo 1983,* 2 vols. (Milan: Comune di Milano e Archivio Storico Civico-Biblioteca Trivulziana, 1983), vol. 1, 29–56, esp. 32, 41.

126. Delaborde, *L'expédition de Charles VIII,* 269–270.

127. Sanuto, *La spedizione,* 675; Cuomo, *Ambrogio Varese,* 73. The same is recounted by the Gonzaga ambassador in Milan, Donato de Pretis, to Isabella. See ASMa, AG 1630, fol. 233r, Milan, October 22, 1494, and fol. 234r. Gian Galeazzo was buried in the Duomo the following day, after his body was exposed in the Duomo for everybody to see (ASMa, AG 1630, fol. 234r).

128. Sanuto, *La spedizione,* 117; Cuomo, *Ambrogio Varese,* 73. Cf. ASMa, AG 1630, fol. 275v, Donato de Pretis to Francesco Gonzaga, November 29, 1494.

129. ASMa, AG 1630, fol. 236v, Donato de Pretis to Francesco Gonzaga, Milan, October 25, 1494: "Hozi matina ad hore 16 per ponte [sic] de astrologia la excellentia sua insiema cum la illustrissima sua consorte è partita de qui per andare a Castello Santo Donino de Piasentina dove se ritrova il Re de Franza"; fol. 264r, Donato de Pretis to Francesco Gonzaga, Milan, November 13, 1494: "Sua signoria volse intrare per ponto de astrologia per la prima intrata doppo l'assumption del ducato."

130. See Renzo Negri, "Birago, Francesco," *DBI* 10 (1968), 581–584.

131. ASMi, *Autografi*, Medici 219, Ambrogio Varesi to Ludovico, Milan, November 15, 1495.

132. Ibid., Milan, August 15, 22, 26 and 27, 1498. An earlier letter dated March 25, 1498, similiarly reported that Varesi and Ascanio's physician (possibly Luigi Marliani) had decided to let Ascanio's blood. See ASMi, SPS 1468, Birago (?) to Ludovico, Castello di Porta Giovia (Milan), March 25, 1498.

133. Ibid., especially August 14, 1498, where he marks the fourth and the seventh day of the illness as particularly significant. On this theory, see the Introduction and Chapter 4.

134. ASMi, *Autografi*, Medici 219, Ambrogio Varesi to Ludovico, Vespolate (Novara), October 18, 1494. Nicolò Orsini da Pitigliano was made captain of the papal troops in May 1489. The position was renewed at the ascension of Rodrigo Borgia to the papacy in 1492, and he was still holding that position in 1494 when he was wounded. He was captured by Charles VIII in Naples in February 1495 but escaped at the battle of Fornovo in July, soon after becaming Venice's captain-general. See Christine Shaw, *The Political Role of the Orsini Family from Sixtus IV to Clement VIII: Barons and Factions in the Papal States* (Rome: Istituto Storico Italiano per il Medio Evo, 2007), 178–182. According to Sanuto, Nicolò Orsini himself was "homo [che] segue molto astrologi et hore" and asked to receive the command of the Venetian army at the elected time. See Sanuto, *La spedizione*, 549 and 656.

135. ASMi, *Autografi*, Medici 219, Giovanni Bassino and Giovanni da Rosate to Ludovico, October 24, 1494.

136. ASMi, *Autografi*, Medici 219, Varesi to Bartolomeo Calco, ex domo (Milan?), September 26 and 30, 1497, and ex castro (Porta Giovia?), January 23, 1498.

137. The best moment for the "passage of the baton" to the Count of Caiazzo was determined astrologically: "El sig. Ludovico a hore 9 1/2 , la qual hora volse aver astro-logica dal suo maistro Ambrosio de Cerato [sic], ottimo astrologo, senza il consiglio dil quale non faria alcuna cosa, et molto varda tal hore e ponti, dette el stendardo et baston al conte de Caiazo capitano de 500 homeni d'arme." See Sanuto, *La spedizione*, 59. The day and time to issue the "condotta" of the Marquis of Mantua, Francesco Gonzaga, was also established astrologically. See ASMi, *Autografi*, Medici 219, Ambrogio Varesi to Ludovico, June 19, 1498 (for its transcription, see Cuomo, *Ambrogio Varese*, 82–83). Ludovico employed similar "astrological precautions" when his banner and baton were passed to Francesco Gonzaga on December 19, 1498. See Leon-G. Pélissier, *Recherches dans les archives Italiennes. Louis XII et Ludovic Sforza (8 Avril 1498–23 Juillet 1500)*, 2 vols. (Paris: Thorin et fils, 1896), vol. I, 210, n. 1. On passing the baton, see also Casanova, "L'astrologia e la consegna del bastone," 134–143. The examples cited include Nicolò Orsini, Count of Pitigliano, Costanzo Sforza, lord of Pesaro, and Ercole d'Este.

138. ASMo, Ambasciatori, Milano 7, Giacomo Trotti to Ercole d'Este, Vigevano, October 14, 1492, on Varesi's letter to Ludovico advising on the best day for his ambassadors' departure for Rome.

139. ASMi, *Autografi,* Medici 219, Ambrogio Varesi to Ludovico, s.l. and s.d. (but sometime around April 19, 1495). Also the earlier league between Ludovico and Charles VIII had been signed on a day and time decided astrologically. See Delaborde, *L'expédition de Charles VIII,* 228. On the anti-French Italian league of 1495, see also Michele Jacoviello, "La lega anti-francese del 31 Marzo 1495 nella fonte veneziana del Sanudo," *ASI* 143 (1985): 39–90.

140. Benedetto Capilupi to Francesco Gonzaga, Milan, January 23, 1495, as quoted in *COM,* vol. 15 (2005), 80–81. See also Cuomo, *Ambrogio Varese,* 73–74.

141. Delaborde, *L'expédition de Charles VIII,* 540–562.

142. Sanuto, *La spedizione,* 352.

143. "Or questo disse al Duca, come a dì 29 Zugno el Re de Franza harebbe una gran rotta, la qual cossa poco radegò; che a dì 6 Luio seguite la battaglia, et fo fugato dal nostro esercito." Sanuto, *La spedizione,* 412.

144. Ibid., 438.

145. Ibid.

146. Ibid., 474.

147. Guicciardini, *The History of Italy,* Book II, esp. 98.

148. On Charles VIII's retreat and the battle of Fornovo, see also the famous account by Guicciardini in ibid., esp. Book II.

149. ASMi, *Autografi,* Medici 219, Varesi to Ludovico, Milan, July 18, 1496: "La Illustrissima Signoria Vostra vedarà il bono successo del Reame quale gli ho predito questo anno tante volte, et insiema avarà intesso la sperduta dela Regina de Francya, dela quale disse quando morte il primo fiollo avaria pochi fiolli, che questa altra venì a vero, et così li addirmo e non averà o vero avendoli morirano et ala fine farà di pochi fiolli et questo per la natività del Christianissimo Re."

150. Ibid.: "Similemente ho dito più volta ala Illustrissima et Excellentissima Signoria chel fine de Fiorentini sarya triste, et che remanaryano inganati in mane del Re de Francya."

151. Letter of Augusto Somenzi to Ludovico, from Überlingen (on the lake of Konstanz), quoted in Leon-G. Pélissier, *Documents relatifs au règne de Louis XII et à sa politique en Italie* (Montpellier: Imprimerie Générale du Midi, 1912), 150: "Alla parte me scrisse l'E. V., de quello haveva predicto el magnifico maestro Ambrosio de Roxato in le cose de Pisa et in molte altre occorrentie sue, e de quello concludeva dela gloriosa vittoria haveria a conseguire la Maestà Cesarea contra li Suiceri et maxime fin ali 22 del presente, io lo notificai alla prefata Maestà con le parole che la me scrive; la quale ne rise et hebe grandissimo piacere intenderlo, demonstrando bel non dare troppa fede ad astronomi; pur ho inteso da alcuni che la sua Maestà, a tavola più fiate e altramente, a molti de questi signori e zentilhomeni ha dicto con grandissima alegreza l'Excellentia Vostra haverli dato aviso chel suo astronomo ha dicto che S.M. sarà victoriosa in questa impresa, talmente che ho conosciuto la S.M. haverne havuto piacere; e così nel advenire quando l'E.V. ne scriva qualche cosa de simile natura,

credo li sarà piacere e grato." I wish to thank Isabella Lazzarini for providing this reference. The Swiss war is recounted in detail in Sanuto, where the historian recounts how the Milanese ambassador Angelo da Firenze had reached Konstanz and offered 16,000 ducats (of the 33,000 promised) in exchange for the Emperor's peace with the Swiss and Ludovico's inclusion in the terms of the peace. See Marino Sanuto, *I Diarii di Marino Sanuto (MCCCCXCVI–MDXXXIII) dall'autografo Marciano ital. cl. VII cod. CDXIX–CDLXXVII,* ed. Guglielmo Berchet, Federico Stefani, Nicolò Barozzi, Rinaldo Fulin, and Marco Allegri, 58 vols. (Venice: Visentini, 1879–1903), vol. 2, 1180–1181.

152. See Guicciardini, *The History of Italy,* Book IV.

153. On the financial difficulties of the duchy at the end of the century, see Franca Leverotti, "La crisi finanziaria del ducato di Milano alla fine del Quattrocento," in *Milano nell'età di Ludovico il Moro,* vol. 2, 585–632.

154. ASMo, Ambasciatori, Milano 15, Antonio Constabili to Ercole d'Este, Milano, August 8, 1499: "Ad hore xv & meza questo illustrissimo signore se fece incontra fuora de Melegnano dui terzi de miglio al Reverendissimo Monsignor Aschanio suo fratello perché così fu parere delo astrologo d'epso Monsignor & de quelli de sua excellentia." See also Pélissier, *Recherches,* vol. 1, 430 and 451. That Constabili either followed basic astrological rules himself or, more likely, knew how Ludovico's days were planned astrologically is indicated in another letter where he informs Ercole that he had hastened to see Ludovico "inanti che sopragiungesseno le hore combuste." See ASMo, Ambasciatori, Milano 15, Antonio Constabili to Ercole d'Este, Milan, August 5, 1499.

155. Sanuto, *Diarii,* vol. 2, 1138. See also Pélissier, *Recherches,* vol. 2, 41.

156. Sanuto, *Diarii,* vol. 2, 1138.

157. Ibid., 1187. See also Pélissier, *Recherches,* vol. 2, 41.

158. Sanuto, *Diarii,* vol. 2, 1103: "l'cardinal Ascanio ha menato uno astrologo con lui, el qual à ditto queste parole al prefato cardinal e concluso: s'il ducha non perde il stato vol lo fazi impichar." See Pèlissier, *Recherches,* vol. 2, 41.

159. Sanuto, *Diarii,* vol. 2, 1187. This could be the episode involving Veronica da Binasco and Ludovico that took place in 1492 and is now recounted in BAM, MS I 179 inf., *Vita della beata Veronica del monastero di Santa Marta,* fols. 213v–215r. On this occasion, Veronica predicted to the duke that, unless the "great abominations" happening at his court ceased, God would punish him and his court with His wrath. I wish to thank John Gagné for pointing me to this text. On Veronica da Binasco, who died in 1497, see Zarri, *Le sante vive,* 52, 95–96.

160. Sanuto, *Diarii,* vol. 2, 1183. The events of the summer of 1499, including Ludovico's negotiations with a reluctant Maximilian, are recounted in Constabili's dispatches in ASMo, Ambasciatori, Milano 14 and 15. For these events, and those of September, see also Pélissier, *Recherches,* vol. 2, 1–106.

161. ASMo, Ambasciatori, Milano 15, Antonio Constabili to Ercole d'Este, Milan, August 18 and 20, 1499.

162. Ibid., August 22, 1499: "<Li> astrologi li quali questo illustrissimo ha havuti questa sira tutti uniti <qui>, similmente concludeno che seben sua excellentia non fosse principe di questo stato niente di mancho le influencie sono de sorte che <ad> ogni modo questo stato conveneria patire. Ma affir<mano> che alla fine sua celsitudine se defenderà & restarà superiore. Et Magistro Ambroso è in disfavore grande, perché disputa<ndo> cum questi altri astrologi el consente quello che loro di<cono>. Et sua excellentia dice haverli prima preditto el contrario. Per quanto se comprende potria essere chel sia stato bona causa de non lassare fare a questo illustrissimo signore quelle provisione che < . . . > < . . . >lo haveva fatte & che seriano state necessarie et se <a> celsitudine vostra parvesse havere a se Magistro Petro Bono <per> <inte>rrogarlo del parere suo, credo che <sua> excellentia seria se non grato, benché però epsa non me ne habia ditto niente. Alchuni de questi astrologi dicono questa influentia durarà tutto questo anno, alchuni per tutto Octob<re>, alchuni per insin a 27 de Septembre. [. . .]."

163. Ibid., August 23, 24, 25, and 26, 1499.

164. Ibid., September 3 1499.

165. Sanuto, *Diarii*, vol. 2, 1191–1198. Pélissier, *Recherches*, vol. 2, 48–53.

166. Sanuto, *Diarii*, vol. 2, 1255 and 1275.

167. On the loss of his possessions and the lengthy legal cause that followed, see Cuomo, *Ambrogio Varese*, 28–38. On Ermodoro, see the document quoted in ibid., 32; and also Pélissier, *Recherches*, vol. 2, 162.

168. ASMo, Ambasciatori, Milano 16, Ettore Bellingeri to Ercole d'Este, Milan, September 18, 1499: "La illustrissima Duchessa Isabella gli disse che maistro Ambrosio da Roxate havea confessato havere dato il veneno al quodam illustrissimo signore suo consorte in uno syroppo cum saputa del spetiale ad instantia del signore Ludovico, e che sua signoria era per far formare uno processo per mandarlo al Re de Romani, ma dubitando non sia disturbata la cosa, faceva conto de aspectare la venuta del Christianissimo Re, al quale dice voleva andare incontro insime col figliolino suo a notificarli il tuto." Varesi was imprisoned in the house of Luigi Trivulzio, accused of Gian Galeazzo's murder. ASMa, AG 1633, fol. 277v, Rozono *eques* to Francesco Gonzaga, Milan, September 14, 1499. Pélissier, *Recherches*, vol. 2, 206 recounts this story but erroneously indicates the name of Nicolò Bianchi as the Ferrarese ambassador in Milan. At the time, however, Bianchi was in charge of following the French king approaching the city. In 1492, Ludovico had accused Isabella of Aragon of attempting to poison him and Galeazzo Sanseverino with the help of Bernardino da Cotignola. ASMo, Ambasciatori, Milano 7, Giacomo Trotti to Ercole d'Este, Vigevano, October 15, 16, 17, and 18, 1492.

169. On the French occupation of Milan, see now Letizia Arcangeli, ed., *Milano e Luigi XII. Ricerche sul primo dominio francese in Lombardia (1499–1512)* (Milan:

FrancoAngeli, 2002); Stefano Meschini, *Luigi XII duca di Milano: gli uomini e le istituzioni del primo dominio francese, 1499–1512* (Milan: FrancoAngeli, 2004); idem, *La Francia nel ducato di Milano: La politica di Luigi XII (1499–1512),* 2 vols. (Milan: FrancoAngeli, 2006); and Philippe Contamine and Jean Guillaume, eds., *Louis XII en Milanais. XLI° colloque international d'études humanistes, 30 juin–3 juillet 1998* (Paris: H. Champion, 2003).

Epilogue

1. Black, *Absolutism in Renaissance Milan,* 183–184.
2. Grendler, *The Universities of the Italian Renaissance,* 90–93.
3. Ibid., 164–65. On these events, see also Cardano, *De libris propriis,* 102, 281.
4. Grafton, *Cardano's Cosmos,* 72–73; Azzolini, "Refining the Astrologer's Art," 5.

Bibliography

Primary Sources

MANUSCRIPTS

Belgium
Bruxelles
Bibliothèque Royale (BRB)
    II.1685
France
Paris
Bibliothèque Nationale de France (BNF)
    Italien 1524, 1586, 1589
    Lat. 7038, 7214, 7258, 7267, 7307, 7336, 7363, 7400, 7409, 10133, 16021
    Nouv. acq. Lat. 315
Bibliothèque Mazarine (BMP)
    3893
Germany
Munich
Bayerische Staatsbibliothek (BSB)
    Clm 51

Great Britain
Cambridge
Cambridge University Library (CUL)
  li.III.3
  li.I.15
London
British Library (BL)
  Arundel 88
Oxford
Bodleian Library (BLO)
  Canon. Misc. 23, 24
Italy
Corbetta (Milan)
Archivio Pisani-Dossi (APD)
  401
Florence
Archivio di Stato (ASF)
  Mediceo Avanti il Principato (MAP) 12, 50
Biblioteca Laurenziana (BLF)
  Plut. 30. 30
  Ashburnham 134
Biblioteca Nazionale Centrale (BNCF)
  II, V, 140
Mantua
Archivio di Stato (ASMa)
  Archivio Gonzaga (AG) 1232, 1623, 1630, 1632, 1633, 2991
Milan
Archivio di Stato (ASMi)
  Autografi, Medici 212, 213, 217, 219
  Autografi, Uomini Celebri, Scienziati, e Letterati 215
  Diplomatico, Diplomi e Dispacci Sovrani, Milano 6
  Sforzesco, Carteggio Interno 903, 927
  Sforzesco, Miscellanea 1569
  Sforzesco, Potenze Estere (SPE), Francia 549
  —Mantova 390, 394.
  —Napoli 218, 227, 252
  —Roma 106
  —Romagna 183
  Sforzesco, Potenze Sovrane (SPS) 1457, 1464, 1466, 1468, 1469, 1470
  Studi, Parte Antica 390
Biblioteca Ambrosiana (BAM)
  I 179 inf.

M 28 sup.

S 54 sup.

Biblioteca Trivulziana (BT)

    Triv. 1329

Modena

Archivio di Stato (ASMo)

    Ambasciatori, Milano 6, 7, 14, 15, 16

    Materie, Medicina

Rome

Biblioteca Alessandrina (BAR)

    102

Biblioteca Corsiniana (BC)

    232 (36 D 28)

Venice

Biblioteca Marciana (BMV)

    Lat. 3506

United States of America

New Haven

Beineke Library, Yale University (BLYU)

    Mellon 13

Vatican City

Biblioteca Apostolica Vaticana (BAV)

    Vat. Lat. 3379, 4080

PRINTED SOURCES

Abraham ibn Ezra. *Liber coniunctionum.* In *Abrahe Avenaris Judei Astrologi peritissimi in re iudiciali opera,* translated by Pietro D'Abano. Fols. LXXVIr–LXXXVr. Venice: Peter Lichtenstein, 1507.

Albumasar. *De magnis coniunctionibus, annorum revolutionibus, ac eorum profectionibus.* Augsburg: Erhard Ratdolt, March 31, 1489.

———*Introductorium in astronomiam Albumasaris abalachi octo continens libros partiales.* Augsburg: Erhard Ratdolt, 1489.

———*Abu Masar. The Abbreviation of the Introduction to Astrology.* Edited by Charles Burnett, Keiji Yamamoto, and Michio Yano. Leiden and New York: Brill, 1994.

Alcabitius. *Liber introductorius.* Venice: Erhard Ratdolt, 1482.

———*Al-Qabisi (Alcabitius): The Introduction to Astrology: Editions of the Arabic and Latin Texts and an English Translation.* Edited by Charles Burnett, Keiji Yamamoto, and Michio Yano. London and The Warburg Institute and Nino Aragno Editore, 2004.

Annius of Viterbo. *De futuris Christianorum triumphis in Saracenos, seu Glossa super Apocalypsin* [*sic*]. Genoa: Battista Cavallo, 1480.

Arquato, Antonio. *Astrorum fata.* Ferrara: Lorenzo de Rubeis da Valenza, after September 21, 1491.

Ashenden, John of (Johannes Eschuid). *Summa astrologiae iudicialis.* Venice: Johannes Lucilius Santritter for Francesco Bolano, July 7, 1489.

Barzizza, Cristoforo. *Introductorium ad opus practicum medicinae. Liber IX Almansoris.* Pavia: Antonio Carcano for Ottaviano Scoto, 1494.

Bembo, Pietro. *History of Venice.* Edited and translated by Robert W. Ulery, Jr. Cambridge, MA: Harvard University Press, 2007.

Calco, Tristano. *Nuptiae Mediolanensium Ducum, sive Ioannis Galeacii cum Isabella Aragona, Ferdinandi Neapolitanorum Regis.* In *Tristani Chalci Mediolanensis Historiographi, Residua e Bibliotheca Patricii Nobilissimi Lucii Hadriani Cottae.* Milan: Giovanni Battista and Giulio Cesare Malatesta, 1644.

————*Nuptiae Mediolanensium et Estensium Principium scilicet Ludovici Mariae cum Beatrice, Alphonsi Estensis sorore, vicissimque Alphonsi cum Anna, Ludovici nepote.* In Lopez, *Festa di Nozze,* in Latin with Italian translation, 118–142.

Carteggio degli oratori mantovani alla corte Sforzesca (1450–1500). Coordinated and directed by Franca Leverotti. Rome: Ministero per i beni e le attività culturali, 1999– *(COM).*

—vol. 1 (1999): *(1450–1459)* edited by Isabella Lazzarini.

—vol. 2 (2000): *(1460)* edited by Isabella Lazzarini.

—vol. 3 (2000): *(1461)* edited by Isabella Lazzarini.

—vol. 4 (2002): *(1462)* edited by Isabella Lazzarini.

—vol. 5 (2003): *(1463)* edited by Marco Folin.

—vol. 6 (2001): *(1464–1465)* edited by Maria Nadia Covini.

—vol. 15 (2003): *(1495–1498)* edited by Antonella Grati and Arturo Pacini.

Corio, Bernardino. *Storia di Milano.* 2 vols. Turin: UTET, 1978.

Corradi, Alfonso. *Memorie e documenti per la storia dell'Università di Pavia e degli uomini più illustri che v'insegnarono.* Pavia: Stabilimento Tipografico-Librario Successori Bizzoni, 1877–1878.

————*Cronica gestorum in partibus Lombardie et reliquis Italie.* Edited by Giuliano Bonazzi, *RIS*[2,] t. 22, pt. III. Città di Castello: S. Lapi, 1904.

Decembrio, Pier Candido. *De genitura hominis.* Rome: Stephan Plannck, about 1495.

————*Vita di Filippo Maria Visconti.* Milan: Adelphi, 1983.

D'Anghiera, Pietro Martire. *Opus epistolarum [Petri Martyris Anglerii Mediolanensis, Protonotarii Apostolici . . .]* Amsterdam: Daniel Elzevier, 1670.

Del Mayno, Giasone. *Epithalamion in nuptiis Maximiliani et Blancae Mariae.* Milan: Leonardo Pachel, after April 8, 1494.

De Paganica, Nicolaus. *Compendium medicinalis astrologiae.* Edited by Giuseppe Dell'Anna. Galatina: Congedo, 1990.

Dondi Dall'Orologio, Giovanni. *Tractatus astrarii*. Edited and translated by Emmanuel Poulle. Geneva: Droz, 2003.

Ferrarini, Girolamo. *Memoriale estense (1476–1489)*. Edited by Primo Griguolo. Rovigo: Minelliana, 2006.

Ficino, Marsilio. *Three Books on Life: A Critical Edition and Translation with Introduction and Notes*. Edited by Carol V. Kaske and John R. Clark. Tempe, AZ: Arizona Center of Medieval and Renaissance Studies and The Renaissance Society of America, 2002. First edition Binghamton, NY: Medieval and Renaissance Text and Studies in Conjunction with the Renaissance Society of America, 1989.

Galen. *De diebus decretoriis*. In Karl Gottlob Kühn, ed., *Claudii Galeni opera omnia*, vol. 9, 769–941. Hildesheim: Georg Olms, 1964–1965. First edition Leipzig: C. Cnobloch, 1821–1833.

———*Selected Works*. Translated with an introduction and notes by P. N. Singer. Oxford: Oxford University Press, 1997.

Gaurico, Luca. *Tractatus astrologicus in quo agitur de praeteritis multorum hominum accidentibus per proprias eorum genituras ad unguem examinatis*. Venice: Curzio Troiano Nàvo, 1552.

Giovio, Paolo. *Le vite de i dodeci Visconti, e di Sforza prencipi di Milano . . . tradotte per M. Lodovico Domenichi*. Venice: Gabriel Giolito de' Ferrari, 1558.

Giustiniani, Pietro. *Rerum Venetarum ab urbe condita historia*. Venice: Comino Tridino da Monferrato, 1560.

Guicciardini, Francesco. *The History of Italy*. Translated and edited with notes and an introduction by Sidney Alexander. Princeton, NJ: Princeton University Press, 1984. First printed New York: Macmillan, 1969.

Hippocrates. *Nature of Man. Regimen in Health. Humours. Aphorisms. Regimen 1–3. Dreams*. Translated by W. H. S. Jones, Loeb Classical Library, 150. London: W. Heinemann/Cambridge, MA: Harvard University Press, 1931.

Hyginus. *Poetica astronomica*. Edited by Jacobus Sentinus and Johannes Lucilius Santritter. Venice: Erhard Ratdolt, 1485.

Infessura, Stefano. *Diario della città di Roma*. Rome: Forzani, 1890.

Judicium cum tractatibus planetariis. Milan: Filippo Mantegazza, December 20, 1496.

Ketham, Johannes de. *Fasiculo de medicina*. Venice: Giovanni and Gregorio De Gregori, 1493.

Lübeck, Joahnnes de. *Prognosticon super Antichristi adventu Judaeorumque Messiae*. [Padua]: Bartholomeo Valdezocco, [not before April 1474].

Machiavelli, Niccolò. *Opere*. 4 vols. Turin: UTET, 1984–1999.

———*Istorie Fiorentine*. Edited by Alessandro Montevecchi. In Machiavelli, *Opere*, vol. 2 (1986).

———*Il Principe*. Edited by Rinaldo Rinaldi. In Machiavelli, *Opere*, vol. 1 (1999).

Medici, Lorenzo de'. *Lettere.* 13 vols. Florence: Giunti-Barbèra, 1977–2010.

—vol. 2 (1977): *(1474–1478)* edited by Riccardo Fubini.

—vol. 4 (1981): *(1479–1480)* edited by Nicolai Rubinstein.

Messahallah. *Epistola [Messahallah] in rebus eclipsis lune et in coniunctionibus pleanetarum ac revolutionbus annorum breviter elucidata.* In *Opera astrologica,* fols. 148r–149r.

———*De revolutionibus annorum mundi.* In *Opera astrologica,* fols. 149r–152r.

[Opera astrologica]. Liber quadripartiti Ptholemei. Centiloquium eiusdem. Centiloquium hermetis. Eiusdem de stellis beibenijs. Centiloquium bethem. [et] de horis planeta[rum]. Eiusdem de significatione triplicitatum ortus. Centu[m] quinquaginta p[ro]po[sitio]nes Almansoris. Zahel de interrogationibus. Eiusdem de electionibus. Eiusde[m] de te[m]po[rum] significationib[us] in iudicijs. Messahallach de receptionibus planetaru[m]. Eiusdem de interrogationibus. Epistola eiusde[m] cu[m] duodecim capitulis. Eiusdem de reuolutionibus anno[rum] mundi. Venice: Bonetto Locatelli for Ottaviano Scoto, December 20, 1493.

Pacioli, Luca. *De divina proportione.* Introduction by Augusto Marinoni. Milan: Silvana, 1982.

Peurbach, Georg von. *Theorica nova planetarum.* Venice: Simone Bevilacqua da Pavia, 1495.

Pico della Mirandola, Giovanni. *Disputationes adversus astrologiam divinatricem.* Edited by Eugenio Garin. 2 vols. Turin: Aragno, 2004. First printed Florence: Vallecchi, 1946–1952.

Ps.-Albertus Magnus. *Liber aggregationis, seu liber secretorum de virtutibus herbarum, lapidum et animalium quorundam. De mirabilibus mundi.* [Ferrara: Severino Ferrarese, about 1477].

———*De secretis mulierum cum commento [Henrici de Saxonia].* [Rome?] Venice: Simone Bevilacqua da Pavia, July 8, 1499.

———*The Book of Secrets of Albertus Magnus of the Virtues of Herbs, Stones and Certain Beasts: Also, A Book of the Marvels of the World.* Edited by Michael R. Best and Frank H. Brightman. Oxford: Clarendon Press, 1973.

Ptolemy. *Liber quadripartiti Ptholomei.* In *Opera astrologica,* fols. 2r–106v.

———*Tetrabiblos.* Translated by F. E. Robins, Loeb Classical Library, 435. London: W. Heinemann/Cambridge, MA: Harvard University Press, 1940.

———*Cosmographia: Bologna, 1477.* Introduction by Raleigh A. Skelton. Amsterdam: N. Israel, Meridian Publishing Co., 1963.

Regiomontanus, Johannes. *Ephemerides 1475–1506.* [Nuremberg]: Johannes Regiomontanus, 1474. (Digital edition Bayrische StaatsBibliothek: http://daten .digitale-sammlungen.de/~db/0003/bsb00031043/image_1).

———*Kalendar.* Venice: Erhard Ratdolt, August 9, 1482.

———*Opera Collectanea. Faksimiledrucke von neun Schriften Regiomontans und einer von ihm gedruckten Schrift seines Lehrers Purbach. Zusammengestellt und mit einer Einleitung hrsg. von Felix Schmeidler.* Osnabrück: Otto Zeller, 1972.

Rolewinck, Werner. *Fasciculum temporum*. Venice: Erhard Ratdolt, 1480; reprinted 1481.

Sacrobosco, Johannes. *Sphaera*. Venice: Simone Bevilacqua da Pavia, 1499.

Sanuto, Marino. *La spedizione di Carlo VIII in Italia*. Edited by Riccardo Fulin. Venice: Tipografia del Commercio di Marco Visentini, 1883.

———*I Diarii di Marino Sanuto (MCCCCXCVI–MDXXXIII) dall'autografo Marciano ital. cl. VII cod. CDXIX–CDLXXVII*. Edited by Guglielmo Berchet, Federico Stefani, Nicolò Barozzi, Rinaldo Fulin, and Marco Allegri. 58 vols. Venice: Visentini, 1879–1903.

—vol. 2 (1879): edited by Guglielmo Berchet.

Savonarola, Michele. *Practica canonica de febribus*. Venice: Giunta, 1552.

Schöner, Johannes. *De iudiciis nativitatum libri tres*. Nuremberg: Johannes Montanus and Ulricus Neuber, 1545.

Suetonius. *The Lives of the Caesars*. Translated by G. C. Rolfe. Loeb Classical Library, 38. London: W. Heinemann/Cambridge, MA: Harvard University Press, 1914.

Zambotti, Bernardino. *Diario ferrarese dall'anno 1476 sino al 1504*. Edited by Giuseppe Pardi. *RIS²*, XXIV/7. Bologna: Zanichelli, 1934.

Secondary Sources

Abulafia, David, ed. *The French Descent into Renaissance Italy, 1494–95. Antecedents and Effects*. Aldershot: Ashgate, 1995.

———"The Inception of the Reign of King Ferrante I of Naples: the Events of the Summer 1458 in the Light of the Documentation from Milan." In Abulafia, *The French Descent*, 71–90.

———"Ferrante of Naples: The Statecraft of a Renaissance Prince." *History Today* 45.2 (February 1995): 19–25.

Ady, Cecilia M. *Milan under the Sforza*. London: Methuen & Co., 1907.

Agrimi, Jole. *Tecnica e scienza nella cultura medievale. Inventario dei manoscritti relativi alla scienza e alla tecnica medievale (secc . XI–XV). Biblioteche di Lombardia*. Florence: La Nuova Italia, 1976.

———*Ingeniosa scientia nature. Studi sulla fisiognomica medievale*. Florence: SISMEL-Edizioni del Galluzzo, 2002.

Agrimi, Jole, and Chiara Crisciani. *Medicina del corpo, medicina dell'anima: note sul sapere medico fino all'inizio del secolo XIII*. Milan: Episteme Editrice, 1978.

Akasoy, Anna, Charles Burnett, and Ronit Yoeli-Tlalim, eds. *Astro-Medicine: Astrology and Medicine, East and West*. Florence: SISMEL-Edizioni del Galluzzo, 2008.

Albala, Ken. *Eating Right in the Renaissance*. Berkeley: University of California Press, 2002.

Albertini Ottolenghi, Maria Grazia. "La biblioteca dei Visconti e degli Sforza: gli inventari del 1488 e del 1490." *Studi Petrarcheschi* 8 (1991): 1–238.

Arcangeli, Letizia. "Gian Giacomo Trivulzio marchese di Vigevano e il governo francese nello stato di Milano." In eadem, *Gentiluomini di Lombardia. Ricerche sull'aristocrazia padana nel Rinascimento*, 3–70. Milan: Unicopli, 2003.

———ed. *Milano e Luigi XII. Ricerche sul primo dominio francese in Lombardia (1499–1512)*. Milan: FrancoAngeli, 2002.

Aubert, Alberto. *La crisi degli antichi Stati Italiani (1492–1521)*. Florence: Le Lettere, 2003.

Azzolini, Monica. "Anatomy of a Dispute: Leonardo, Pacioli, and Scientific Entertainment in Renaissance Milan." *Early Science and Medicine* 9.2 (2004): 115–135.

———"Reading Health in the Stars: Prognosis and Astrology in Renaissance Italy." In Oestmann, Rutkin, and Stuckrad, *Horoscopes and Public Spheres*, 183–205. Berlin and New York: Walter de Gruyter, 2005.

———"Annius of Viterbo Astrologer: Predicting the Death of Ferrante of Aragon, King of Naples." *Bruniana & Campanelliana* 14.2 (2008): 619–632.

———"The Politics of Prognostication: Astrology, Political Conspiracy and Murder in Fifteenth-Century Italy." *History of Universities* 23.2 (2009): 6–34.

———"The Political Uses of Astrology: Predicting the Illness and Death of Princes, Kings and Popes in the Italian Renaissance." *Studies in the History and Philosophy of Biological and Biomedical Sciences* 41 (2010): 135–145.

———"Refining the Astrologer's Art: Astrological diagrams in Bodleian MS Canon. Misc. 24 and Cardano's *Libelli quinque* (1547)." *Journal for the History of Astronomy* 42 (2011): 1–25.

———"Exploring Generation: A Context to Leonardo's Anatomies of the Female and Male Bodies." In Domenico Laurenza and Alessandro Nova, eds., *Leonardo da Vinci's Anatomical World: Language, Context, and "Disegno"*, 79–97. Venice: Marsilio Editore, 2011.

Barnes, Robin Bruce. *Prophecy and Gnosis: Apocalypticism in the Wake of the Lutheran Reformation*. Stanford: Stanford University Press, 1988.

Barton, Tamsyn. *Ancient Astrology*. London and New York: Routledge, 1994.

———*Power and Knowledge: Astrology, Physiognomics, and Medicine under the Roman Empire*. Ann Arbor: University of Michigan Press, 2002; originally published 1994.

Battioni, Gianluca. *La diocesi parmense durante l'episcopato di Sacramoro da Rimini (1476–1482)*. In Chittolini, ed., *Gli Sforza, la Chiesa lombarda, la corte di Roma*, 115–210.

———"Indagini su una famiglia di 'officiali' fra tardo medioevo e prima età moderna: I Sacramoro da Rimini (fine secolo XIV–inizio secolo XVII)." *Società e Storia* 52 (1991): 271–295.

Bedini, Silvio A., and Francis R. Maddison. "Mechanical Universe: The Astrarium of Giovanni de' Dondi." *Transactions of the American Philosophical Society* 56 (October 1966): 1–69.

Belotti, Bortolo. *Il dramma di Gerolamo Olgiati*. Milan: Cogliati, 1929. Later republished as *Storia di una congiura (Olgiati)*, Milan, 1965.

Beltrami, Luca. "L'annullamento del contratto di matrimonio fra Galeazzo Maria Sforza e Dorotea Gonzaga (1463)." *ASL* (1889): 126–132.

Benjamin, Francis S., and G. J. Toomer, eds. *Campanus of Novara and Medieval Planetary Theory. Theorica planetarum*. With an introduction, English translation, and commentary. Madison: University of Wisconsin Press, 1971.

Bentley, Jerry H. *Politics and Culture in Renaissance Naples*. Princeton, NJ: Princeton University Press, 1987.

Bernstein, Jane A. *Music Printing in Renaissance Venice*. Oxford: Oxford University Press, 1998.

Bertoni, Giulio. *La biblioteca estense e la coltura ferrarese ai tempi del Duca Ercole I (1471–1505)*. Turin: Loescher, 1903.

Bertozzi, Marco. *La tirannia degli astri: Aby Warburg e l'astrologia di Palazzo Schifanoia*. Bologna: Cappelli, 1985. Republished Livorno: Sillabe, 1999.

———*Lo Zodiaco del Principe. I decani di Schifanoia di Maurizio Bonora*. Ferrara: Tosi, 1992.

Biagioli, Mario. *Galileo, Courtier. The Practice of Science in the Culture of Absolutism*. Chicago: The University of Chicago Press, 1993.

Bietenholz, Peter G., and Thomas B. Deutscher, eds. *Contemporaries of Erasmus: A Biographical Register of the Renaissance and Reformation*. 3 vols. Toronto and Buffalo: University of Toronto Press, 1985–1987.

Biller, Peter, and Joseph Ziegler, eds. *Religion and Medicine in the Middle Ages*. York: York Medieval Press, 2001.

Bisaha, Nancy. *Creating East and West: Renaissance Humanists and the Ottoman Turks*. Philadelphia: University of Pennsylvania Press, 2004.

Black, Jane. *Absolutism in Renaissance Milan. Plenitudo of Power under the Visconti and the Sforza, 1328–1535*. Oxford: Oxford University Press, 2009.

Blume, Dieter. *Regenten des Himmels: astrologische Bilder im Mittelalter und Renaissance*. Berlin: Akademie Verlag, 2000.

Boillet, Danielle, and Marie-Françoise Piéjus, eds. *Les guerres d'Italie. Histoire, pratique, représentations*. Paris: Université de Paris III Sorbonne Nouvelle, 2002.

Bònoli, Fabrizio, and Daniela Piliarvu. *I lettori di astronomia presso lo Studio di Bologna dal XII al XX secolo*. Bologna: CLUEB, 2001.

Borris, Kenneth, and George S. Russeau, eds. *The Sciences of Homosexuality*. London: Routledge, 2007.

Borsa, Mario. "Pier Candido Decembrio e l'umanesimo in Lombardia." *ASL* 20 (1893): 5–75, 358–441.

Bouché-Leclercq, Auguste. *L'astrologie grecque*. Paris: E. Leroux, 1899.

Boudet, Jean-Patrice. "Simon de Phares et les rapports entre astrologie et prophétie à la fin du Moyen Âge." *MEFRM* 102 (1990): 617–648.

————"Les astrologues et le pouvoir sous le règne de Louis XI." In Bernard Ribémont, ed., *Observer, lire, écrire le ciel au Moyen Age. Actes du Colloque d'Orléans (22–23 avril 1989)*, 7–61. Paris: Klincksieck, 1991.

————ed. *Le recueil des plus celebres astrologues de Simon de Phares*. 2 vols. Paris: Honoré Champion, 1999.

————"Un jugement astrologique en français sur l'année 1415." In Jacques Paviot and Jacques Verger, eds., *Guerre, pouvoir et noblesse au Moyen Age. Mélanges en l'honneur de Philippe Contamine*, 111–120. Paris: Presses de l'université de Paris-Sorbonne, 2000.

————"Astrology." In Thomas F. Glick, Stephen J. Livesey, and Faith Wallis, eds., *Medieval Science, Technology and Medicine: An Encyclopedia*, 61–64. New York: Rutledge, 2005.

————*Entre science et nigromance. Astrologie, divination et magie dans l'Occident médiéval (XIIe–XVe siècle)*. Paris: Publications de la Sorbonne, 2006.

Boudet, Jean-Patrice, and Therese Charmasson. "Une consultation astrologique princière en 1427." In *Comprendre et maîtriser la nature au Moyen Age. Mélanges d'histoire des sciences offerts à Guy Beaujouan*, 255–278. Geneva: Droz, 1994.

Boudet, Jean-Patrice, and Emmanuel Poulle. "Les jugements astrologiques sur la naissance de Charles VII." In Françoise Autrand, Claude Gauvard, and Jean-Marie Moeglin, eds., *Saint-Denis et la royauté. Études offertes à Bernard Guenée*, 170–178. Paris: Publications de la Sorbonne, 1999.

Bowd, Stephen. *Venice's Most Loyal City: Civic Identity in Renaissance Brescia*. Cambridge, MA: Harvard University Press, 2010.

Broecke, Steven Vanden. *The Limits of Influence: Pico, Louvain, and the Crisis of Renaissance Astrology*. Leiden and Boston: Brill, 2003.

Brosseder, Claudia. *Im Bann der Sterne. Caspar Peucer, Philipp Melanchthon und andere Wittenberger Astrologen*. Berlin: Akademie Verlag, 2004.

Bueno de Mesquita, Daniel M. *Giangaleazzo Visconti, Duke of Milan (1351–1402): A Study in the Political Career of an Italian Despot*. Cambridge: Cambridge University Press, 1941.

Bühler, Curt. *The University and the Press in Fifteenth-Century Bologna*. Notre Dame, IN: The Medieval Institute, 1958.

Burnett, Charles S. F. "The Earliest Chiromancy in the West." *JWCI* 50 (1987): 189–195.

————"The Planets and the Development of the Embryo." In G. R. Dunstan, *The Human Embryo: Aristotle and the Arabic and European Traditions*, 95–135. Exeter: University of Exeter Press, 1990.

————"Astrology." In Frank A. C. Mantello and Arthur G. Rigg, eds., *Medieval Latin: An Introduction and Bibliographical Guide,* 369–382. Washington, DC: Catholic University of America, 1996.

————"The Coherence of the Arabic-Latin Translation Program in Toledo in the Twelfth Century." *Science in Context* 14 (2001): 249–288.

Burkhardt, Jacob. *The Civilization of the Renaissance in Italy.* London: Phaidon Press, 1960.

Bynum, William F., and Vivian Nutton, eds. *Theories of Fever from Antiquity to the Enlightenment.* Medical History Suppl. N. 1. London: The Wellcome Institute, 1981.

Byrne, James Steven. "A Humanist History of Mathematics? Regiomontanus's Padua Oration in Context." *Journal of the History of Ideas* 67 (2006): 41–61.

Canestrini, Giuseppe. "Lettera di Ludovico il Moro all'Imperatore Massimiliano (30 Settembre 1495)." *ASI* Appendice III (1846): 113–124.

Canetta, Carlo. "Le sponsalie di Casa Sforza con Casa d'Aragona (Giugno-Ottobre 1455)." *ASL* 9 (1882): 136–144.

————"Le 'sponsalie' di Casa Sforza con Casa d'Aragona." *ASL* 10 (1883): 769–782.

Capodieci, Luisa. Medicæa Medæa: *Art, astres et pouvoir à la cour de Catherine de Médicis.* Geneva: Droz, 2011.

Capp, Bernard. *Astrology and the Popular Press: English Almanacs, 1500–1800.* London: Faber, 1979.

Carey, Hilary M. *Courting Disaster: Astrology at the English Court and University in the Later Middle Ages.* London: Macmillan, 1992.

————"Judicial Astrology in Theory and Practice in Later Medieval Europe." *Studies in History and Philosophy of Biological and Biomedical Sciences* 41 (2010): 90–98.

Carmody, Francis J. *Arabic Astronomical and Astrological Sciences in Latin Translation: A Critical Bibliography.* Berkeley and Los Angeles: University of California Press, 1965.

————ed. *Theorica planetarum Gerardi.* Berkeley: [s.n.], 1942.

Caroti, Stefano, ed. "Quaestio contra divinatores horoscopios." *Archives d'histoire doctrinale et littéraire du moyen age* 43 (1976): 201–310.

————"La critica contro l'astrologia di Nicole Oresme e la sua influenza nel Medioevo e nel Rinascimento." *Atti della Accademia Nazionale dei Lincei (Classe di Scienze Morali, Storiche e Filologiche)* 23 (1979): 545–685.

————"Nicole Oresme's Polemic against Astrology in his *Quodlibeta.*" In Curry, *Astrology, Science and Society: Historical Essays,* 75–93.

Cartwright, Julia. *Beatrice d'Este, Duchess of Milan, 1475–1497: A Study of the Renaissance.* London: J. M. Dent, 1912.

Casanova, Eugenio. "L'astrologia e la consegna del bastone al capitano generale della Repubblica fiorentina." *ASI* 7 (1891): 134–143.

Castagnola, Raffaella. "Un oroscopo per Cosimo I." *Rinascimento* 29 (1989): 125–189.

———ed. *I Guicciardini e le scienze occulte. L'oroscopo di Francesco Guicciardini, lettere di alchimia, astrologia e cabala a Luigi Guicciardini.* Florence: Olschki, 1990.

Castelli, Patrizia, ed. *La rinascita del sapere: libri e maestri dello studio ferrarese,* 293–306. Venice: Marsilio, 1991.

———ed. *«In supreme dignitatis». Per la storia dell'Università di Ferrara (1391–1991). Atti del Convegno dell'Università di Ferrara, 15–19 ottobre 1991.* Florence: Olschki, 1995.

———"L'oroscopo di Pico." In Patrizia Castelli, ed., *Giovanni e Gianfrancesco Pico: L'opera e la fortuna di due studenti ferraresi,* 17–43. Florence: Olschki, 1998.

———"Leonardo, i due Pico e la critica alla divinazione." In Frosini, *Leonardo e Pico,* 131–172.

Castelnuovo, Guido. "Offices and Officials." In Andrea Gamberini and Isabella Lazzarini, eds., *The Italian Renaissance State,* 368–384. Cambridge: Cambridge University Press, 2012.

Catalano, Franco. "La Nuova Signoria: Francesco Sforza." In *Storia di Milano,* vol. 7 (1956): 1–224.

———"La pace di Lodi e la Lega Italica." In *Storia di Milano,* vol. 7 (1956): 3–81

———"La politica Italiana dello Sforza." In *Storia di Milano,* vol. 7 (1956): 82–172.

———"La fine della signoria sforzesca." In *Storia di Milano,* vol. 7 (1956): 431–508.

Cavagna, Anna Giulia. "«Il libro desquadernato: la carta rosechata da rati». Due nuovi inventari della libreria Visconteo-Sforzesca." *Bollettino della Società Pavese di Storia Patria* 41 (1989): 29–97.

Cerrini, Simonetta. "Libri dei Visconti-Sforza. Schede per una nuova edizione degli inventari." *Studi Petrarcheschi* 8 (1991): 239–281.

Chabás, José, and Bernard R. Goldstein. *The Alfonsine Tables of Toledo.* Dordrecht and London: Kluwer Academic, 2003.

———*The Astronomical Tables of Giovanni Bianchini.* Leiden and Boston: Brill, 2009.

Chiabò, Maria, Silvia Maddalo, Massimo Miglio, and Anna Maria Oliva, eds. *Roma di fronte all'Europa al tempo di Alessandro VI. Atti del convegno, Città del Vaticano-Roma, 1–4 dicembre 1999.* 3 vols. Rome: Roma nel Rinascimento, 2001.

Chittolini, Giorgio, ed. *Gli Sforza, la Chiesa lombarda, la corte di Roma. Strutture e pratiche beneficiarie nel Ducato di Milano (1450–1535).* Naples, Liguori Editore, 1989.

Cirrani, Rosalba, Laura Giusti, and Gino Fornaciari. "Prostatic Hyperplasia in the Mummy of an Italian Renaissance Prince." *Prostate* 45.4 (2000): 320–322.

Cirrani, Rosalba, Valentina Giuffra, and Gino Fornaciari, "Ergonomic Pathology of Pandolfo III Malatesta." *Medicina nei secoli* 15.3 (2003): 581–594.

Clark, Charles W. *The Zodiac Man in Medieval Medical Astrology.* Ph.D. dissertation. University of Colorado, 1979.

——— "The Zodiac Man in Medieval Medical Astrology." *Journal of the Rocky Mountain Medieval and Renaissance Association* 3 (1982): 13–38.

Cognasso, Francesco. "L'unificazione della Lombardia sotto Milano." In *Storia di Milano,* vol. 5 (1955): 1–567.

——— "Il Ducato Visconteo da Gian Galeazzo a Filippo Maria." In *Storia di Milano,* vol. 6 (1955): 1–383 (includes: "Milano sotto Giovanni Maria Visconti," 108–152; and "Il crollo dell'egemonia Milanese," 193–247).

——— "Istituzioni comunali e signorili di Milano sotto i Visconti." In *Storia di Milano,* vol. 6 (1955): 451–544.

——— *I Visconti.* [Milan]: Dall'Oglio Editore, 1966.

Colombo, Alessandro. "Abbozzo dell'alleanza tra lo Sforza ed il Gonzaga in previsione di una guerra con Venezia (ottobre–novembre 1450)." *Nuovo Archivio Veneto* 65 (1907): 143–151.

——— "Nuovo contributo alla storia del contratto di matrimonio fra Galeazzo Maria Sforza e Susanna Gonzaga." *ASL* 12 (1909): 204–211.

——— "Il 'grido di dolore' di Isabella d'Aragona duchessa di Milano." In *Studi di storia napoletana in onore di Michelangelo Schipa,* 331–346. Naples: Industrie tipografiche e affini, 1926.

Contamine, Philippe and Jean Guillaume, eds. *Louis XII en Milanais. XLIe colloque international d'études humanistes, 30 juin–3 juillet 1998.* Paris: H. Champion, 2003.

Cooper, Glen M. *Galen, De diebus decretoriis, from Greek into Arabic: A Critical Edition, with Translation and Commentary, of Hunayn ibn Ishaq, Kitab ayyam al-buhran.* Farnham: Ashgate, 2011.

——— "Galen and Astrology: A Mésalliance?" *Early Science and Medicine* 16 (2011): 120–146.

Coopland, George W. *Nicole Oresme and the Astrologers: A Study of His* Livre de divinacions. Liverpool: Liverpool University Press, 1952.

Corbo, Anna Maria. *Paolo II Barbo dalla mercatura al papato (1464–1471).* Rome: Edilazio, 2004.

Cosmacini, Giorgio. "La malattia del duca Francesco." *COM* vol. 3 (2000): 23–26.

Covini, Maria Nadia. "Introduzione." *COM* vol. 6 (2001): 7–31.

Cox-Rearick, Janet. *Dynasty and Destiny in Medici Art: Pontormo, Leo X, and the Two Cosimos.* Princeton, NJ: Princeton University Press, 1984.

Crisciani, Chiara. "Fatti, teorie, 'narratio' e i malati a corte. Note su empirismo in medicina nel tardo-medioevo." *Quaderni Storici* 3 (2001): 695–718.

Crisciani, Chiara, and Gabriella Zuccolin, eds. *Michele Savonarola: Medicina e cultura di corte*. Florence: SISMEL-Edizioni del Galluzzo, 2011.

Crum, Roger J. "'Cosmos, the World of Cosimo': The Iconography of the Uffizi Façade." *The Art Bulletin* 71 (1989): 237–253.

Cumont, Franz. *Astrology and Religion among the Greeks and Romans*. New York and London: Putnam, 1912.

Cuomo, Alberto M. *Ambrogio Varese: un rosatese alla corte di Ludovico il Moro*. Rosate: Amministrazione comunale, 1987.

Curry, Patrick, ed. *Astrology, Science and Society: Historical Essays*. Woodbridge: Boydell, 1987.

Cusin, Fabio. "Le aspirazioni straniere sul ducato di Milano e l'investitura imperiale (1450–1454)." *ASL* (1936): 277–369.

———"L'impero e la successione degli Sforza ai Visconti." *ASL* (1936): 3–116.

———"Le relazioni tra l'impero ed il ducato di Milano dalla pace di Lodi alla morte di Francesco Sforza." *ASL* (1938): 3–110.

De Ferrari, Augusto. "Conti, Nicolò." *DBI* 28 (1983): 461–462.

Delaborde, Henri Francois. *L'expédition de Charles VIII en Italie: histoire diplomatique et militaire*. Paris: Firmin-Didot, 1888.

D'Elia, Anthony. *The Renaissance of Marriage in Fifteenth-Century Italy*. Cambridge, MA: Harvard University Press, 2005.

Dell'Anna, Giuseppe. *Dies critici: La teoria della ciclicità delle patologie nel XIV secolo*. 2 vols. Galatina: Mario Congedo Editore, 1999.

Denis, Anne. *Charles VIII et les Italiens: Histoire et mythe*. Geneva: Droz, 1979.

De' Reguardati, Fausto M. *Benedetto de' Reguardati da Norcia «medicus tota Italia celeberrimus»*. Trieste: Lint, 1977.

Dessì, Rosa Maria. "'Quanto di poi abia a bastare il mondo . . . ' Tra predicazione e recezione. Apocalittica e penitenza nelle 'reportationes' dei sermoni di Michele Carcano da Milano." *Florensia. Bollettino del Centro Internazionale di Studi Gioachimiti*, 3–4 (1989): 71–90. Republished in French as "Entre prédication et réception: le thèmes eschatologiques dans les 'reportationes' des sermons de Michele Carcano de Milan (Florence, 1461–1466)." *MEFRM* 102 (1990): 457–479.

De' Vecchi, Bindo. "I libri di un medico umanista fiorentino del sec. XV." *La Bibliofilia* 34 (1932): 293–301.

Di Giorgio, Michela, and Christiane Klapisch-Zuber, eds. *Storia del matrimonio*. 3 vols. Rome: Laterza, 1996.

Dilg, Peter. "The Antiarabism in the Medicine of Humanism." In *La diffusione delle scienze islamiche nel Medio Evo Europeo*, 269–289. Rome: Accademia Nazionale dei Lincei, 1987.

Dillon Bussi, Angela. "I Benzi a Ferrara." In Castelli, *«In supreme dignitatis»*, 431–439.

Dina, Achille. "Ludovico il Moro prima della sua venuta al governo." *ASL* 3 (1886): 737–776.

———"Qualche notizia su Dorotea Gonzaga." *ASL* 4 (1887): 562–567.

———"Isabella d'Aragona duchessa di Milano e di Bari." *ASL* 48 (1921): 269–457.

Dispacci sforzeschi da Napoli. 4 vols. Salerno-Naples-Battipaglia: Carlone-Laveglia&Carlone, 1997–2009.

—vol. 1 (Salerno: Carlone, 1997): (1444–2 Luglio 1458) edited by Francesco Senatore.

Documenti per la storia dell'Università di Pavia nella seconda metà del '400. 3 vols. Bologna and Milan: Cisalpino, 1994–2010.

—vol. 1 (1994): *(1450–1455)* edited by Agostino Sottili.

—vol. 2 (2002): *(1456–1560)* edited by Agostino Sottili and Paolo Rosso.

Dover, Paul M. "Royal Diplomacy in Renaissance Italy: Ferrante d'Aragona (1458–1494) and his Ambassadors." *Mediterranean Studies* 14 (2005): 57–94.

Draelants, Isabelle. *Le «Liber de virtutibus. Herbarum lapidum et animalium (liber aggregationis)». Un texte à succès attribué à Albert Le Grand.* Florence: SISMEL, 2008.

Duranti, Tommaso. *Il carteggio di Gerardo Cerruti, oratore sforzesco a Bologna (1470–1474).* 2 vols. Bologna: CLUEB, 2007.

———*Mai sotto Saturno: Girolamo Manfredi medico e astrologo.* Bologna: CLUEB, 2008.

Dykes, Benjamin N. "Introduction." In Guido Bonatti, *Book of Astronomy,* translated by Benjamin N. Dykes. 2 vols. Golden Valley, MN: The Cazimi Press, 2007.

Eade, C. J. *The Forgotten Sky: A Guide to Astrology in English Literature.* Oxford: Clarendon Press, 1984.

Ernst, Germana. *Religione, ragione e natura.* Milan: FrancoAngeli, 1991.

Ettlinger, Helen. "Visibilis et Invisibilis: The Mistress in Italian Renaissance Court Society." *RQ* 47.4 (1994): 770–792.

Eubel, Konrad. *Hierarchia catholica medii aevi sive Summorum pontificum, S. R. E. cardinalium, ecclesiarum antistitum series e documentis tabularii praesertim Vaticani collecta, digesta, edita.* 8 vols. Monasterii: sumptibus et typis librariae regensbergianae, 1898–1979.

Falcioni, Anna. "Malatesta, Malatesta." *DBI* 68 (2007): 77–81.

Fantoni, Giuliana. "Un carteggio femminile del sec. XV. Bianca Maria Visconti e Barbara di Hohenzollern-Brandeburgo." *Libri e documenti: Archivio storico civico e Biblioteca Trivulziana* 7 (1981): 6–29.

Fattori, Daniela. "Il dottore padovano Alessandro Pellati, la sua biblioteca e *l'editio princeps* del *De medicorum astrologia.*" *La Bibliofilía* 110 (2008): 117–137.

Federici Vescovini, Graziella. *Astrologia e scienza: la crisi dell'aristotelismo sul cadere del Trecento e Biagio Pelacani da Parma.* Florence: Vallecchi, 1979.

————*Il "Lucidator dubitabilium astronomiae" di Pietro d'Abano: opere scientifiche inedite.* Padua: Programma e 1+1 Editori, 1988.

————"L'astrologia all'università di Ferrara nel Quattrocento." In Castelli, *La rinascita del sapere,* 293–306.

————"I programmi degli insegnamenti del collegio di medicina, filosofia e astrologia dello statuto dell'università di Bologna del 1405." In Hamesse, *Roma, magistra mundi,* vol. 1, 193–223.

————"La medicina astrologica dello «Speculum phisionomie» di Michele Savonarola." In Castelli, *«In supreme dignitatis»,* 415–429.

————"Note di commento a alcuni passi del «Libro di Pittura». «L'astrologia che nulla fa senza la prospettiva . . . »." In Frosini, *Leonardo e Pico,* 99–129.

Ferorelli, Nicola. "Il ducato di Bari sotto Sforza Maria Sforza e Ludovico il Moro." *ASL* 41 (1914): 389–433.

Ferrari, Henri-Maxime. *Une chaire de médecine au XV* siècle. Un Professeur à l'Université de Pavie de 1432–1472.* Paris: Félix Alcan, 1899.

Ferrari, Monica. *"Per non mancare in tuto del debito mio". L'educazione dei bambini Sforza nel Quattrocento.* Milan: FrancoAngeli, 2000.

Fornaciari, Gino. "Malignant Tumor in the Mummy of Ferrante Ist of Aragon, King of Naples (1431–1494)." *Medicina nei Secoli,* 6.1 (1994): 139–146.

————"Renaissance Mummies in Italy." *Medicina nei Secoli,* 11.1 (1999): 85–105.

Fornaciari, Gino, Antonio Marchetti, Silvia Pellegrini, and Rosalba Cirrani. "K-ras Mutation in the Tumor of King Ferrante I of Aragon (1431–1494) and Environmental Mutagens at the Aragonese Court of Naples." *International Journal of Osteoarcheology* 9 (1999): 302–306.

Fornaciari, Gino, Rosalba Cirrani, and Luca Ventura. "Paleoandrology and Prostatic Hyperplasia in Italian Mummies (XV–XIX Century)." *Medicina nei secoli* 13.2 (2001): 269–284.

Fossati, Felice. "Ludovico Sforza avvelenatore del nipote? (Testimonianza di Simone Del Pozzo)." *ASL* 2, serie 4 (1904): 162–171.

Fournel, Jean-Louis, and Jean-Claude Zancarini, eds. *Les guerres d'Italie. Des batailles pour l'Europe (1494–1559).* Paris: Gallimard, 2003.

French, Roger K. "Astrology in Medical Practice." In Ballester, French, Arrizabalaga, and Cunningham, *Practical Medicine,* 30–59. Reprinted in Roger French, *Ancients and Moderns in the Medical Sciences: From Hippocrates to Harvey,* 30–59. Aldershot: Ashgate, 2000.

Freudenthal, Gad. "The Medieval Astrologization of Aristotle's Biology: Averroes on the Role of the Celestial Bodies in the Generation of Animate Beings." *Arabic Sciences and Philosophy* 12 (2002): 111–137. Reprinted in idem, *Science in the Medieval Hebrew and Arabic Traditions,* 111–137. Aldershot: Ashgate, 2005.

Frosini, Fabio, ed. *Leonardo e Pico. Analogie, contratti, confronti. Atti del Convegno di Mirandola (10 Maggio 2003).* Florence: Olschki, 2005.

Fubini, Riccardo. "L'assassinio di Galeazzo Maria Sforza nelle sue circostanze politiche." In Medici, *Lettere*, vol. 2, 247–250 and 523–535.

———"Osservazioni e documenti sulla crisi del ducato di Milano nel 1477 e sulla riforma del consiglio segreto ducale di Bona Sforza." In Sergio Bertelli and Gloria Ramakus, eds., *Essays Presented to Myron P. Gilmore*, 2 vols. Vol. 1, 47–103. Florence: La Nuova Italia, 1978.

———"Lega italica e 'politica dell'equilibrio' all'avvento di Lorenzo de' Medici al potere." In idem, *Italia Quattrocentesca. Politica e diplomazia nell'età di Lorenzo il Magnifico*, 185–219. Milan: Franco Angeli, 1994. Published in English in *Journal of Modern History* 67 suppl. (1995): 166–199.

Fumagalli, Edoardo. "Aneddoti della vita di Annio da Viterbo O.P. (I-II-III)." *Archivum Fratrum Praedicatorum* 50 (1980): 189–199; and 52 (1982): 197–218.

Furdell, Elizabeth Lane. *The Royal Doctors, 1485–1714: Medical Personnel at the Tudor and Stuart Courts*. Rochester, NY: University of Rochester Press, 2001.

Gabotto, Ferdinando. "L'astrologia nel Quattrocento in rapporto colla civiltà: osservazioni e documenti storici." *Rivista di filosofia scientifica* 8 (June/July 1889): 377–413.

———*Nuove ricerche e documenti sull'astrologia alla corte degli Estensi e degli Sforza*. Torino: La Letteratura, 1891.

———*Bartolomeo Manfredi e l'astrologia alla corte di Mantova: ricerche e documenti*. Torino: La Letteratura, 1891.

———*Lettere inedite di Joviano Pontano in nome de' reali di Napoli*. Bologna: Romagnoli Dall'Acqua, 1893.

Gabriel, Astrik L. *A Summary Catalogue of Microfilms of One Thousand Scientific Manuscripts in the Ambrosiana Library, Milan*. Notre Dame, IN: Mediaeval Institute, University of Notre Dame, 1968.

Galasso, Giuseppe, and Carlos José Hernando Sánchez, eds. *El reino de Nápoles y la monarquía de España: Entre agregación y conquista (1458–1535)*. Rome: Real Academia de España en Roma, 2004.

Galimberti, Paolo M. "Il testamento e la biblioteca di Ambrogio Griffi, medico milanese, protonotario apostolico e consigliere sforzesco." *Aevum* 72 (1998): 447–483.

García-Ballester, Luis, Roger French, Jon Arrizabalaga, and Andrew Cunningham, eds. *Practical Medicine from Salerno to the Black Death*. Cambridge: Cambridge University Press, 1994.

Garin, Eugenio. *Astrology in the Renaissance: The Zodiac of Life*. London: Routledge & Kegan Paul, 1983.

Gasparrini Leporace, Tullia. "Due biblioteche mediche del Quattrocento." *La Bibliofilía* 52 (1950): 205–220.

Geneva, Ann. *Astrology and the Seventeenth-Century Mind: William Lilly and the Language of the Stars*. Manchester: Manchester University Press, 1995.

Gil-Sotres, Pedro. "Derivation and Revulsion: The Theory and Practice of Medieval
    Phlebotomy." In Garcìa-Ballester, French, Arrizabalaga, and Cunningham,
    Practical Medicine, 110–155.
Gingerich, Owen, and Jerzy Dobrzycki, eds. *Astronomy of Copernicus and its
    Background.* Colloquia Copernicana III Wroclaw-Warszawa-Krakow-Gdansk:
    Ossolineum, 1975.
Giordano, Luisa, ed. *Beatrice d'Este, 1475–1497.* Pisa: ETS, 2008.
Girguolo, Primo. "Professori di medicina senesi tra Ferrara e Padova: notizie dei
    Benzi e di Alessandro Sermoneta." *Quaderni per la storia dell'Università di
    Padova* 41 (2008): 135–150.
Grafton, Anthony. "Girolamo Cardano and the Tradition of Classical Astrology:
    The Rothschild Lecture, 1995." *Proceedings of the American Philosophical Society*
    142, no. 3 (1998): 323–354.
———*Cardano's Cosmos: The Worlds and Works of a Renaissance Astrologer.*
    Cambridge, MA: Harvard University Press, 1999.
Grafton, Anthony, and Nancy Siraisi. "Between the Election and My Hopes:
    Girolamo Cardano and Medical Astrology." In Newman and Grafton, *Secrets of
    Nature,* 69–131.
Grant, Edward. *A Source Book in Medieval Science.* Cambridge, MA: Harvard
    University Press, 1974.
———"The influence of the celestial region on the terrestrial." In idem, *Planets,
    Stars, and Orbs: The Medieval Cosmos, 1200–1687,* 569–617. Cambridge and New
    York: Cambridge University Press, 1994,
Grendler, Paul. "New Scholarship on Renaissance Universities." *RQ* 53 (2000):
    1174–1182.
———*The Universities of the Italian Renaissance.* Baltimore and London: The Johns
    Hopkins University Press, 2001.
Grossi Turchetti, Maria Luisa. "La dotazione libraria di un collegio universitario del
    Quattrocento." *Physis* 22 (1980): 463–475.
Grössing, Helmut, and Franz Graf-Stuhlhofer. "Versuch einer Deutung der Rolle
    der Astrologie in den persönlichen und politischen Entscheidungen einiger
    Habsburger des Spätmittelalters." *Anzeiger der Österreichischen Akademie der
    Wissenschaften,* phil.-hist. Kl., 117 (1980): 267–283.
Hamesse, Jaqueline, ed. *Roma, magistra mundi. Itineraria culturae medievalis.
    Mélanges offerts au Père L. E. Boyle à l'occasion de son 75ᵉ anniversaire.* 3 vols.
    Louvain-la-Neuve: Féderation Internationale des Instituts d'Études
    Médiévales, 1998.
Hankins, James. *Plato in the Italian Renaissance.* 2 vols. Leiden and New York: Brill,
    1991 (second impression with addenda and corrigenda).
Hanson, Ann Ellis. "The Eight Months' Child and The Etiquette of Birth: *Obsit
    Omen!*" *Bulletin of the History of Medicine* 61 (1987): 589–602.

Harwell, Gregory. "Ruled by Mars: Astrology and the Habsburg-Sforza Alliance of 1494." Unpublished paper. Cambridge (UK): Renaissance Society of America, 2005.

Hauser, Henri. *Les Sources de l'histoire de France: XVIe siècle (1494–1610)*. 4 vols. Paris: A. Picard, 1906–1915.

Hayton, Darin. *Astrologers and Astrology in Vienna during the Era of Emperor Maximilian I (1493–1519)*. Ph.D. dissertation. Notre Dame University, 2004.

———"Joseph Grünpeck's Astrological Explanation of the French Disease." In Kevin Siena, ed., *Sins of the Flesh: Responding to Sexual Disease in Early Modern Europe*, 81–106. Toronto: Centre for Reformation and Renaissance Studies, 2005.

———"Astrology as Political Propaganda: Humanist Responses to the Turkish Threat in Early Sixteenth-Century Vienna." *Austrian History Yearbook* 38 (2007): 61–91.

———"Martin Bylica at the Court of Matthias Corvinus: Astrology and Politics in Renaissance Hungary." *Centaurus* 49 (2007): 185–198.

———"Expertise ex Stellis: Comets, Horoscopes, and Politics in Renaissance Hungary." In Erik H. Ash, ed., *Expertise and the Early Modern State* (Osiris, vol. 25), 27–46. Chicago: The University of Chicago Press, 2010.

Hegedus, Tim. *Early Christianity and Ancient Astrology*. New York: Peter Lang, 2007.

Heiberg, J. L. *Beiträge zur Geschichte Georg Valla's und seiner Bibliothek*. Leipzig: Otto Harrassowitz, 1896.

Herlihy, David, and Christiane Klapisch-Zuber. *Tuscans and Their Families: A Study of the Florentine Catasto of 1427*. New Haven: Yale University Press, 1978. Originally published as *Les Toscans et leurs familles. Une étude du Catasto florentin de 1427*. Paris: Fondation Nationale des Sciences Politiques, 1978.

Hersey, George L. *Alfonso II and the Artistic Renewal of Naples, 1485–1495*. New Haven: Yale University Press, 1969.

Hilfstein, Erna, Pawel Czartoryski, and Frank D. Grande, eds. *Science and History: Studies in Honor of Edward Rosen*. Studia Copernicana XVI. Wroclaw-Warszawa-Krakow-Gdansk: Ossolineum, 1978.

Hollegger, Manfred. *Maximilian I (1459–1519). Herrscher und Mensch einer Zeitenwende*. Stuttgard: W. Kohlhammer, 2005.

Ianziti, Gary. *Humanistic Historiography under the Sforzas: Politics and Propaganda in Fifteenth-Century Milan*. Oxford: Clarendon Press, 1988.

Ilardi, Vincent. "The Assassination of Galeazzo Maria Sforza and the Reaction of Italian Diplomacy." In Lauro Martines, ed., *Violence and Civil Disorder in Italian Cities 1200–1500*, 72–103. Berkeley-Los Angeles-London: University of California Press, 1972.

———"Towards the Tragedia d'Italia: Ferrante and Galeazzo Maria Sforza, Friendly Enemies and Hostile Allies." In Abulafia, *The French Descent*, 91–122.

Il Rinascimento Italiano e l'Europa. 6 vols. Treviso: Fondazione Cassamarca & Angelo Colla Editore, 2005–2010.

Jacoviello, Michele. "La lega anti-francese del 31 Marzo 1495 nella fonte veneziana del Sanudo." *ASI* 143 (1985): 39–90.

Jacquart, Danielle. "De *crasis* à *complexio:* Note sur le vocabulaire su tempérament en Latin médiéval." In eadem, *La science médicale occidentale entre deux renaissances,* 71–76. Aldershot: Ashgate, 1997. Originally published in Guy Sabbah, ed., *Textes médicaux latins antiques,* 71–76. Saint-Etienne: Université de Saint-Étienne, 1984.

———"Theory, Everyday Practice, and Three Fifteenth-Century Physicians." In eadem, *La science médicale occidentale,* Ch. XIII: 142–153. Originally published in McVaugh and Siraisi, *Renaissance Medical Learning:* 140–160.

Jervis, Jane L. *Cometary Theory in Fifteenth-Century Europe.* Dordrecht-Boston-Lancaster: Reidel Publishing, 1985.

Johnson, John Noble. *The Life of Thomas Linacre: Doctor in Medicine, Physician to King Henry VIII.* London: Edward Lumley, 1835.

Juste, David, ed. *Les manuscrits astrologiques latins conservés à la Bayerische Staatsbibliothek de Munich.* Paris: CNRS Éditions, 2011.

Kemp, Martin. *Leonardo da Vinci: The Marvellous Works of Nature and Man.* Revised edition. Oxford: Oxford University Press, 2006.

Kibre, Pearl. "*Astronomia* or *Astrologia Ypocratis.*" In Hilfstein, Czartoryski, and Grande, *Science and History,* 133–156.

———*Hippocrates Latinus: Repertorium of Hippocratic Writings in the Latin Middle Ages.* Revised edition with additions and corrections. New York: Fordham University Press, 1985.

Knecht, Robert J. *The Rise and Fall of Renaissance France, 1483–1610.* Oxford: Blackwell, 2001. First edition 1996.

Krauss, Samuel. "Le Roi de France, Charles VIII, et les espérances messianiques." *Revue des études juives* 51 (1906): 87–95.

Kristeller, Paul O. "Giovanni Pico della Mirandola and his Sources." In *L'opera e il pensiero di Giovanni Pico della Mirandola nella storia dell'Umanesimo.* 2 vols. Vol. 1, 35–142. Florence: Istituto Nazionale di Studi sul Rinascimento, 1965.

Kunitzsch, Paul. *Mittelalterliche astronomisch-astrologische Glossare mit arabischen Fachausdrücken.* Munich: Bayerische Akademie der Wissenschaften, 1977.

Kuntz, Marion Leathers. *The Anointment of Dionisio: Prophecy and Politics in Renaissance Italy.* University Park, PA: Pennsylvania State University Press, 2001.

Kurze, Dietrich. *Johannes Lichtenberger: Eine Studie zur Geschichte der Prophetie und Astrologie.* Lübeck and Hamburg, 1960.

———"Popular Astrology and Prophecy in the Fifteenth and Sixteenth Centuries: Johannes Lichtenberger." In Zambelli, *"Astrologi Hallucinati,"* 177–193.

Langermann, Y. Tzvi. "The Astral Connections of Critical Days. Some Late Antique Sources Preserved in Hebrew and Arabic." In Akasoy, Burnett, and Yoeli-Tlalim, *Astro-Medicine,* 99–118.

Lauree pavesi nella seconda metà del '400. 3 vols. Bologna and Milan: Cisalpino, 1995–2008.

———vol. 1 (1995): (1450–1475) edited by Agostino Sottili.

Lazzarini, Isabella. "L'informazione politico-diplomatica nell'età della pace di Lodi: raccolta, selezione, trasmissione. Spunti di ricerca dal carteggio Milano-Mantova nella prima età sforzesca (1450–1466)." *Nuova Rivista Storica* 83 (1999): 247–280.

———"Gonzaga, Dorotea." *DBI* 57 (2001): 707–708.

Lemay, Helen. "The Stars and Human Sexuality: Some Medieval Scientific Views." *Isis* 71.1 (1980): 127–137.

———*Women's Secrets: A Translation of Pseudo-Albertus Magnus'* De secretis mulierum *with Commentaries.* Albany, NY: State University of New York Press, 1992.

Lemay, Richard. *Abû Ma'shar and Latin Aristotelianism in the Twelfth Century: The Recovery of Aristotle's Natural Philosophy through Arabic Astrology.* Beirut: American University of Beirut, 1962.

———"The Teaching of Astronomy in Medieval Universities, Principally at Paris in the Fourteenth Century." *Manuscripta* 20 (1976): 197–217.

———"Gerard of Cremona." In *Dictionary of Scientific Biography,* vol. 15, 173–192. New York: Schribner's Sons, 1978.

Leverotti, Franca. "La crisi finanziaria del ducato di Milano alla fine del Quattrocento." In *Milano nell'età di Ludovico il Moro,* vol. 2, 585–632.

———*Diplomazia e Governo dello Stato: I 'famigli cavalcanti' di Francesco Sforza (1450–1466).* Pisa: GISEM-ETS Editrice, 1992.

———"Lucia Marliani e la sua famiglia: il potere di una donna amata." In Letizia Arcangeli and Susanna Peyronel, eds., *Donne di potere nel Rinascimento,* 281–311. Rome: Viella, 2008.

Lippincott, Kristen. "Two Astrological Ceilings Reconsidered: The Sala di Galatea in the Villa Farnesina and the Sala del Mappamondo at Caprarola." *JWCI* 53 (1990): 185–207.

Lonie, Iain M. "Fever Pathology in the Sixteenth-Century: Tradition and Innovation." In William F. Bynum and Vivian Nutton, eds., *Theories of Fever from Antiquity to the Enlightenment,* Medical History Suppl. N. 1, 19–44. London: The Wellcome Institute, 1981.

Lopez, Guido. *Festa di Nozze per Ludovico il Moro: Fasti nuziali e intrighi di potere alla corte degli Sforza, tra Milano, Vigevano e Ferrara.* Milan: Mursia, 2008.

Lowry, Martin. *Nicolas Jenson and the Rise of Venetian Publishing in Renaissance Europe.* Cambridge, MA: Blackwell, 1991.

Lubkin, Gregory P. *A Renaissance Court: Milan under Galeazzo Maria Sforza*. Berkeley: University of California Press, 1994.

Lucentini, Paolo, and Vittoria Perrone Compagni. *I testi e i codici di Ermete nel Medioevo*. Florence: Edizioni Polistampa, 2001.

Luzio, Alessandro. "Isabella d'Este e Francesco Gonzaga promessi sposi." *ASL* 35 (1908): 34–69.

Luzio, Alessandro, and Rodolfo Renier. "Delle relazioni di Isabella d'Este Gonzaga con Ludovico e Beatrice Sforza." *ASL* 17 (1890): 74–119, 346–399, 619–674.

MacDonald, Alasdair A., Zweder R. W. M. von Martels, and Jan R. Veenstra, eds., *Christian Humanism: Essays in Honour of Arjo Vanderjagt*. Leiden and Boston: Brill, 2009.

Maclean, Ian. *Logic, Signs and Nature in the Renaissance: The Case of Learned Medicine*. Cambridge: Cambridge University Press, 2002.

———"A Chronology of the Composition of Cardano's Works." In Girolamo Cardano, *The Libris propriis: The Editions of 1544, 1550, 1557, 1562, with supplementary material*, edited by Ian Maclean, 43–111. Milan: FrancoAngeli, 2004.

———"The Reception of Medieval Practical Medicine in the Sixteenth Century: The Case of Arnau de Vilanova." In idem, *Learning and the Market Place: Essays in the History of the Early Modern Book*, 89–106. Leiden and Boston: Brill, 2009.

Magenta, Carlo. *I Visconti e gli Sforza nel castello di Pavia, e loro attinenze con la certosa e la storia cittadina*. 2 vols. Milan: Hoepli, 1883.

Magistretti, Pietro. "Galeazzo Maria Sforza e la caduta di Negroponte." *ASL* I (1884): 79–120 (fasc. I) and 337–356 (fasc. II).

Majocchi, Rodolfo. *Codice diplomatico dell'Università di Pavia*. 2 vols. Bologna: Forni editore, 1971. Originally published Pavia: Fratelli Fusi, 1905–1915.

Malagola, Carlo. *Statuti delle università e dei collegi dello Studio bolognese*. Bologna: Nicola Zanichelli, 1888.

Malaguzzi Valeri, Francesco. *La corte di Ludovico il Moro*. 4 vols. Milan: Hoepli, 1913–1923.

Mallett, Michael. "Personalities and Pressures: Italian Involvement in the French Invasion of 1494." In Abulafia, *The French Descent*, 151–163.

———"Venezia e la politica italiana: 1454–1530." In *Storia di Venezia*. 12 vols. Rome: Enciclopedia Italiana Treccani, 1991–2002. Vol. 4: *Il Rinascimento: politica e cultura*, Edited by Alberto Tenenti and Ugo Tucci. (1996): 245–310.

Margaroli, Paolo. "Bianca Maria e Galeazzo Maria Sforza nelle ultime lettere di Antonio da Trezzo (1467–1469)." *ASL* III (1985): 327–377.

Mari, Francesco, Aldo Polettini, Donatella Lippi, and Elisabetta Bertol. "The Mysterious Death of Francesco I de' Medici and Bianca Cappello: An Arsenic Murder?" *British Medical Journal* 333 (December 23, 2006): 1299–1301.

Marinozzi, Silvia, and Gino Fornaciari. *Le mummie e l'arte medica: per una storia dell'imbalsamazione artificiale dei corpi umani nell'evo Moderno*. Rome: Università la Sapienza, 2005.

Martin, Craig. *Renaissance Meteorology: Pomponazzi to Descartes*. Baltimore, MD: The Johns Hopkins University Press, 2011.

Martorelli Vico, Romana. *Medicina e filosofia. Per una storia dell'embriologia medievale nel XIII e XIV secolo*. Milan: Guerini e associati, 2002.

Mattiangeli, Paola. "Annio astrologo e alchimista." In *Annio da Viterbo. Documenti e ricerche*, 269–275. Rome: Consiglio Nazionale delle Ricerche, 1981.

Maxwell-Stuart, P. G. "Representations of Same-Sex Love in Early Modern Astrology." In Kenneth Borris and George S. Rousseau, eds., *The Sciences of Homosexuality*, 165–182. London: Routledge, 2007.

Mazzatinti, Giuseppe. "Inventario delle carte dell'Archivio Sforzesco contenute nei codd. Ital. 1583–1593 della Biblioteca Nazionale di Parigi." *ASL* 10 (1883): 222–326.

McCall, Timothy. "Traffic in Mistresses: Sexualized Bodies and Systems of Exchange in the Early Modern Court." In Allison Levy, ed., *Sex Acts in Early Modern Italy: Practice, Performance, Perversion, Punishment*, 125–136. Farnham: Ashgate, 2010.

McVaugh, Michael. "The Nature and Limits of Certitude at Early Fourteenth-Century Montpellier." In McVaugh and Siraisi, *Renaissance Medical Learning*, 62–84.

———*Medicine Before the Plague: Practitioners and Their Patients in the Crown of Aragon, 1285–1345*. Cambridge: Cambridge University Press, 1993.

McVaugh, Michael and Nancy Siraisi, eds., *Renaissance Medical Learning: The Evolution of a Tradition* (Osiris, vol. 6). Philadelphia: History of Science Society, 1990.

Mentgen, Gerd. *Astrologie und Öffentlichkeit im Mittelalter*. Stuttgart: Anton Hiersemann, 2005.

Merkley, Paul A., and Lora L. M. Merkley. *Music and Patronage in the Sforza Court*. Turnhout: Brepols and Pietro Antonio Locatelli Foundation, 1999.

Meschini, Stefano. *Luigi XII duca di Milano: gli uomini e le istituzioni del primo dominio francese, 1499–1512*. Milan: FrancoAngeli, 2004.

———*La Francia nel ducato di Milano: La politica di Luigi XII (1499–1512)*. 2 vols. Milan: FrancoAngeli, 2006.

Meserve, Margaret. "News from Negroponte: Politics, Popular Opinion, and Information Exchange in the First Decade of the Italian Press." *RQ* 59 (2006): 440–480.

Midelfort, H. C. Erik. *Mad Princes of Renaissance Germany*. Charlottesville and London: University of Virginia Press, 1999. First edition 1994.

Mikkeli, Heikki. *Hygiene in the Early Modern Medical Tradition*. Helsinki: Finnish Academy of Science and Letters, 1999.

Milano nell'età di Ludovico il Moro. Atti del convegno internazionale, 28 febbraio–4 Marzo 1983. 2 vols. Milan: Comune di Milano e Archivio Storico Civico-Biblioteca Trivulziana, 1983.

Moreton, Jennifer. "John of Sacrobosco and the Calendar." *Viator* 25 (1994): 229–244.

Morpurgo, Piero. "Scot, Michael (*d.* in or after 1235)." *ODNB,* http://www .oxforddnb.com/view/article/24902, last accessed July 28, 2011.

Motta, Emilio. "I Sanseverino feudatari di Lugano e Balerna, 1434–1484." *Periodico della società storica comense* 2 (1882): 155–310.

Moulinier-Brogi, Laurence. *Guillaume l'Anglais le frondeur de l'uroscopie médiévale (XIIIe siècle). Édition commentée et traduction du* De urina non visa. Geneva: Droz, 2011.

Mugnai Carrara, Daniela. *La biblioteca di Nicolò Leoniceno. Tra Aristotele e Galeno: cultura e libri di un medico umanista.* Florence: Olschki, 1991.

———"Fra causalità astrologica e causalità naturale. Gli interventi di Nicolò Leoniceno e della sua scuola sul morbo gallico." *Physis* 21 (1979): 37–54.

Nanni, Romano. "Le «disputationes» pichiane sull'astrologia e Leonardo." In Frosini, *Leonardo e Pico,* 53–98.

Narducci, Enrico, ed. *Catalogo di manoscritti ora posseduti da D. Baldassarre Boncompagni.* Rome: Tipografia delle scienze matematiche e fisiche, 1892. First edition 1862.

Negri, Paolo. "Milano, Ferrara e l'Impero durante l'impresa di Carlo VIII in Italia." *ASL* 44 (1917): 423–571.

Negri, Renzo. "Birago, Francesco." *DBI* 10 (1968): 581–584.

Neugebauer, Otto, and H. B. van Hoesen. *Greek Horoscopes.* Philadelphia: American Philosophical Society, 1959.

Newman, William R., and Anthony Grafton, eds. *Secrets of Nature: Astrology and Alchemy in Early Modern Europe.* Cambridge, MA: The MIT Press, 2001.

Niccoli, Ottavia. "Il diluvio del 1524 fra panico collettivo e irrisione carnevalesca." In *Scienze, credenze occulte, livelli di cultura. Convegno Internazionale di Studi (Firenze, 26–30 giugno 1980),* 369–392. Florence: Olschki, 1982.

———*Prophecy and People in Renaissance Italy.* Translated by Lydia C. Cochrane. Princeton, NJ: Princeton University Press, 1990.

Nicoud, Marilyn. "Expérience de la maladie et échange épistolaire: les derniers moments de Bianca Maria Visconti (mai–octobre 1468)." *MEFR* 112. 1 (2000): 311–458.

———"Les pratiques diététiques à la cour de Francesco Sforza." In Bruno Laurioux and Laurence Moulinier-Brogi, eds., *Scrivere il Medioevo. Lo spazio, la santità, il cibo. Un libro dedicato ad Odile Redon,* 393–404. Rome: Viella, 2001.

———"La médecine à Milan a la fin du Moyen Âge: les composantes d'un milieu *professional.*" In Frank Collard and Evelyne Samama, eds., *Mires, physiciens, barbiers et charlatans. Les marges de la médecine de l'Antiquité au XVIe siècle,* 101–131. Langres: D. Gueniot, 2004.

————"Les médecins à la cour de Francesco Sforza ou comment gouverner le Prince." In Odile Redon, Line Sallmann, and Sylvie Steinberg, eds., *Le Désir et le Goût. Une autre histoire (XIIIᵉ–XVIIIᵉ siècle)*, 201–217. Paris: Presses universitaires de Vincennes, 2005.

————*Les régimes de santé au Moyen Âge: Naissance et diffusion d'une écriture médicale, XIIIᵉ–XVᵉ siècle.* 2 vols. Rome: École française de Rome, 2007.

North, John D. "Astrology and the Fortune of Churches." *Centaurus* 24 (1980): 181–211.

————*Horoscopes and History.* London: Warburg Institute, 1986.

————"Celestial Influence—The Major Premise of Astrology." In Zambelli, *Astrologi Hallucinati*, 45–100.

————"Types of Inconsistency in the Astrology of Ficino and Others." In MacDonald, Alasdair A., Zweder R. W. M. von Martels, and Jan R. Veenstra, eds. *Christian Humanism: Essays in Honour of Arjo Vanderjagt*, 281–302. Leiden and Boston: Brill, 2009.

Nuovo, Angela. *Il commercio librario nell'Italia del Rinascimento*. Milan: FrancoAngeli, 1998.

Nuovo, Angela, and Christian Coppens. *I Giolito e la stampa nell'Italia del XVI secolo*. Geneva: Droz, 2005.

Nutton, Vivian. "Greek Medical Astrology and the Boundaries of Medicine." In Akasoy, Burnett, and Yoeli-Tlalim, *Astro-Medicine*, 17–31.

O'Boyle, Cornelius. *Medieval Prognosis and Astrology: A Working Edition of the Aggregationes de crisi et creticis diebus, with Introduction and English Summary.* Cambridge: Wellcome Unit for the History of Medicine, 1991.

————"An Updated Survey of the Life and Works of Bartholomew of Bruges († 1356)." *Manuscripta* 40.2 (1996): 67–95.

Oestmann, Günther, H. Darrel Rutkin, and Kocku von Stuckrad, eds. *Horoscopes and Public Spheres.* Berlin: Walter de Gruyter, 2005.

Pack, Roger A. "Pseudo-Aristoteles: Chiromantia." *Archives d'histoire doctrinale et littéraire du Moyen Age* 39 (1972): 289–320.

Page, Sophie. "Richard Trewythian and the Uses of Astrology in Late Medieval England." *JWCI* 64 (2001): 193–228.

Panconesi, Emiliano, and Lorenzo Marri Malacrida. *Lorenzo il Magnifico in salute e in malattia.* With an introduction by Eugenio Garin. Florence: Alberto Bruschi, 1992.

Pannella, Liliana. "Anselmi, Giorgio, Senior." *DBI* 3 (1961): 377–378.

Parel, Anthony. *The Machiavellian Cosmos.* New Haven: Yale University Press, 1992.

Park, Katharine. *Doctors and Medicine in Early Renaissance Florence.* Princeton, NJ: Princeton University Press, 1985.

————"The Organic Soul." In Charles B. Schmitt, Quentin Skinner, and Eckhard Kessler, eds., *The Cambridge History of Renaissance Philosophy,* 464–484. Cambridge: Cambridge University Press, 1988.

————"Natural Particulars: Medical Epistemology, Practice, and the Literature of Healing Springs." In Anthony Grafton and Nancy Siraisi, eds., *Natural Particulars: Nature and the Disciplines in Renaissance Europe*, 347–367. Cambridge, MA: The MIT Press, 1999.

————*Secrets of Women: Gender, Generation, and the Origins of Human Dissection.* New York: Zone Books, 2006.

Pedersen, Olaf. "The *Theorica Planetarum*-literature of the Middle Ages." *Classica et Medievalia* 23 (1962): 225–232.

————"The *Corpus Astronomicum* and the Traditions of Medieval Latin Astronomy." In Gingerich and Dobrzycki, *Astronomy of Copernicus and Its Background*, 57–96.

————"The Decline and Fall of the *Theorica Planetarum*: An Essay in Renaissance Astronomy." In Hilfstein, Czartoryski, and Grande, *Science and History*, 157–185.

————"The Origins of the *Theorica Planetarum*." *Journal for the History of Astronomy* 12 (1981): 113–123.

Pedralli, Monica. "Il medico ducale milanese Antonio Bernareggi e i suoi libri." *Aevum* 70 (1996): 307–350.

————*Novo, grande, coverto, e ferrato: gli inventari di biblioteca e la cultura a Milano nel Quattrocento.* Milan: Vita e Pensiero, 2002.

Pedretti, Carlo. *Studi Vinciani: documenti, analisi e inediti leonardeschi.* Geneva: Droz, 1957.

Pélissier, Leon-G. *Recherches dans les archives Italiennes. Louis XII et Ludovic Sforza (8 Avril 1498–23 Juillet 1500).* 2 vols. Paris: Thorin et fils, 1896. Athens.-Ecole Française d'Athènes. Bibliothèque des Écoles Françaises d'Athènes et de Rome. fasc. 75, 76.

————*Documents relatifs au règne de Louis XII et à sa politique en Italie.* Montpellier: Imprimerie Générale du Midi, 1912.

Pellegrin, Elisabeth. *La Bibliothèque des Visconti et des Sforza, ducs de Milan.* Paris: Vente au Service des publications du C.N.R.S., 1955.

Pellegrini, Marco. "Ascanio Maria Sforza: La creazione di un cardinale 'di famiglia'." In Chittolini, *Gli Sforza, la chiesa lombarda, e la corte di Roma*, 215–298.

————*Ascanio Sforza: La parabola politica di un cardinale-principe del Rinascimento,* 2 vols. Rome: Istituto Storico Italiano per il Medio Evo, 2002.

————*Le guerre d'Italia, 1494–1530.* Bologna: Il Mulino, 2009.

Pennuto, Concetta. "The Debate on Critical Days in Renaissance Italy." In Akasoy, Burnett, and Yoeli-Tlalim, *Astro-Medicine*, 17–31.

Pesenti Marangon, Tiziana. "La miscellanea astrologica del prototipografo padovano Bartolomeo Valdizocco e la diffusione dei testi astrologici e medici tra i lettori padovani del '400." *Quaderni per la storia dell'università di Padova* 11 (1978): 87–106.

———*Marsilio Santasofia tra corti e università. La carriera di un «Monarcha Medicinae» del Trecento.* Treviso: Antilia, 2003.

Petrella, Giancarlo. *La* Pronosticatio *di Johannes Lichtenberger. Un testo profetico nell'Italia del Rinascimento.* Udine: Forum, 2010.

Piana, Celestino. "Documenti intorno al B. Marco Fantuzzi da Bologna († 1479)." *Studi Francescani* 25 (1953): 224–235.

———"Il beato Marco da Bologna e il suo convento di S. Paolo in Monte nel Quattrocento." *Atti e Memorie della Deputazione di Storia Patria per le Provincie di Romagna* II, n.s. 22 (1971): 85–265.

Pomata, Gianna. *Contracting a Cure: Patients, Healers, and the Law in Early Modern Bologna.* Baltimore and London: The Johns Hopkins University Press, 1998.

Pompeo Faracovi, Ornella. *Gli oroscopi di Cristo.* Venice: Marsilio, 1999.

Porro, Giulio. "Nozze di Beatrice d'Este e di Anna Sforza. Documenti copiati dagli originali esistenti nell'Archivio di Stato di Milano." *ASL* 9 (1882): 483–534.

Portigliotti, Giuseppe. "G. B. Boerio alla Corte d'Inghilterra." *Illustrazione medica Italiana* 5 (Gennaio 1923): 8–10.

Portioli, Attilio. "Nascita di Massimiliano Sforza." *ASL* 9 (1882): 325–334.

Potestà, Gian Luca, ed. *Il profetismo gioachimita tra Quattrocento e Cinquecento: atti del III Congresso internazionale di studi gioachimiti, San Giovanni in Fiore, 17–21 Settembre 1989.* Genoa: Marietti, 1991.

Potter, David. *Prophets and Emperors: Human and Divine Authority from Augustus to Theodosius.* Cambridge, MA: Harvard University Press, 1994.

———*A History of France, 1460–1560: The Emergence of a Nation-State.* Basingstoke: Macmillan, 1995.

Poulle, Emmanuel. *La Bibliothèque scientifique d'un imprimeur humaniste au XVe siecle: Catalogue des manuscrits d'Arnaud de Bruxelles a la Bibliothèque Nationale de Paris.* Geneva: Droz, 1963.

———"Horoscopes princiers des XIVe et XVe siècles." *Bulletin de la société nationale des antiquaires de France* (1969): 63–69.

———*Les Tables Alphonsines avec les Canons de Jean de Saxe.* Paris: Éditions du centre national de la recherche scientifique, 1984.

Poulle, Emmanuel, and Owen Gingerich, eds. *Les positiones des planètes au moyen âge: application du calcul électronique aux tables alphonsines.* Paris: Klincksieck, 1968.

Préaud, Maxime. *Les astrologues à la fin du Moyen Age.* Paris: Lattès, 1984.

Quinlan-McGrath, Mary. "The Astrological Vault of the Villa Farnesina: Agostino Chigi's Rising Sign." *JWCI* 47 (1984): 91–105.

———"The Villa Farnesina: Time-telling Conventions and Renaissance Astrological Practice." *JWCI* 58 (1995): 52–71.

Raponi, Nicola. "Arcimboldi, Nicolò." *DBI* 3 (1961): 779–781.

Reeves, Marjorie. *The Influence of Prophecy in the Later Middle Ages: A Study of Joachimism.* Notre Dame, IN: University of Notre Dame, 1993. First edition Clarendon Press, 1969.

Reti, Ladislao. "The Two Unpublished Manuscripts of Leonardo da Vinci in the Biblioteca Nacional of Madrid-II." *The Burlington Magazine* 110 (1968): 81–89.

Richardson, Brian. *Printing, Writers, and Readers in Renaissance Italy.* Cambridge: Cambridge University Press, 1999.

Rider, Catherine. *Magic and Impotence in the Middle Ages.* Oxford: Oxford University Press, 2006.

Robinson, A. Mary F. "The Claim of the House of Orleans to Milan." *English Historical Review* 3.9 (1988): 34–62.

———"The Claim of the House of Orleans to Milan (continued)." *English Historical Review* 3.10 (1988): 270–291.

Rodriguez Salgado, Maria-Jose. "Terracotta and Iron. Mantuan Politics (ca. 1450–ca. 1550)." In Cesare Mozzarelli, Robert Oresko, and Leandro Ventura, eds., *La corte di Mantova nell'età di Andrea Mantegna: 1450–1550,* 15–61. Rome: Bulzoni, 1997.

Rogledi Manni, Teresa. *La tipografia a Milano nel XV secolo.* Florence: Olschki, 1980.

Rosmini, Carlo de'. *Dell'Istoria di Milano.* 4 vols. Milan: Tip. Manini e Rivolta, 1820.

———*Dell'istoria intorno alle militari imprese e alla vita di Gian-Jacopo Trivulzio detto il Magno.* 2 vols. Milan: Tip. Gio. Giuseppe Destefanis, 1815.

Rosso, Paolo. "Presenze studentesche e collegi pavesi nella seconda metà del Quattrocento." *Schede umanistiche* 2 (1994): 24–42.

Rusconi, Roberto. *Profezia e profeti alla fine del Medioevo.* Rome: Viella, 1999.

Rutkin, H. Darrel. *Astrology, Natural Philosophy and the History of Science, c.1250–1700: Studies Toward an Interpretation of Giovanni Pico della Mirandola's Disputationes adversus astrologiam divinatricem.* Ph.D. dissertation. Bloomington: Indiana University, 2002.

———"Astrology." In *The Cambridge History of Science.* 5 vols. Cambridge: Cambridge University Press, 2002–2009. Vol. 3: *Early Modern Science.* Edited by Katharine Park and Lorraine Daston. (2006): 541–561.

———"Astrological Conditioning of Same-Sex Relations in Girolamo Cardano's Theoretical Treatises and Celebrity Genitures." In Kenneth Borris and George S. Russeau, eds., *The Sciences of Homosexuality,* 183–199. London: Routledge, 2007.

Ryan, Michael A. *A Kingdom of Stargazers: Astrology and Authority in the Late Medieval Crown of Aragon.* Ithaca, NY: Cornell University Press, 2011.

Ryder, Alan. "The Angevin Bid for Naples, 1380–1480." In Abulafia, *The French Descent,* 56–69.

Sambin, Paolo. "Barzizza, Cristoforo." *DBI* 7 (1970): 32–34.

Santoro, Caterina. *Gli Uffici del dominio Sforzesco (1450–1500)*. Milan: Fondazione Treccani, 1947.

———*I Registri delle lettere ducali del periodo Sforzesco*. Milan: Castello Sforzesco, 1961.

Scarton, Elisabetta. "La congiura dei baroni del 1485–'87 e la sorte dei ribelli." In Francesco Senatore and Francesco Storti, eds., *Poteri, relazioni, guerra nel regno di Ferrante d'Aragona*, 213–290. Naples: Cliopress, 2011.

Schechner Genuth, Sara. *Comets, Popular Culture, and the Birth of Modern Cosmology*. Princeton, NJ: Princeton University Press, 1997.

Segrè, Arturo. "Ludovico Sforza, detto il Moro, e la Repubblica di Venezia dall'autunno 1494 alla primavera 1495." *ASL* 18 (1902): 249–317, and 20 (1903): 33–109, 368–443.

Seidel Menchi, Silvana. "Boerio, Giovanni Battista." *DBI* 11 (1969): 126–127.

Sela, Shlomo. *Abraham Ibn Ezra and the Rise of Medieval Hebrew Science*. Leiden-Boston: Brill, 2003.

Selby, Talbot R. "Filippo Villani and his Vita of Guido Bonatti." *Renaissance News* 11.4 (1958): 243–248.

Senatore, Francesco. *«Uno mundo de carta». Forme e strutture della diplomazia sforzesca*. Naples: Liguori Editore, 1998.

Setton, Kenneth M. *The Papacy and the Levant (1204–1571)*. 4 vols. Philadelphia: American Philosophical Society, 1978.

Sevesi, Paolo M,. O.F.M. "Corrispondenza Milanese del B. Marco da Bologna." *Archivum Franciscanum Historicum* 48 (1955): 298–323.

Shank, Michael. "L'astronomia nel Quattrocento tra corti e università." *Il Rinascimento Italiano e l'Europa*, 6 vols. Treviso: Fondazione Cassamarca & Angelo Colla Editore, 2005–2010. Vol. 5: *Le Scienze*. Edited by Antonio Clericuzio and Germana Ernst. (2008): 3–20.

Shaw, Christine. *The Politics of Exile in Renaissance Italy*. Cambridge: Cambridge University Press, 2000.

———*The Political Role of the Orsini Family from Sixtus IV to Clement VIII: Barons and Factions in the Papal States*. Rome: Istituto Storico Italiano per il Medio Evo, 2007.

Signorini, Rodolfo. *Fortuna dell'astrologia a Mantova: Arte, letteratura, carte d'archivio*. Mantua: Sometti, 2007.

Silver, Larry. *Marketing Maximilian: The Visual Ideology of a Holy Roman Emperor*. Princeton, NJ: Princeton University Press, 2008.

Simili, Alessandro. "Alcuni documenti inediti intorno a Francesco Benzi." *Rivista di storia delle scienze mediche e naturali* 33 (1942): 81–102.

Simonetta, Marcello. *Rinascimento Segreto: Il mondo del segretario da Petrarca a Machiavelli*. Milan: FrancoAngeli, 2004.

Siraisi, Nancy G. *Arts and Sciences at Padua: The* Studium *of Padua before 1350.*
Toronto: Pontifical Institute for Medieval Studies, 1973.

———*Taddeo Alderotti and His Pupils: Two Generations of Italian Medical Learning.* Princeton, NJ: Princeton University Press, 1981.

———*Avicenna in Renaissance Italy: The Canon and Medical Teaching in Italian Universities after 1500.* Princeton, NJ: Princeton University Press, 1987.

———*Medieval and Early Renaissance Medicine: An Introduction to Knowledge and Practice.* Chicago: The University of Chicago Press, 1990.

———"Disease and Symptom as Problematic Concepts in Renaissance Medicine." In Eckhard Kessler and Ian Maclean, eds., *Res et verba in der Renaissance,* 217–240. Wiesbaden: Harrassowitz Verlag, 2002.

Smithuis, Renate. "Abraham Ibn Ezra's Astrological Works in Hebrew and Latin: New Discoveries and Exhaustive Listing." *Aleph: Historical Studies in Science and Judaism* 6 (2006): 239–338.

Smoller, Laura Ackerman. *History, Prophecy, and the Stars: The Christian Astrology of Pierre D'Ailly, 1350–1420.* Princeton, NJ: Princeton University Press, 1994.

———"Of Earthquakes, Hail, Frogs, and Geography: Plague and the Investigation of the Apocalypse in the Later Middle Ages." In Caroline Walker Bynum and Paul H. Freedman, eds., *Last Things: Death and the Apocalypse in the Middle Ages,* 156–187. Philadelphia, PA: University of Pennsylvania Press, 2000.

———"Teste Albumasare cum Sibylla: Astrology and the Sibyls in Medieval Europe." *Studies in History and Philosophy of Biological and Biomedical Sciences* 41.2 (2010): 76–89.

Soldi Rondinini, Gigliola. "Ambasciatori e ambascerie al tempo di Filippo Maria Visconti." *Nuova Rivista Storica* 49 (1965): 313–344.

———"Ludovico il Moro nella storiografia coeva." In *Milano nell'età di Ludovico il Moro. Atti del convegno internazionale, 28 Febbraio–4 Marzo 1983,* vol. 1, 29–56. Milan: Comune di Milano e Archivio Storico Civico-Biblioteca Trivulziana, 1983.

———"Milano, il Regno di Napoli e gli Aragonesi (secoli XIV–XV)." In *Gli Sforza a Milano e in Lombardia e i loro rapporti con gli Stati italiani ed europei (1450–1535). Convegno internazionale, Milano 18–21 maggio 1981,* 229–290. Milan: Cisalpino-Goliardica, 1992. Reprinted in eadem, *Saggi di storia e storiografia visconteo-sforzesca,* 83–129. Bologna: Cappelli, 1984.

Somaini, Francesco. "Processi costitutivi, dinamiche politiche e strutture istituzionali dello Stato visconteo-sforzesco." In *Storia d'Italia.* Edited by Giancarlo Andenna, Renato Bordone, Francesco Somaini, and Massimo Vallerani. 25 vols. Turin: UTET, 1976–2008. Vol. 6: *Comuni e signorie nell'Italia settentrionale: La Lombardia.* (1998): 681–786 and 809–825.

Sottili, Agostino. "Università e cultura a Pavia in età visconteo-sforzesca." In *Storia di Pavia,* vol. 3, t. 2, 359–451.

Storia di Milano. 17 vols. Milan: Fondazione Treccani, 1953–1962.

—vol. 5 (1955): *La signoria dei Visconti (1310–1392)*.

—vol. 6 (1955): *Il Ducato visconteo e la Repubblica ambrosiana (1392–1450)*.

—vol. 7 (1956): *L'età sforzesca dal 1450 al 1500*.

Storia di Pavia. 5 vols. Milan: Banca del Monte di Lombardia, 1990.

Sudhoff, Karl. "Zur Geschichte der Lehre von den kritischen Tagen im Krankenhaitsverlaufe." *Wiener medizinische Wochenschrift* 52 (1902): 210–275.

————"Daniels von Morley *Liber de Naturis Inferiorum et Superiorum*.", *Archiv für Geschichte der Naturwissenschaften und der Technik* 8 (1918): 1–40.

Tester, S. Jim, *A History of Western Astrology*. Woodbridge, Suffolk: Boydell Press, 1987.

Thomas, Keith. *Religion and the Decline of Magic*. London: Weidenfeld & Nicolson, 1971.

Thorndike, Lynn. "Alfodhol and Almadel: Hitherto Unnoted Mediaeval Books of Magic in Florentine Manuscripts." *Speculum* 2 (1927): 326–331.

————"Alfodhol de Merengi Again." *Speculum* 4 (1929): 90.

————"Alfodhol and Almadel Once More." *Speculum* 20 (1945): 88–91.

————*The* Sphere *of Sacrobosco and its Commentators*. Chicago: The University of Chicago Press, 1949.

————*University Records and Life in the Middle Ages*. New York: Norton, 1975; reprint of the 1944 edition.

————"The Three Latin Translations of the Pseudo-Hippocratic Tract on Astrological Medicine." *Janus* 49 (1960): 104–129.

————*Michael Scot*. London: Nelson, 1965.

Thorndike, Lynn, and Pearl Kibre, eds. *A Catalogue of Incipits of Mediaeval Scientific Writings in Latin*. Revised and augmented. London: Mediaeval Academy of America, 1963.

Vaglienti, Francesca M. "Galeazzo Maria Sforza." *DBI* 51 (1998): 398–409.

————"Gian Galeazzo Maria Sforza, duca di Milano." *DBI* 54 (2000): 391–397.

————"Anatomia di una congiura. Sulle tracce dell'assassinio del duca Galeazzo Maria Sforza tra scienza e storia." *Atti dell'Istituto Lombardo. Accademia di Scienze e Lettere di Milano* 136 (2002): 237–273.

Valeri, Elena. *Italia dilacerata. Girolamo Borgia nella cultura storica del Rinascimento*. Milan: FrancoAngeli, 2007.

Vasoli, Cesare. "Avogaro (Dell'Avogaro, Avogari, Avogario, Arvogari, Advogaro), Pietro Buono." *DBI* 6 (1962): 709–710.

————"Bedemandis, Prosdocimo." *DBI* 7 (1965): 551–554.

————"L'astrologia a Ferrara tra la metà del Quattrocento e la metà del Cinquecento." In *Il Rinascimento nelle corti padane. Società e cultura*, 469–567. Bari: De Donato, 1977.

————*I miti e gli astri*. Naples: Guida, 1977.

————"Il mito della monarchia francese nelle profezie fra 1490 e 1510." In Dario Cecchetti, Lionello Sozzi, and Louis Terreaux, eds., *L'aube de la Renaissance: pour le dixième anniversaire de la disparition de Franco Simone.* Geneva: Editions Slatkine, 1991.

Veen, Henk Th. van. *Cosimo I de' Medici and his Self-Representation in Florentine Art and Culture.* Cambridge: Cambridge University Press, 2006.

Veenstra, Jan R. *Magic and Divination at the Courts of Burgundy and France: Text and Context of Laurens Pignon's* Contre les devineurs (1441). Leiden: Brill, 1998.

Villani, Filippo. *De origine civitatis Florentie et de eiudsem famosis civibus.* Edited by Giuliano Tanturli. Padua: Antenore, 1997.

Villard, Renaud. *Du bien commun au mal nécessaire. Tyrannies, assassinats politiques et souveraineté en Italie.* Rome: École française de Rome, 2008.

Viti, Paolo, ed. *Pico, Poliziano e l'Umanesimo di fine Quattrocento.* Florence: Olschki, 1994.

————"Decembrio, Pier Candido." *DBI* 33 (1987): 488–498.

Volpati, Carlo. "Gli Scotti di Monza tipografi-editori in Venezia." *ASL* 59 (1932): 365–382.

Warburg, Aby. "Pagan-Antique Prophecy in Words and Images in the Age of Luther." In *The Renewal of Pagan Antiquity: Contributions to the Cultural History of the European Renaissance,* with an introduction by Kurt W. Forster and translated by David Britt, 597–697. Los Angeles: Getty Research Institute, 1999. Originally published as Aby Warburg, *Heidnisch-Antike Weissagung in Wort und Bild zu Luthers Zeiten,* Heidelberg: C. Winter, 1920.

Weill-Parot, Nicolas. *Les "images astrologiques" au Moyen Âges et à la Renaissance: Spéculations intellectuelles et pratiques magiques, XIIème–XVème siècles.* Paris: Honoré Champion, 2002.

Weiss, Roberto. "Traccia per una biografia di Annio da Viterbo." *Italia medioevale e umanistica* 5 (1962): 425–441.

Welch, Evelyn. *Art and Authority in Renaissance Milan.* New Haven: Yale University Press, 1995.

————"Between Milan and Naples: Ippolita Maria Sforza, Duchess of Calabria." In Abulafia, *The French Descent,* 123–136.

————*Shopping in the Renaissance: Consumer Cultures in Italy, 1400–1600.* New Haven and London: Yale University Press, 2005.

Welch, Evelyn, and Michelle O'Malley. *The Material Renaissance.* Manchester: Manchester University Press, 2007.

Westman, Robert S. *The Copernican Question.* Berkeley, CA and London: The University of California Press, 2011.

Wiesflecker, Hermann. *Maximilian I: das Reich, Österreich und Europa an der Wende zur Neuzeit.* 5 vols. Munich: Oldenbourg, 1971–1986.

Yearl, Mary Katherine Keblinger Hague. *The Time of Bloodletting*. Ph.D. dissertation. Yale University, 2005.

Zaccaria, Vittorio. "Sulle opere di Pier Candido Decembrio." *Rinascimento* 7 (1956): 13–74.

Zambelli, Paola. "Fine del mondo o inizio della propaganda? Astrologia, filosofia della storia e propaganda politico-religiosa nel dibattito sulla congiunzione del 1524. In *Scienze, credenze occulte, livelli di cultura. Convegno Internazionale di Studi (Firenze, 26–30 giugno 1980)*, 291–368. Florence: Olschki, 1982.

——ed. *"Astrologi Hallucinati": Stars and the End of the World in Luther's Time.* Berlin and New York: Walter de Gruyter, 1986.

——"Astrologi consiglieri del principe a Wittenberg." *Annali dell'Istituto storico italo-germanico in Trento/Jahrbuch des italienisch-deutschen historischen Instituts in Trient* 18 (1992): 497–543.

——*Una reincarnazione di Pico ai tempi di Pomponazzi*. Milan: Il Polifilo, 1994.

Zarri, Gabriella. *Le Sante vive: profezie di corte e devozione femminile tra '400 e '500.* Turin: Rosenberg & Sellier, 1990.

Ziegler, Joseph. *Medicine and Religion c. 1300: The Case of Arnau de Vilanova.* Oxford: Clarendon Press, 1998.

Zutshi, Patrick. "An Unpublished Letter of Isabella of Aragon, Duchess of Milan." *Renaissance Studies* 20 (2006): 494–501.

Acknowledgments

This book would have never been conceived without the support of Sachiko Kusukawa. A beacon of professionalism and high scholarly standards, she served as my "informal" supervisor during my doctoral years, saving me from bleak despair on more than one occasion. She was the first one to have confidence in my abilities and my work. Over the years, she has been a treasured friend and a source of constant support. I owe her more than I can say. This book is dedicated to her. My work has been influenced immensely by the work of two other scholars, now friends and mentors, to whom this book is also dedicated: Nancy Siraisi and Katharine Park. Their scholarship is no less impressive than their unfailing generosity towards young scholars. I am deeply indebted to them for being such models of scholarship and for providing constant encouragement and guidance over the years. Katharine Park was the first one to read the whole manuscript in draft and to suggest that I submit it to this series. This book has benefited enormously from her generous encouragement and, above all, her rigorous and thoughtful criticism. She cannot be thanked enough for this. A few other scholars have been with me since the beginning: I am enormously grateful to Ian Maclean and Bill

Newman for their gracious support and advice and for their appreciation of the merits of archival research.

This book travelled with me across three continents and four countries. In writing it, therefore, I have accumulated a great many debts. It seems only fair that I thank explicitly those people and institutions that have enriched my knowledge and helped along the way. My greatest debt goes to the community of Villa I Tatti, the Harvard University Center for Italian Renaissance Studies, where I spent a blissful year in 2005/6 as Ahmanson Fellow in History. This book was first conceived while I was there. I am grateful to my fellow Tattiani that year, particularly Janie Cole, Brian Curran, Stefano Dall'Aglio, Alison Frazier, John Gagné, Marco Gentile, Miguel Gotor, David Lines, Agata Pincelli, and Darrel Rutkin, for providing delightful social and intellectual companionship during my time in Florence. We have remained friends since. Special thanks go to Joe Connors, Director of Villa I Tatti from 2002 to 2010, for welcoming all of us so warmly in this *locus amoenus*. My two Heads of School at the University of New South Wales, John Gascoigne and Rae Frances, deserve my sincerest thanks: the former for letting me go to Florence barely a year after my appointment and the latter for welcoming me back so warmly at my return. Australia holds a big place in my heart. I wish to thank in particular Nick Doumanis, Nick Eckstein, Robert Forgacs, Gary Ianziti, Maria Cristina Mauceri, Frances Muecke, Nerida Newbigin, David Juste, Peter and Elle Schrijvers, and John O. Ward for making my time there so memorable. Kate Colleran has been a most cheerful research assistant, getting me books and articles I could not get from my own university with her customary efficiency. I would also like to thank her, Luke, and little Matilda for offering me last-minute hospitality when I was leaving Sydney. Robyn and Sean Dunne merit special mention and heartfelt thanks for being the warmest Australian family one could hope for.

Earlier on in the project, support came in the form of a Travelling Research Fellowship from the Australian Academy of the Humanities in 2004, a Young Scholar Award from the Renaissance Society of America, and a series of research grants from the Faculty of Arts and Social Sciences at the University of New South Wales in 2005. This support allowed me to return to the Milanese archives and facilitated my research at the Bibliothèque Nationale in Paris. A Frances Yates Fellowship at the Warburg Institute, London, from November 2006 to February 2007, provided the ideal setting in which to refine my skills and learn more about the intricacies of Renaissance astrology.

My thanks go to Charles Burnett and Dorian Greenbaum for their constant guidance and for sharing their unique expertise. Thanks also to Guido Giglioni and Colin Homiski for helping me track down difficult material and for being such a constant source of companionship and good humor. Finally, a generous grant from Trinity College allowed me to spend an idyllic month in Cambridge almost a decade after my doctorate. Chapter 1 and part of Chapter 2 were written in such scholarly paradise during an unusually hot July of 2010.

I hold a double debt towards Villa I Tatti, as much of the writing and re-writing of this book took place there during the academic year 2008/9, when I returned to Florence as the recipient of a twelve-month Leverhulme Research Fellowship. My thanks thus go to the Leverhulme Trust for providing crucial support for my next project while also allowing me to finish this one. The Berenson Library once again provided a privileged place in which to work with the necessary continuity. Special thanks go to Michael Rocke and all the library staff for welcoming me back and for their unfailing support with my many requests. Last but not least, I wish to thank Professor Lino Pertile, VIT's current director, Professor Ed Muir, Editor of I Tatti Studies in Italian Renaissance History, and C. Ian Stevenson, Assistant Editor for the Humanities at Harvard University Press, for welcoming this book in their splendid series.

Milan holds pride of place in inspiring this book: not only is its history the subject of this work, but is also the place where I was born and grew up. Few Anglo-American scholars have dared to enter the secluded space of Renaissance studies on Milan. I thus wish to thank especially Evelyn Welch and Gary Ianziti for showing me the way so capably and for encouraging me to pursue research on Renaissance Milan at various stages in my research. My work in the Archivio di Stato di Milano was greatly facilitated by its helpful staff. I wish to thank particularly Maria Pia Bortolotti, Giovanni Liva, and Marina Valori for patiently answering my many questions, and especially Alba Osimo for sharing her unique knowledge of the Sforza collection and teaching me the ropes of archival research and Italian palaeography when I first entered the *Archivio* rather unprepared for the task. My thanks go also to the fellow historians who accompanied my researches and were generous with their advice: Massimiliano Ferri, Luca Fois, Arnaldo Ganda, Ada Grossi, Charles Morscheck, and Marilyn Nicoud. Each of them taught me something. If ever a granting body wants to fund something close to a science lab in the humanities, its answer is certainly the archives. In the humanities, no other place breeds so much camaraderie, encourages teamwork, and allows for collaboration to flourish.

My thanks are due also to the many other libraries and archives where I conducted my research. I must make special mention of the Bibliothèque Nationale de France, Paris, the Biblioteca Nazionale, Florence, the British Library and the Warburg Institute Library, London, Cambridge University Library and the Whipple Library, Cambridge, the Bodleian Library, Oxford, the Crawford Library, Royal Observatory, Edinburgh, Edinburgh University Library, and the State Archives of Florence, Mantua, and Modena.

Many of the colleagues mentioned above, and many others more, have been instrumental in offering the support and encouragement needed to finish the book. Among those yet to be mentioned I wish to thank my colleagues at the University of Edinburgh: fellow historians of science John Henry and Jane Ridder-Patrick, many of the members of the Centre for Medieval and Renaissance Studies, and my colleagues Thomas Ahnert, Pertti Ahonen, Steve Boardman, Stephen Bowd, Jill Burke, Sarah Cockram, Douglas Cairns, Tom Devine, James Fraser, Alvin Jackson, and Stana Nenadic. Thanks to all. A special thanks goes to fellow Italian expatriate Isabella Lazzarini, who has been unfailingly generous in sharing her profound knowledge of late medieval and Renaissance Italian diplomatic and institutional history with me for the past four years. Among fellow historians of astrology/astronomy, a number of colleagues have been particularly generous with their time and knowledge: together with Darrel Rutkin, I wish to thank particularly Jean-Patrice Boudet, Darin Hayton, David Juste, and Adam Mosley. I am immensely grateful to Stefano Dall'Aglio, Serena Ferente, Marco Gentile, David Juste, Isabella Lazzarini, Tim McCall, Katy Park, Darrel Rutkin, and the anonymous reader of Harvard University Press for reading part or the whole of the manuscript at various stages. Their keen eyes, constructive advice, and benign criticism have improved this book enormously. David Juste's superb knowledge of medieval and Renaissance astrology and his unique familiarity with the manuscript tradition have saved me more than once from embarrassing mistakes. Needless to say, any errors and idiosyncrasies in this book remain my own.

Portions of this book have appeared in print elsewhere. An earlier version of most of Chapter 3 appeared as "The Politics of Prognostication: Astrology, Political Conspiracy and Murder in Fifteenth-Century Italy" in *History of Universities* 23/2 (2009), while another part of the same chapter appeared as

"Annius of Viterbo Astrologer: Predicting the Death of Ferrante of Aragon, King of Naples" in *Bruniana & Campanelliana* 14.2 (2008). I am grateful to Oxford University Press, and Fabrizio Serra Editore for permission to reuse their material here.

My family has been consistently supportive of my work, even when, especially early on, they could not see much sense in it. I owe a particular debt to my mother for encouraging me to study and fulfill my intellectual aspirations as a woman. I also wish to thank my parents for putting up with me writing in a language that is not their own. Last but not least, I wish to thank my partner, Stefano, who at times took time out from his own work to follow my trail through Italian and European archives and libraries and read this book in its many unfolding drafts. Without his unflinching optimism, his enduring *buonsenso,* his eye for detail, and boundless faith in this project and my abilities, this book may have never appeared. Having suffered for years academic birth-pangs, it seems time to let go and see this book to publication. I am sure many of the people I have thanked here will be delighted and not least relieved to see this book appear in print.

A number of books came out too late to be properly integrated in this study. These include: Glen M. Cooper, *Galen,* De diebus decretoriis, *from Greek into Arabic* (Farnham: Ashgate, 2011), Robert S. Westman, *The Copernican Question* (Berkley, CA and London: The University of California Press, 2011), and Michael A. Ryan, *A Kingdom of Stargazers: Astrology and Authority in the Late Medieval Crown of Aragon* (Ithaca, NY: Cornell University Press, 2011).

Index

Abraham Ibn Ezra, 34, 45
   Liber de nativitatibus, 34, 45
   De mundo vel seculo, 34
Albertus Magnus, 76
   De secretis mulierum, 76
Albumasar, 23, 34, 40, 45, 51, 62, 87, 109,
   144, 148
   Flores astrologiae, 51
   Introductorium maius, 23, 45, 144, 148
   De magnis coniunctionibus, 23, 34,
     51, 62
   De revolutionibus annorum mundi (or *De*
     experimentis), 40
Alcabitius, 23, 28, 34, 40, 45–47, 51, 58, 87,
   103, 107
   Introductorius, 23, 28, 34, 45–46, 51,
     103
alchocoden (giver of the years), 60, 108–109,
   113
Alexander VI (Rodrigo Borgia), pope, 176,
   183, 197–199

Alexander of Aphrodisia, 161
   Problemata, 161
Alfodhol, 51
   Liber iudiciorum et consiliorum, 51–52
Alfonso II of Aragon, Duke of Calabria and
   King of Naples, 96–97, 101, 120–122
Alfonso V of Aragon (the Magnificent),
   King of Aragon and Naples, 120, 139,
   143, 150, 200, 202
Alfraganus, 47, 51
   De aggregationibus scientiae stellarum,
     47, 51
almuten, 110
Alphonsine tables, 27, 29, 42, 46, 51, 105
anaereta, 108, 112
Angeli, Teodora, 175
animodar, 105, 111
Anne of Britanny, 179, 182
Anne of France, 182
Annius of Viterbo (Giovanni Nanni), 7, 20,
   116, 118, 122–123, 125–126, 192

annual prognostications (or *iudicia*), 1, 4–5, 7, 10, 19–20, 30, 49, 67–71, 116–117, 126, 128–129, 131–133

Anselmi, Giorgio (da Parma), 47, 149
   Astronomia, 47, 149

Antiquario, Jacopo, 160

Antonio da Trezzo, 121, 128

Arcimboldi, Nicolò, 74

Ariosto, Francesco, 70

Aristotle (natural philosophical works), 78

armillary sphere, 45

Artoni, Giovanni, 131

Ascletarion, 113

aspects, 12, 54, 111

astrolabe, 28, 105, 175

astrological elections, 2, 17–19, 47, 52, 60, 63, 66, 79, 119, 132, 149, 155, 169, 174, 176–178, 180, 185, 200, 202–203

astrological embryology, 33, 74, 76, 78, 88

astrological interrogations, 17–20, 33, 47, 52, 60, 62–63, 96, 118–119, 121, 123, 125, 168–169, 184–185, 192–194, 197, 204

astrological talismans, 149–150

astrological techniques, 3–4, 50, 53–64, 66, 212

astrology
   and astronomy, 3, 10
   and death, 17–18, 21, 156–159
   and marriage, 17–18, 20–21, 60, 89–93, 95, 149–150, 167, 170–176, 185–190, 212
   and medicine (*see* medicine)
   and prophecy (*see* prophecy)
   and travel, 2, 18, 21, 60, 66, 73, 167, 176–178
   and war, 2, 17–18, 20–21, 60, 66, 73, 167, 199–207, 212
   at university, 3, 9–10, 20–21, 24–29, 39–42, 46–47, 77
   circulation of, 5, 102, 117, 128–129, 133–134
   in chronicles, 4, 165
   judicial, 11, 23–24, 47, 52, 66, 79, 86–87
   mundane, 47, 52

   natural, 11, 79
   political uses of, 4–5, 8, 10, 16–21, 63, 66, 68–71, 93–98, 102–103, 115, 118–119, 125–135, 166–167, 170–175, 178–180, 184–191, 204–208, 212
   Roman emperors' use of, 10, 113

astronomy, 23–24, 26–27, 42, 45–46, 50–51, 53

Attendoli, Bolognino degli, 84

Avicenna, 158
   Canon, 158

Avogario, Pietro Buono, 5, 115, 130–132, 207–208

Baghdad, 63

Barbarigo, Agostino, 176

Barbiano, Carlo (di Belgioioso), 179, 183, 200

Barbo, Marco, 191

Bardone, Lanfranco (da Parma), 73, 79

Bartholomew of Bruges (*see* ps.-Galen)

Barzizza, Cristoforo, 42
   Introductorium ad opus practicum medicinae, 42

Beldomandi, Prosdocimo de', 40, 46
   De electionibus, 40
   Algorismi, 42

Belleto, Lorenzo, 131

Bellingeri, Ettore, 208

Bembo, Pietro, 141, 165
   History of Venice, 141, 165

Bencora, Thebit, 42, 51
   Tractatus de motu octavae sphaerae, 42, 51

Benivieni, Girolamo, 45–46

Bentivoglio, Giovanni II, 132

Benzi, Francesco, 49

Bergamo, 121

Bernareggi, Antonio, 46–47, 73, 78–87, 90, 92, 98, 149

Bianchini, Giovanni, 42

bloodletting (*see* phlebotomy)

Boerio, Bartolomeo, 29

Boerio, Giovanni Battista, 16, 29–30, 33–34, 36, 39–40, 45, 48–49, 52, 62, 146, 148

Boioni, Giovanni, 71

Bologna, 42, 47, 52, 68–69, 127, 129–131, 148, 211

   curriculum (in arts and medicine), 20, 27, 29, 34, 40, 48, 50

Bona of Savoy, 92, 96, 101, 136–141, 150, 161, 171–172, 192

Bonatti, Guido, 47, 51, 86, 167–168

   Liber astronomiae, 47, 51, 86

Bontemps, Jean, 185

Borgia, Girolamo, 149, 165

Bossis Polonio, Giovanni de (da Polonia), 115

Brasca, Erasmo, 181

Brescia, 121

Burckhardt, Jacob, 119

Bylica, Martin, 115

Calco, Agostino, 151–152, 179–180

Calco, Bartolomeo, 151–152

Calco, Tristano, 173–174, 179, 189

Calixtus III (Alfonso Borgia), pope, 119

Camera, Antonio da, 65–69, 71

Campanus of Novara, 27, 42, 51

Campofregoso, Battista II, Doge of Genoa, 49

Carafa, Oliviero, 197

Cardano, Fazio, 168

Cardano, Girolamo, 18, 112–114, 118, 159, 168–169, 178, 187–188, 191, 209, 211

   Liber de exemplis centum geniturarum, 18, 209

   Liber duodecim geniturarum, 112

Casati, Francesco, 171

Cavalli, Gianmarco, 190

Cecco d'Ascoli, 28

celestial influence, 2, 10, 48, 53, 76–79, 88, 101, 118, 146, 185, 207

Centiloquium Hermetis, 29, 33, 39–41

Cerruti, Gerardo, 129

Charles V of Habsburg, Holy Roman Emperor, 210–211

Charles VI of Valois, King of France, 120

Charles VIII of Valois, King of France, 22, 96, 120, 154, 161–164, 178–179, 181–183, 190–191, 199–206

Chiromancy, 23, 30, 45

   ps.-Aristotle (*Chiromantia*), 30

Christ, 67

Ciccinello, Antonio, 122

Colis, Giovanni Battista of Cremona, 49

combustion (of the Moon), 146–147, 179–180, 193–194, 179–180, 193

Confalonieri, Dionisio, 152–156, 164

Constabili, Antonio, 206–208,

Conti, Nicolò de', 40, 42, 69–70,

   De motu octave sphere, 40, 69

Copernicus, Nicholaus, 42

Corio, Bernardino, 100–101, 108, 113, 128, 133, 141, 149, 190

   Historia di Milano, 100, 108

Crema, 121

Cremona, 121, 176

Crisciani, Chiara, 87

Cristoforo da Soncino, 82

critical days

   theory of, 11–12, 16, 40, 48, 51, 135–136, 144, 147–148, 152, 155, 202

Crivelli, Boldrino, 140

Cusano, Nicolò, 156–157, 178

Dactilo, Lazzaro (da Piacenza), 155–156

Daniele da Birago, 201

Decembrio, Pier Candido, 72–79, 88

   Cosmographia, 74

   De genitura hominis et de signis conceptionis, 74, 76, 78

   De muneribus romanae rei publicae, 74

   Historia peregrina, 74, 76

   Life of Filippo Maria Visconti, 73, 77, 80

della Porta, Ardicino, 197–198

della Rovere, Giuliano, 191–192, 198

Del Pozzo, Simone, 165

dignities (astrological), 53, 57, 60

diplomatic sources, 5, 7

Doge of Venice (Barbarigo, Agostino), 176, 183

Domitian, Roman Emperor, 113

Dondi, Giovanni, 73

earthquake, 70, 78–79,

Edward VI, King of England, 112

Eleonor of Portugal (wife of Frederick III), 186–187

Eleonora of Aragon (wife of Ercole d'Este), 120, 170, 173, 176

Elia (Jewish physician), 73

Erasmus of Rotterdam, 30, 33

Ernst the Iron, Archduke of Austria, 188

Este, Alfonso d', Duke of Ferrara, Moderna, and Reggio, 170, 172–173

Este, Beatrice d', Duchess of Milan, 150, 170–173, 175–178, 183, 185, 201

Este, Borso d', Duke of Ferrara, Modena and Reggio, 69, 102

Este, Ercole I d', Duke of Ferrara, Modena and Reggio, 49, 102, 115, 129–130, 150, 170–174, 176, 208

Este, Isabella d', Duchess of Ferrara, Modena, and Reggio, 171, 175–176, 185

Este, family, 70, 171, 173–174, 176

Euclid, 22, 27–28

    Elementa geometriae, 22, 27–28, 51

Fantucci, Stefano (da Faenza), 73

Ferdinand II of Aragon, King of Naples, 202

Ferrante (Ferdinand I) of Aragon, King of Naples, 7, 96, 118–123, 125–126, 136, 139, 150–151, 161, 170–171, 180–181, 191–192, 197–200

    illness, 118–119, 122–123

Ferrara, 69, 127, 130–131, 171, 176–177, 208, 211

Ferrari, Giovanni Matteo (da Grado), 84

Ferrarini, Girolamo, 5

fevers, 12

Ficino, Marsilio, 45, 149

    De vita coelitus comparanda, 149

Finé, Oronce, 28

Florence, 20, 22, 67, 121, 140, 154, 171, 174, 197, 200, 205

Florentine republic, 2, 68, 205–206

Francis I of Valois, King of France, 49, 210–211

Frederick III of Habsburg, Holy Roman Emperor, 89, 162, 185–186

Gabotto, Ferdinando, 8–9

Galen, 11, 16, 51, 144

    De crisibus, 51, 148

    De diebus criticis, 11, 16, 28, 48, 51, 148

    On Different Kinds of Fevers, 158

Gaspare da Pesaro (*see* Venturelli, Gaspare)

Gaurico, Luca, 18, 112

Gazio, Antonio, 132–133, 137, 159–160

genethlialogy (*see* horoscope, natal chart)

Geneva, Ann, 63

geniture (*see* horoscope, natal chart)

Genoa, 30, 119, 123, 125, 178, 198

Gerard of Cremona, 51

Giacomo da Palazzo, 93

Giovanni Battista da Cotignola, 130–132

Giovanni de Fundis, 69, 77

Giovanni de Lucoli, 131

Giovio, Paolo, 161–162

Giorgio Albanese, 131

Giustiniani, Pietro, 161–162

globe (celestial or terrestrial), 50

Gonnel, William, 33

Gonzaga, Barbara (Barbara Hohenzollern Gonzaga), 90–93, 103, 125

Gonzaga, Dorotea, 86, 89–93, 95, 103

Gonzaga, family, 20, 72, 89, 97, 149
Gonzaga, Francesco II, 94–95, 171, 176
Gonzaga, Giovanni, 208
Gonzaga, Ludovico, Marquis of Mantua, 65,
    67–69, 84, 89, 91–96, 102, 125
Gonzaga, Susanna, 89, 92
Grafton, Anthony, 63
Grendler, Paul, 25, 28
Guainieri (or Guarnieri), Antonio, 149
Guarnieri, Teodoro (da Pavia), 165
Guicciardini, Francesco, 162, 165, 204

Haly (Ali Ibn Ridwan), 28, 45, 47
Haly Abenragel, 45–47, 86
    De iudiciis astrorum, 45–47, 86
Harwell, Gregory, 190
Hayton, Darin, 115
Henry VII Tudor, King of England, 16, 30,
    180
Henry VIII Tudor, King of England, 16,
    30, 49
horologium (large clock), 50, 73
horoscope (celestial chart), 17–18, 29, 34, 49,
    93–94, 101, 103, 105, 111, 114–116, 125,
    132–133, 137, 159, 173, 186–187, 189–190,
    205, 209, 211
    casting of, 9, 17–18, 28, 49, 58, 63, 65, 91,
        93, 96, 101, 103, 105, 132
    collections, 18, 132–133, 159, 209, 211
    decumbiture, 13, 123, 148
    natal chart (geniture or nativity), 16–19,
        49, 53, 58, 60, 62, 65–67, 90–91, 93–94,
        98, 101, 103–105, 107, 110, 112–116, 125,
        132–133, 137, 148, 158, 173, 186–187,
        189–190, 205, 211
houses (mundane), 54, 57–58, 111
humanism, 10, 30
Hyginus, 45
    Poeticon astronomicon, 45
hyleg (giver of life), 60, 107–109, 112

Ilardi, Vincent, 119–120
Infessura, Stefano, 198

Innocent VIII (Gianbattista Cibo), pope,
    191–194, 197–200
Innsbruck, 186, 190
Isabella of Aragon, Duchess of Milan, 97,
    121, 136, 139, 142–143, 148, 150–153,
    162, 170–172, 175, 178, 200, 208
Israeli, Ysaac, 158
    Book of Fevers, 158
iudicium (*see* annual prognostications or
    horoscope)

Jean de Linières (Johannes de Lineris), 27,
    29, 51
    Canons, 27, 29, 51
Johannes of Bruges, 34
    De veritate astronomie, 34, 36, 39
Johannes of Lübeck, 42
John of Ashenden (Johannes Eschuid),
    45, 47
    *Summa astrologiae iudicialis de
        accidentibus mundi,* 45
John of Saxony, 27

Ketham, Johannes de, 12
    Fasiculo de medicina, 12–13

Lampugnani, Giovanni Andrea, 126, 132
Lanfranco da Parma (*see* Bardone,
    Lanfranco)
Lapini, Pietro (da Montalcino), 79
Leonardo da Vinci, 22–23, 30, 45
    Paragone, 23
Leopold III, Duke of Austria, 188
Lichtenberger, Johannes, 39
    Prognosticatio, 39
Linacre, Thomas, 30
Locatelli, Boneto, 41
Louis I of Valois, Duke of Orléans, 120
Louis XI of Valois, King of France, 90,
    92–96, 121, 129, 136
Louis XII of Valois, Duke of Orléans and
    King of France, 120, 161, 183, 199, 202,
    205–206, 210

Lowry, Martin, 41

Lubkin, Gregory, 102

Lucian, 33

   De astrologia, 33

Lunar nodes (*caput and cauda draconis*), 54,
   57, 60, 109–110, 113

Luther, 36, 62

Machiavelli, Niccolò, 140, 206

   The Prince, 206

Mahomet, 67

Maino, Agnese del, 73, 80, 82, 88, 90–91, 93,
   98, 101, 103

Malatesta, Giovanni, 84

Malatesta, Malatesta, 69

Malatesta, Sigismondo Pandolfo, 65, 68

Maletta, Francesco, 122

maleficium, 149–150

Manfredi, Bartolomeo, 91–97, 103

Manfredi, Girolamo, 52, 131–132

Mantegazza, Filippo, 34

Mantua, 20, 22, 65–67, 72, 84, 92

Marco da Bologna, 114–115

Margaret of Habsburg, 182

Maria of Savoy (wife of Filippo Maria
   Visconti), 84

Marliani, Giovanni, 160

Marliani, Lucia, 100–101

Marliani, Luigi, 177–178

Marliani, Pietro Antonio, 156

Marsilio da Bologna, 131–132

Marsilio da Santa Sophia, 52

Mary of Burgundy, 181, 186

Matthias Corvinus, King of Hungary,
   115–116

Maximilian I of Habsburg, Holy Roman
   Emperor and King of the Romans,
   162–163, 170–171, 178, 180–183, 185–190,
   199–201, 205–207, 210

Mayno, Giasone del, 188–198

Medici, Francesco (da Busto), 114, 117

Medici, Piero de', 200

Medici, Lorenzo de', 150, 152

medicine, 7, 30, 42, 46, 49–50, 58, 74–75,
   82–84, 139–148, 151–160, 177, 201–202

   and astrology, 4–5, 10–17, 19–21, 26–27,
      33, 48–49, 51–52, 87, 91, 135–136,
      144–148, 157–160, 201–202, 212

   at court, 16–17, 19, 49, 75, 82–83, 87–88,
      139–147, 151–160

Melchior von Meckau, bishop of Brixen
   (Bressanone), 163–165, 185

melothesia, 12, 133

Messahallah, 30, 26, 39–41, 45, 47, 51

   De interrogationibus, 51

   De receptionibus planetarum, 41

   De revolutionibus annorum mundi, 30, 36,
      39–41, 47

   Epistola in rebus eclipsis (or *Epistula de
      eclipsi lunae*), 41, 51

Meteorology (weather prediction), 4, 11, 53,
   62, 66, 68–69

Midelfort, Erik H. C., 4

Milan, 8, 11, 20, 22, 24–26, 34, 63, 67,
   69–70, 72, 74–75, 78, 81, 84, 86, 96–97,
   113, 119–121, 128, 130, 135–136, 140, 142,
   150, 152–153, 160–161, 171–174, 178,
   180–182, 184–185, 191, 198–199,
   201–202, 206–211

Modena, 20

Montano, Cola, 132

Moses, 67

Nanni, Giovanni (*see* Annius of Viterbo)

Naples, 78–79, 96–97, 120–123, 150–151, 154,
   171, 176, 178, 182–183, 192, 199–200,
   202

Negroponte, 121

Nicholas V (Tommaso Parentuccelli),
   pope, 75

Nicolò da Arsago, 116

Nicoud, Marilyn, 7, 87

North, John D., 38

Novara, Domenico Maria, 42

O'Boyle, Cornelius, 148
Olgiati, Gerolamo, 126, 132
Orfeo da Ricavo, 141
Orléans, family, 120
Orsini, Nicolò, Count of Pitigliano, 202
Oxford, 49
    university of, 49

Pacioli, Luca, 22–23,
    De divina proportione, 23
Padua, 25, 52, 69, 211
    University of (*Studium*), 25, 52
Paganica, Nicolò, 40
    Compendium medicinalis astrologie, 40
Paleologo, Bonifacio III, Marquis of
    Montferrat, 178–180
Paleologo, Guglielmo VIII, Marquis of
    Montferrat, 139
Pandolfini, Pietro Filippo, 150–152
Parati, Guido (da Crema), 75, 84, 138
Paris, 24, 28, 148, 179–180, 183
Pars fortunae, 60
patronage, 9–10, 17, 72, 80–82, 84–85, 98,
    102, 107, 113, 115, 117–118, 169, 189,
    207–208, 211–212
Pavia, 16, 24, 26, 29–30, 34, 39, 41–42,
    46–50, 52, 80–81, 84–86, 136, 142, 146,
    148–150, 152–153, 160, 163–164, 172–173,
    178, 211
    University of (*Studium*), 5, 7, 10, 20,
        23–27, 29, 40, 46, 49–50, 52, 62,
        84–86, 105, 117, 149, 160, 211
    curriculum (in arts and medicine), 24–25,
        52, 85, 146
    teaching of astrology and astronomy,
        24–29, 39–40, 46–50, 52, 85, 105,
        146, 148
Pedersen, Olef, 24, 28
Pelacani, Biagio da Parma, 52
Pellati, Alessandro, 42, 45–46
Pellegrin, Elizabeth, 50

Peroni, Luca, 8
Pesenti, Tiziana Marangon, 40
Petrucci, Antonello, 123
Peurbach, Georg, 28, 42
    Theoricae novae planetarum, 28, 42
Phares, Simon de, 20
phlebotomy, 12–13
physiognomy, 23, 33, 45
Piasio, Battista, 49, 69, 115
Pico della Mirandola, Giovanni, 77, 86
    *Disputationes adversus astrologiam
    divinatricem,* 77
Pietro d'Abano, 52
Pierre d'Ailly, 36
Pietro da Montalcino (*see* Lapini,
    Pietro)
Pirovano, Gabriele, 7, 30, 49, 117, 154–157,
    202–203
Pisa, 28, 140
    University of, 28
Pius II (Enea Silvio Piccolomini), pope,
    94, 96
plague, 30, 33, 36, 62, 68, 70, 79, 91,
    115
planetary conjunctions, 34, 36, 39, 54, 60,
    62, 65, 79, 91, 95, 145–146, 186–190,
    202
Pontano, Giovanni, 192
prodiges (and omens), 10, 66, 73, 198–199,
    204
profection, 133,
prophecy, 10, 36, 39, 66, 114–115, 207, 211
prorogation, 107, 111–112
ps.-Galen, 48, 148, 153
    Aggregationes de crisi et creticis diebus,
    48–49, 51, 148, 153
ps.-Hippocrates, 12, 48, 148
    Astronomia or *Astrologia Ypocratis,* 12, 45,
    48, 148
ps.-Ptolemy, 28–29, 33, 39, 41, 45–46, 148
    Centiloquium, 28–29, 33–34, 39–41,
    45–46, 51, 148

Ptolemy, 28–29, 41, 45–47, 51, 87, 103,
    107–108, 132, 143
    Almagest, 28, 47, 51
    Cosmographia (or *Geography*), 50, 132
    Planispherium, 51
    Quadripartitum (Tetrabiblos), 28–29, 41,
        45–46, 107, 143

quadrant, 22–23, 27, 45–46

Ratdolt, Erhard, 41
regimen sanitatis, 11, 154
Regiomontanus, Johannes, 23, 42, 69,
    187
    Kalendar, 23, 45
Reguardati, Benedetto (da Norcia), 84, 88,
    92–93
    Regimen sanitatis, 88
revolutions, 17–18, 60, 62, 65–66, 70, 114,
    133, 193–194
    of the natal chart, 17–18, 60, 65, 70, 114,
        133
    of the year, 62, 193–194
Rhasis, 42
    Liber IX Almansoris, 42
Rhazes, 158
    Almansor, 158
Robertus Anglicus, 50
    Tractatus quadrantis, 50
Rolewinck, Werner, 45
    Fasciculus temporum, 45
Rome, 67, 74–75, 136, 145, 182–183, 199,
    202

Sacrobosco, Johannes, 22, 24, 28, 30, 42,
    46, 51, 77
    Algorismus (de minutiis), 27, 42, 51
    Computus, 30, 42, 46
    Sphaera mundi, 22–23, 27–28, 42, 46,
        51, 77
Sacramoro, Sacramoro (da Rimini), 123,
    129
Saint Thomas of Pavia (convent), 46–47, 86

Salio, Girolamo da Faenza, 42
Sallust, 127
    The Conspiracy of Catiline, 127
Sanseverino, Gianfrancesco, Count of
    Caiazzo, 179–180
Sanseverino, Roberto, 140–141, 192
Santoseverino, Margherita da, 82
Sanuto, Marino, 149, 190–191, 200–201,
    203–204
Savelli, Giovanni Battista, 199
Savonarola, Girolamo, 45
Savonarola, Michele, 52
Schöner, Johannes, 187
Scot, Michael, 23, 51
    Liber introductorius, 51
    Liber particularis (or *Astronomia*), 51
    Physiognomia, 23, 45
Scoto, Ottaviano, 41–42
Sententiae (or *Capitula*) *Almansoris,*
    39–41, 51
Serigatti, Francesco, 23, 65
Sforza, Anna, 89, 128, 172–173
Sforza, Ascanio, 88, 98, 136, 140, 142,
    145–147, 182, 191–192, 197–199, 202,
    206–208
Sforza, Bianca Maria (daughter of Galeazzo
    Maria), 89, 128, 162, 170, 181, 183–188,
    190, 199
Sforza, Carlo, 90
Sforza, court, 5, 7–9, 16, 19–20, 23–24, 48,
    63, 77–78, 80, 89, 97, 102, 105, 180, 185,
    192, 197, 211
Sforza, Elisabetta, 82, 139
Sforza, Ercole (later Massimiliano), Duke of
    Milan, 174, 176, 178, 209–210
Sforza, family, 1, 3, 5, 16, 18, 19, 21, 63,
    73–74, 80, 88, 160, 173–177, 209–210,
    212
Sforza, Filippo, 82, 140
Sforza, Francesco I, Duke of Milan, 21,
    65–71,74, 80–85, 88–89, 92–98,
    103–104, 119–120, 136, 138, 173,
    209

Sforza, Francesco II, Duke of Milan, 157, 211

Sforza, Galeazzo Maria, Duke of Milan, 5, 21, 69, 72–73, 80–82, 86, 89–91, 97–98, 100–101, 104, 107–110, 113–120, 122–123, 125–134, 136–140, 160, 162, 170, 173, 185
   and Annius of Viterbo, 7, 20, 116, 126
   and Dorotea Gonzaga, 86, 89–93, 95–96, 101, 103, 173
   death, 21, 102, 107–111, 113, 117, 126–134, 140

Sforza, Gian Galeazzo, Duke of Milan, 16, 21, 48, 88, 97, 116, 127–128, 135–142, 144–167, 169–170, 172, 174, 178–179, 181, 192, 200, 208
   illness and death, 16, 21, 48, 88, 97, 127, 135–137, 139–142, 144–148, 151–166, 192, 200, 208

Sforza, Hermes, 128

Sforza, Ippolita, 82, 96–97, 101, 122, 139

Sforza, library, 25–26, 41, 46–48, 50–52, 73, 76, 86

Sforza, Ludovico Maria, Duke of Milan, 2, 7, 16, 21–23, 48, 72–73, 82, 88–89, 97–98, 102, 117, 119, 125, 127–128, 132, 136, 140–142, 145–147, 149–157, 159–182, 184–186, 188–194, 197–212

Sforza, Ottaviano, 86, 140

Sforza, Sforza, 120, 128, 140, 162

Sforza, Tristano, 138

Sigismund of Habsburg, Archduke of Austria, 186

Simonetta, Cicco, 7, 75, 78–80, 114–116, 129, 131, 140–141, 161

Simonetta, Giovanni, 129, 141

Siraisi, Nancy, 25

Sixtus IV (Francesco della Rovere), pope, 121, 129

Somenzi, Augusto, 205–206

Stefano da Faenza (see Fantucci, Stefano)

Stuhlhofer, Franz, 186, 190

Suetonius, 113

Tabule Toletane (Toledan tables), 46, 51

Taggia (Genoa), 29–30, 33, 36

Tassino, Antonio, 141

Terzaghi, Luigi, 73

Theorica planetarum, 27–28, 42, 46

Thorndike, Lynn, 28

Tizio, Sigismondo, 165
   History of Siena, 165

Torrella, Girolamo, 149
   Opus praeclarum de imaginibus astrologicis, 149

triplicity, 34, 36, 57, 60, 62

Trivulzio, Gian Giacomo, 159–161, 208, 210

Trotti, Giacomo, 150, 171–174

Turks, 68, 96, 121, 126, 162–163, 188, 207

Tuttavilla, Girolamo, 179

Valdizocco, Bartolomeo, 40, 48

Valenza (Alessandria), 30

Valla, Giorgio, 159–160, 211

Vanden Broecke, Steven, 111

Varesi, Ambrogio da Rosate, 7, 16–17, 21, 23, 88, 117, 149, 156, 167–171, 175–179, 184–185, 189–194, 197, 200–208, 212

Venice, 22, 26, 41, 45, 67, 70, 92, 121, 161, 174, 176–178, 181–183, 197, 205–206, 208, 210

Venturelli, Gaspare (da Pesaro), 83

Vercelli, 84

Vimercati, Gaspare, 118

Vimercati, Pietro Paolo, 117–118

Vimercati, Raffaele, 103–104, 107–111, 113–118, 126, 132–133

Visconti, Antonio, 151

Visconti, Bernabò, 188

Visconti, Bianca Maria, 65, 72, 75, 78, 80–85, 88, 93, 98, 121, 127, 138
   and Antonio Bernareggi, 80–85, 98, 101
   and astrology, 21, 72, 80–81, 98
   illness, 75, 82–84

Visconti, Carlo, 126, 132

Visconti, family, 21, 63, 67, 72, 75, 78, 80–81,
　86, 98, 138
Visconti, Filippo Maria, Duke of Milan,
　46, 72–74, 80–81, 84–85, 88–89,
　120, 138
Visconti, Galeazzo, 179–180, 200
Visconti, Gian Galeazzo, Duke of Milan,
　72, 138
Visconti, Valentina, 120, 210
Visconti, Viridis, 188

Weill-Parot, Nicolas, 149
William of England, 28–29, 48
　De urina non visa, 28–29, 40, 48

Zael, 33, 39, 41, 45, 47, 51, 103
　De electionibus, 41, 51
　De interrogationibus, 41, 45, 47
　Liber temporum, 51
　Quinquaginta praecepta, 33, 39
Zambotti, Bernardino, 5